T0220515

III-Nitride Materials
for Sensing, Energy Conversion and Controlled Light-Matter Interactions

MATERIALS RESEARCH SOCIETY
SYMPOSIUM PROCEEDINGS VOLUME 1202

III-Nitride Materials for Sensing, Energy Conversion and Controlled Light-Matter Interactions

Symposium held November 29–December 3, 2009, Boston, Massachusetts, U.S.A.

EDITORS:

Shangjr Gwo
National Tsing-Hua University
Hsinchu, Taiwan

Joel W. Ager III
Lawrence Berkeley National Laboratory
Berkeley, California, U.S.A.

Fan Ren
University of Florida
Gainesville, Florida, U.S.A.

Oliver Ambacher
Fraunhofer-Institut für Angewandte
Festkörperphysik
Freiburg, Germany

Leo Schowalter
Crystal IS, Inc.
Green Island, New York, U.S.A.

Materials Research Society
Warrendale, Pennsylvania

CAMBRIDGE UNIVERSITY PRESS
Cambridge, New York, Melbourne, Madrid, Cape Town,
Singapore, São Paulo, Delhi, Mexico City

Cambridge University Press
32 Avenue of the Americas, New York NY 10013-2473, USA

Published in the United States of America by Cambridge University Press, New York

www.cambridge.org
Information on this title: www.cambridge.org/9781107408128

Materials Research Society
506 Keystone Drive, Warrendale, PA 15086
http://www.mrs.org

First published 2010
First paperback edition 2012

Single article reprints from this publication are available through
University Microfilms Inc., 300 North Zeeb Road, Ann Arbor, MI 48106

CODEN: MRSPDH

ISBN 978-1-107-40812-8 Paperback

Cambridge University Press has no responsibility for the persistence or
accuracy of URLs for external or third-party internet websites referred to in
this publication, and does not guarantee that any content on such websites is,
or will remain, accurate or appropriate.

This work was supported in part by the Army Research Office under Grant Number W91INF-10-
1-0019. The views, opinions, and/or findings contained in this report are those of the author(s) and
should not be construed as an official Department of the Army position, policy, or decision, unless
so designated by other documentation.

CONTENTS

CHEMICAL AND BIOLOGICAL SENSING

*Invited Paper

ELECTRONIC DEVICES

ENERGY CONVERSION

CONTROLLED LIGHT-MATTER
INTERACTIONS

*Invited Paper

LIGHT EMITTING DIODES

*Invited Paper

GROWTH AND PROCESSING

PREFACE

There has been tremendous progress during the past two decades in the field of III-nitride (GaN, AlN, InN) semiconductors and their device applications. For example, the invention of III-nitride-based blue laser diodes was the key to a new high-density data storage technology, the Blue-ray disc. At the same time, improvements in III-nitride-based solid state light sources have led to their use in an increasing number of illumination applications, moving steadily toward the goal of general lighting. III-nitrides-related symposia have traditionally had excellent attendance at the MRS Fall Meetings. However, as III-nitride technologies have evolved into maturity and become commercialized, several large-scale, dedicated meetings for III-nitride semiconductors have been organized. For this reason, Symposium I, "III-Nitride Materials for Sensing, Energy Conversion and Controlled Light-Matter Interactions," was held November 29–December 3 at the 2009 MRS Fall Meeting in Boston, Massachusetts. Instead of organizing a general III-nitride symposium, we tried to identify three emerging areas for III-nitride applications: chemical and biological sensing, energy conversion, and controlled light-matter interactions. Our selection was based on the potential for these applications to improve our quality of life, as well as the superior material properties III-nitrides offer in terms of high chemical stability, excellent biocompatibility, efficient light absorption from the complete solar spectrum, and strong light-matter coupling. It turned out that with these focused areas in mind, we were able to have a four-day symposium with over 100 oral and poster presentations. Additionally, we organized a half-day tutorial for III-nitride-based sensors for gas, chemical, and medical applications with lectures given by two leading experts in the field, Drs. Volker Cimalla and Fan Ren. A significant number of papers (41) presented during the symposium are collected in this proceedings volume, which we hope can serve to further stimulate related III-nitride research.

Shangjr Gwo
Joel W. Ager III
Fan Ren
Oliver Ambacher
Leo Schowalter

February 2010

ACKNOWLEDGMENTS

We would like to thank the excellent support from the MRS staff. The symposium was supported in part by DCA Instrument in Finland. The tutorial was supported in part by the U.S. Army Research Office (ARO).

MATERIALS RESEARCH SOCIETY SYMPOSIUM PROCEEDINGS

MATERIALS RESEARCH SOCIETY SYMPOSIUM PROCEEDINGS

Prior Materials Research Society Symposium Proceedings available by contacting Materials Research Society

Chemical and Biological Sensing

Mater. Res. Soc. Symp. Proc. Vol. 1202 © 2010 Materials Research Society   1202-I06-01

Recent Advances in Wide Bandgap Semiconductor Biological and Gas Sensors

S.J.Pearton[1], F.Ren[2], Yu-Lin Wang[2], B.H. Chu[2], K.H. Chen[2], C.Y. Chang[2], Wantae Lim[1], Jenshan Lin[3] and D.P. Norton[1]

[1] Department of Materials Science and Engineering, University of Florida, Gainesville, FL 32611 USA
[2] Department of Chemical Engineering, University of Florida, Gainesville, FL 32611 USA
[3] Department of Electrical and Computer Engineering, University of Florida, Gainesville, FL 32611

ABSTRACT

There has been significant recent interest in the use of surface-functionalized thin film and nanowire wide bandgap semiconductors, principally GaN, InN, ZnO and SiC, for sensing of gases, heavy metals, UV photons and biological molecules. For the detection of gases such as hydrogen, the semiconductors are typically coated with a catalyst metal such as Pd or Pt to increase the detection sensitivity at room temperature. Functionalizing the surface with oxides, polymers and nitrides is also useful in enhancing the detection sensitivity for gases and ionic solutions. The wide energy bandgap of these materials make them ideal for solar-blind UV detection, which can be of use for detecting fluorescence from biotoxins. The use of enzymes or adsorbed antibody layers on the semiconductor surface leads to highly specific detection of a broad range of antigens of interest in the medical and homeland security fields. We give examples of recent work showing sensitive detection of glucose, lactic acid, prostate cancer and breast cancer markers and the integration of the sensors with wireless data transmission systems to achieve robust, portable sensors.

INTRODUCTION

Chemical sensors have gained in importance in the past decade for applications that include homeland security, medical and environmental monitoring and also food safety. In the area of detection of medical biomarkers, many different methods, including enzyme-linked immunsorbent assay (ELISA), particle-based flow cytometric assays, electrochemical measurements based on impedance and capacitance, electrical measurement of microcantilever resonant frequency change, and conductance measurement of semiconductor nanostructures. gas chromatography (GC), ion chromatography, high density peptide arrays, laser scanning quantitiative analysis, chemiluminescence, selected ion flow tube (SIFT), nanomechanical cantilevers, bead-based suspension microarrays, magnetic biosensors and mass spectrometry (MS) have been employed [1-6]. Depending on the sample condition, these methods may show variable results in terms of sensitivity for some applications and may not meet the requirements for a handheld biosensor. While the techniques mentioned above show excellent performance under lab conditions, there is also a need for small, handheld sensors with wireless connectivity that have the capability for fast responses.

A promising sensing technology utilizes AlGaN/GaN high electron mobility transistors (HEMTs). HEMT structures have been developed for use in microwave power amplifiers due to their high two dimensional electron gas (2DEG) mobility and saturation velocity. The conducting 2DEG channel of AlGaN/GaN HEMTs is very close to the surface and extremely sensitive to adsorption of analytes. HEMT sensors can be used for detecting gases, ions, pH values, proteins, and DNA. In this review, we discuss recent progress in the functionalization of these semiconductor sensors for applications in detection of gases, pH measurement, biotoxins and other biologically important chemicals and the integration of these sensors into wireless packages for remote sensing capability.

HYDROGEN SENSING

There is great interest in detection of hydrogen sensors for use in hydrogen-fuelled automobiles and with proton-exchange membrane (PEM) and solid oxide fuel cells for space craft and other long-term sensing applications. These sensors are required to detect hydrogen near room temperature with minimal power consumption and weight and with a low rate of false alarms. Due to their low intrinsic carrier concentrations, GaN- and SiC-based wide band gap semiconductor sensors can be operated at lower current levels than conventional Si-based devices and offer the capability of detection to $\sim$600 °C [7].

Unlike conventional sensors where the changes of the sensing material's conductivity or resistivity are used to detect the gas concentration, by integrating the gas sensing material such as Pd or Pt metal on the gate electrode of the HEMTs, the change of the sensing material's conductivity can be amplified through Schottky diode or FET operation. It is generally accepted that H_2 is dissociated when adsorbed on Pt and Pd at room temperature. The reaction is as follows:

$$H_{2(ads)} \rightarrow 2\,H^+ + e^-$$

Dissociated hydrogen causes a change in the channel and conductance change. This makes the integrated semiconductor device based sensors extremely sensitive and the sensors have a broad dynamic range of the sensing concentration. A surface covered reference semiconductor device can also be easily fabricated side-by-side with sensing device to eliminate the temperature variation of the ambient and the fluctuation of the supplied voltage to the sensors [20].

Figure 1 shows a schematic of a Schottky diode hydrogen sensor on AlGaN/GaN HEMT layer structure and a photograph of packaged sensors. A thin (100Å) Pt Schottky contact was deposited by e-beam evaporation for the Schottky metal. The sensor carrier was then placed in our test chamber. Mass flow controllers were used to control the gas flow through the chamber, and the devices were exposed to either 100% pure N_2, or 1% H_2 in nitrogen. Figure 2 shows the linear (top) and log scale (bottom) forward current-voltage (I-V) characteristics at 25 °C of the HEMT diode, both in air and in a 1%H_2 in air atmosphere. For these diodes, the current increases upon introduction of the H_2, through a lowering of the effective barrier height. The data was fit to the relations for thermionic emission and showed decreases in Schottky barrier height Φ_B of 30-50 mV at 50°C and larger changes at higher temperatures .The decrease in barrier height is completely reversible upon removing the H_2 from the ambient and results from diffusion of atomic hydrogen to the metal/GaN interface, altering the interfacial charge.

4

Ohmic Metal:
1. Ti/Al/Pt/Au
2. Ti/Al/TiB₂/Ti/Au

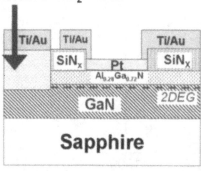
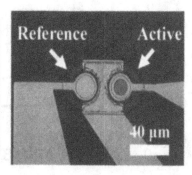

Figure 1. Cross-sectional schematic of completed Schottky diode on AlGaN/GaN HEMT layer structure (top) and plan-view photograph of device (bottom).

Figure 2 shows the linear (top) and log scale (bottom) forward current-voltage (I-V) characteristics at 25 °C of the HEMT diode, both in air and in a 1%H₂ in air atmosphere. For these diodes, the current increases upon introduction of the H₂, through a lowering of the effective barrier height. The data was fit to the relations for thermionic emission and showed decreases in Schottky barrier height Φ_B of 30-50 mV at 50°C and larger changes at higher temperatures .The decrease in barrier height is completely reversible upon removing the H₂ from the ambient and results from diffusion of atomic hydrogen to the metal/GaN interface, altering the interfacial charge.   To achieve the goal of detecting reactions due to hydrogen only and excluding other changes caused by variables such as temperature and moisture, a differential detection interface was used. Several kinds of differential devices have been fabricated and each of its performance has been evaluated to select the most effective solution. These differential devices have two sensors integrated on the same chip. The two sensors are identical except one is designed to react to hydrogen whereas the other one is covered by dielectric protection layer and not exposed to ambient gas. We have also found that nanostructured wide bandgap materials functionalized with Pd or Pt are even sensitive than their thin film counterparts because of the large surface-to-volume ratio.

HEMTs functionalized with polyethyleneimine (PEI) can be used for CO_2 detection, while functionalization with $InGaZnO_4$ can be used for oxygen sensing [8].

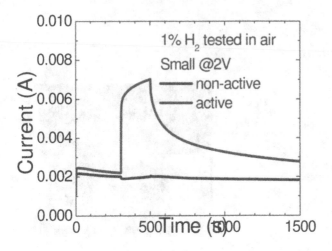

Figure 2. Time response of differential GaN sensor to introduction of 1% H₂ in air ambient.

pH MEASUREMENT

The measurement of pH is needed in many different applications, including medicine, biology, chemistry, food science, environmental science and oceanography. Solutions with a pH less than 7 are acidic and solutions with a pH greater than 7 are basic or alkaline. Ungated AlGaN/ GaN HEMTs also exhibit large changes in current upon exposing the gate region to polar liquids [8]. The polar nature of the electrolyte introduced led to a change of surface charges, producing a change in surface potential at the semiconductor /liquid interface. The use of Sc_2O_3 gate dielectric produced superior results to either a native oxide or UV ozone-induced oxide in the gate region. The ungated HEMTs with Sc_2O_3 in the gate region exhibited a linear change in current between pH 3-10 of 37μA/pH .The HEMT pH sensors show stable operation with a resolution of < 0.1 pH over the entire pH range. The adsorption of polar molecules on the surface of the HEMT affected the surface potential and device characteristics. Figure 3 shows the current at a bias of 0.25V as a function of time from HEMTs with Sc_2O_3 in the gate region exposed for 150s to a series of solutions whose pH was varied from 3-10. The current is significantly increased upon exposure to these polar liquids as the pH is decreased. The change in current was 37 μA/pH. The HEMTs show stable operation with a resolution of ~0.1 pH over the entire pH range, showing the remarkable sensitivity of the HEMT to relatively small changes in concentration of the liquid.

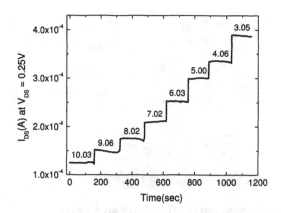

Figure 3. Change in current in gateless HEMT at fixed source-drain bias of 0.25 V with pH from 3-10

BOTULINUM

Antibody-functionalized Au-gated AlGaN/GaN high electron mobility transistors (HEMTs) show great sensitivity for detecting botulinum toxin. The botulinum toxin was specifically recognized through botulinum antibody, anchored to the gate area, as shown in Figure 4. We investigated a range of concentrations from 0.1 ng/ml to 100 ng/ml.
The source and drain current from the HEMT were measured before and after the sensor was exposed to 100 ng/ml of botulinum toxin at a constant drain bias voltage of 500 mV, as shown in Figure 23 (top). Any slight changes in the ambient of the HEMT affect the surface charges on the AlGaN/GaN. These changes in the surface charge are transduced into a change in the concentration of the 2DEG in the AlGaN/GaN HEMTs, leading to the decrease in the conductance for the device after exposure to botulinum toxin [9].

Different concentrations (from 0.1 ng/ml to 100 ng/ml) of the exposed target botulinum toxin in a buffer solution were detected. The sensor saturates above 10ng/ml of the toxin. The experiment at each concentration was repeated four times to calculate the standard deviation of source-drain current response. The limit of detection of this device was below 1 ng/ml of botulinum toxin in PBS buffer solution. The source-drain current change was nonlinearly proportional to botulinum toxin concentration. Figure 5 shows a real time test of botulinum toxin at different toxin concentrations with intervening washes to break antibody-antigen bonds [9]. This result demonstrates the real-time capabilities and recyclability of the chip.

7

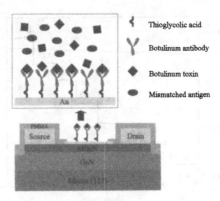

Figure 4. Schematic of functionalized HEMT for botulinum detection.

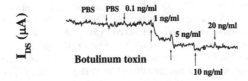

Figure 5. Real-time test from a used botulinum sensor which was washed with PBS in pH 5 to refresh the sensor.

EXHALED BREATH CONDENSATE

There is significant interest in developing rapid diagnostic approaches and improved sensors for determining early signs of medical problems in humans. Exhaled breath is a unique bodily fluid that can be utilized in this regard. Exhaled breath condensate pH is a robust and reproducible assay of airway acidity. For example, the blood pH range that is relevant for humans is 7-8. Humans, even when they are extremely ill, will not have a blood (or interstitial = space between cells in tissue) pH below 7. When they do drift below this value, it almost invariably equals mortality.

While most applications will detect substances or diseases in the breath as a gas or aerosol, breath can also be analyzed in the liquid phase as exhaled breath condensate (EBC). Analytes contained in the breath originating from deep within the lungs (alveolar gas) equilibrates with the blood, and therefore the concentration of molecules present in the breath is closely correlated with those found in the blood at any given time. EBC contains dozens of different biomarkers, such as adenosine, ammonia, hydrogen peroxide, isoprostanes, leukotrienes, peptide, cytokines and nitrogen oxide. Analysis of molecules in EBC is noninvasive and can provide a window on the metabolic state of the human body, including certain signs of cancer, respiratory diseases, liver and kidney functions. In our HEMT devices, the surface is generally functionalized with an antibody or enzyme layer. For use with exhaled breath, the device may include a HEMT bonded on a thermo-electric cooling device, which assists in condensing exhaled breath samples.

AlGaN/GaN HEMTs can be used for measurements of pH in EBC and glucose, through integration of the pH and glucose sensor onto a single chip and with additional integration of the sensors into a portable, wireless package for remote monitoring applications. Figure 6 shows an optical microscopy image of an integrated pH and glucose sensor chip and cross-sectional schematics of the completed pH and glucose device. The gate dimension of the pH sensor device and glucose sensors was 20×50 μm^2.

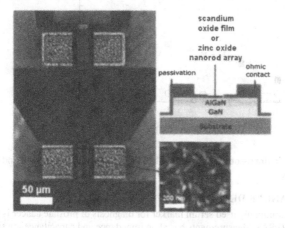

Figure 6. SEM image of an integrated pH and glucose sensor. The insets show a schematic cross-section of the pH sensor and also an SEM of the ZnO nanorods grown in the gate region of the glucose sensor.

For the glucose detection, a highly dense array of 20-30 nm diameter and 2 μm tall ZnO nanorods were grown on the 20×50 μm^2 gate area. The total area of the ZnO was increased significantly with the ZnO nanorods. The ZnO nanorod matrix provides a microenvironment for immobilizing negatively charged GO$_x$ while retaining its bioactivity, and passes charges produced during the GO$_x$ and glucose interaction to the AlGaN/GaN HEMT. The GOx solution was prepared with concentration of 10 mg/mL in 10 mM phosphate buffer saline (pH value of 7.4, Sigma Aldrich). After fabricating the device, 5 μl GO$_x$ (~100 U/mg, Sigma Aldrich) solution was precisely introduced to the surface of the HEMT using a pico-liter plotter. The sensor chip was kept at 4 °C in the solution for 48 hours for GO$_x$ immobilization on the ZnO nanorod arrays followed by an extensively washing to remove the un-immobilized GO$_x$. . Although the response of the HEMT based sensor is similar to that of an electrochemical based sensor, a much

9

lower detection limit of 0.5 nM was achieved for the HEMT based sensor due to this amplification effect. Since there is no reference electrode required for the HEMT based sensor, the amount of sample only depends on the area of gate dimension and can be minimized. The sensors do not respond to glucose unless the enzyme is present, as shown in Figure 7.

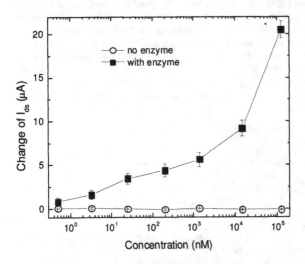

Figure 7.Change in drain-source current in HEMT glucose sensors both with and without localized enzyme.

PROSTATE CANCER DETECTION

The most commonly used serum marker for diagnosis of prostate cancer is prostate specific antigen (PSA). Measurements based on impedance and capacitance are simple and inexpensive but need improved sensitivities for use with clinical samples. Resonant frequency changes of an anti-PSA antibody coated microcantilever enable a detection sensitivity of ~ 10 pg/ml but this micro-balance approach has issues with the effect of the solution on resonant frequency and cantilever damping. Antibody-functionalized nanowire FETs coated with antibody provide for low detection levels of PSA, but the scale-up potential is limited by the expensive e-beam lithography requirements.

Antibody functionalized Au-gated AlGaN/GaN HEMTs were found to be effective for detecting PSA at low concentration levels. The PSA antibody was anchored to the gate area through the formation of carboxylate succinimdyl ester bonds with immobilized thioglycolic acid. The HEMT drain-source current showed a response time of less than 5 seconds when target PSA in a buffer at clinical concentrations was added to the antibody-immobilized surface. The devices could detect a range of concentrations from 1 μg/ml to 10 pg/ml. The lowest detectable concentration was two orders of magnitude lower than the cut-off value of PSA measurements for clinical detection of prostate cancer. Figure 8 shows the real time PSA detection in PBS buffer solution using the source and drain current change with constant bias of 0.5V [10]. No current change can be seen with the addition of buffer solution or nonspecific bovine serum albumin (BSA), but there was a rapid change when10 ng/ml PSA was added to the surface. The abrupt current change due to the exposure of PSA in a buffer solution could be stabilized after the PSA diffused into the buffer solution. The ultimate detection limit appears to be a few pg/ml [10].

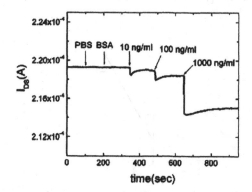

Figure 8. Drain current versus time for PSA detection when sequentially exposed to PBS, BSA, and PSA

BREAST CANCER

Antibody-functionalized Au-gated AlGaN/GaN high electron mobility transistors (HEMTs) show promise for detecting c-erbB-2 antigen. The c-erbB-2 antigen was specifically recognized through c-erbB antibody, anchored to the gate area. We investigated a range of clinically relevant concentrations from 16.7 µg/ml to 0.25 µg/ml.

The Au surface was functionalized with a specific bi-functional molecule, thioglycolic acid. We anchored a self-assembled monolayer of thioglycolic acid, $HSCH_2COOH$, an organic compound and containing both a thiol (mercaptan) and a carboxylic acid functional group, on the Au surface in the gate area through strong interaction between gold and the thiol-group of the thioglycolic acid. The device was incubated in a phosphate buffered saline (PBS) solution of 500 µg/ml c-erbB-2 monoclonal antibody for 18 hours before real time measurement of c-erbB-2 antigen.

After incubation with a PBS buffered solution containing c-erbB-2 antibody at a concentration of 1 µg/ml, the device surface was thoroughly rinsed off with deionized water and dried by a nitrogen blower. The source and drain current from the HEMT were measured before and after the sensor was exposed to 0.25 µg/ml of c-erbB-2 antigen at a constant drain bias voltage of 500 mV. Any slight changes in the ambient of the HEMT affect the surface charges on the AlGaN/GaN. These changes in the surface charge are transduced into a change in the concentration of the 2DEG in the AlGaN/GaN HEMTs, leading to the slight decrease in the conductance for the device after exposure to c-erbB-2 antigen. The source-drain current change was nonlinearly proportional to c-erbB-2 antigen concentration, as shown in Figure 9. Between each test, the device was rinsed with a wash buffer of pH 6.0 phosphate buffer solution containing KCl to strip the antibody from the antigen.

Clinically relevant concentrations of the c-erbB-2 antigen in the saliva and serum of normal patients are 4-6 µg/ml and 60-90 µg/ml respectively. For breast cancer patients, the c-erbB-2 antigen concentrations in the saliva and serum are 9-13 µg/ml and 140-210 µg/ml, respectively. Our detection limit suggests that HEMTs can be easily used for detection of clinically relevant concentrations of biomarkers [11].

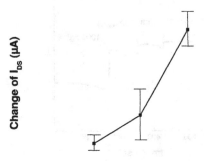

c-erbB-2 antigen concentration (µg/ml)

Figure 9. Change of drain current versus different concentrations from 0.25 µg/ml to 17 µg/ml of c-erbB-2 antigen

LACTIC ACID

Lactic acid can also be detected with ZnO nanorod-gated AlGaN/GaN HEMTs. Interest in developing improved methods for detecting lactate acid has been increasing due to its importance in areas such as clinical diagnostics, sports medicine, and food analysis. An accurate measurement of the concentration of lactate acid in blood is critical to patients that are in intensive care or undergoing surgical operations as abnormal concentrations may lead to shock, metabolic disorder, respiratory insufficiency, and heart failure. Lactate acid concentration can also be used to monitor the physical condition of athletes or of patients with chronic diseases such as diabetes and chronic renal failure. In the food industry, lactate level can serve as an indicator of the freshness, stability and storage quality. For the reasons above, it is desirable to develop a sensor capable of simple and direct measurements, rapid response, high specificity, and low cost. Recent researches on lactate acid detection mainly focus on amperometric sensors with lactate acid specific enzymes attached to an electrode with a mediator. Examples of materials used as mediators include carbon paste, conducting copolymer, nanostructured Si_3N_4 and silica materials. Other methods of detecting lactate acid include utilizing semiconductors and electro-chemiluminescent materials.

A ZnO nanorod array, which was used to immobilize lactate oxidase oxidase (LOx), was selectively grown on the gate area using low temperature hydrothermal decomposition (Figure

10, top). The array of one-dimensional ZnO nanorods provided a large effective surface area with high surface-to-volume ratio and a favorable environment for the immobilization of LOx. The AlGaN/GaN HEMT drain-source current showed a rapid response when various concentrations of lactate acid solutions were introduced to the gate area of the HEMT sensor. The HEMT could detect lactate acid concentrations from 167 nM to 139 Mm, as shown in Figure 10 (bottom) [8].

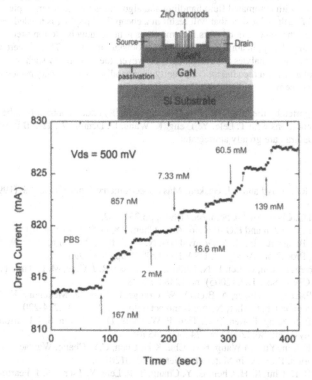

Figure 10. Schematic cross sectional view of the ZnO nanorod gated HEMT for lactic acid detection (top) and plot of drain current versus time with successive exposure to lactate acid from 167 nM to 139 μM (bottom).

CONCLUSION

HEMT sensors show promising results for protein, DNA, prostate cancer, kidney injury molecules, pH values of solutions, mercury ions as well glucose in the exhaled breath condensate. The method relies on an amplification of small changes in antibody-structure due to binding to antigens. The characteristics of these sensors include fast response (liquid phase-5 to

13

10 seconds and gas phase- milli-second), digital output signal, small device size (less than 100 × 100 μm^2) and chemical and thermal stability.

There are still some critical issues. First, the sensitivity for certain antigens (such as prostate or breast cancer) needs to be improved further to allow sensing in body fluids other than blood (urine, saliva). Second, a sandwich assay allowing the detection of the same antigen using two different antibodies (similar to ELISA) needs to be tested. Third, integrating multiple sensors on a single chip with automated fluid handling and algorithms to analyze multiple detection signals, and fourth, a package that will result in a cheap final product is needed. Fourth, the stability of surface functionalization layers in some cases is not conducive to long-term storage and this will limit the applicability of those sensors outside of clinics. There is certainly a need for detection of multiple analytes simultaneously. However, there are many such approaches and acceptance from the clinical community is generally slow for many reasons, including regulatory concerns.

Acknowledgments

This work is supported by the Office of Naval Research (ONR) under Contract Number 00075094. The collaborations with T. Lele, Y. Tseng, K. Wang, D. Dennis, W. Tan, B.P. Gila, W. Johnson and A. Dabiran are greatly appreciated.

REFERENCES

1. A.L. Burlingame, R.K. Boyd and S.J. Gaskell, Mass Spectrometry, Anal. Chem. 68, (1996), pp. 599-611.
2. K.W. Jackson and G. Chen, Anal. Chem. 68, (1996), pp.231-242.
3. J.L. Anderson, E.F. Bowden and P.G. Pickup, Anal. Chem. 68, (1996), pp.379-401
4.   R. J. Chen, S. Bangsaruntip, K. A. Drouvalakis, N. W. S. Kam, M. Shim, Y. Li, W. Kim, P. J. Utz, and H. Dai, Proc. Natl. Acad. Sci. USA 100, (2003), pp4984-4990
5.   C. Li, M. Curreli, H. Lin, B. Lei, F. N. Ishikawa, R. Datar, R. J. Cote, M. E. Thompson, and C. Zhou, J. Am. Chem. Soc. 127, (2005), pp.12484-12498
6.   J. Zhang, H. P. Lang, F. Huber, A. Bietsch, W. Grange, U. Certa, R. Mckendry, H.-J. Güntherodt, M. Hegner and Ch. Gerber, Nature Nanotechnology, 1(2006), pp.214-220.
7.   H. T. Wang, B. S. Kang, F. Ren, L. C. Tien, P. W. Sadik, D. P. Norton, S. J. Pearton, and Jenshan Lin, Appl. Phys. Lett. 86, (2005) pp. 243503-243505
8.   S.J. Pearton, F. Ren, Yu-Lin Wang, B.H. Chu, K.H. Chen, C.Y. Chang, Wantae Lim, Jenshan Lin, D.P. Norton, Progress in Materials Science, 55, 1(2010).
9.   Y. L. Wang, B. H. Chu, K. H. Chen, C. Y. Chang, T. P. Lele, Y. Tseng, S. J. Pearton, J. Ramage, D. Hooten, A. Dabiran, P. P. Chow, and F. Ren, Appl. Phys. Lett. 93, 262101 (2008).
10.   B. S. Kang, H. T. Wang, T. P. Lele, and F. Ren, S. J. Pearton, J.W. Johnson, P. Rajagopal, J.C. Roberts, E.L. Piner, and K.J. Linthicum, Appl. Phys Lett., 91, (2007) pp.112106-112108.
11.   K.H. Shen, B.S.Kang, H.T.Wang, T.P.Lele, F.Ren, Y.L.Wang, C.Y.Chang, S.J.Pearton, J.W. Johnson, P.Rajagopal, J.C.Roberts, E.L.Piner and K.J. Linthicum, Appl. Phys.Lett. 92, 192103 (2008).

Mater. Res. Soc. Symp. Proc. Vol. 1202 © 2010 Materials Research Society   1202-I06-04

*p*H and Biological Sensing of Ultrathin (10nm) InN Based ISFETs

Yen-Sheng Lu[1], Cheng-Yi Lin[1], Yuh-Hwa Chang[2], Yu-Liang Hong[3], Shangir Gwo[3], and J.A Yeh[1,2]

[1]Institute of Electronics Engineering, National Tsing-Hua University,
Hsinchu, 30013, Taiwan
[2]Institute of Nanoengineering and Microsystems, National Tsing-Hua University,
Hsinchu, 30013, Taiwan
[3]Department of Physics, National Tsing-Hua University,
Hsinchu, 30013, Taiwan

ABSTRACT

Ultrathin (~10 nm) InN ion selective field effect transistors (ISFETs) show a current variation ratio of 3.5 % per *p*H decade with a response time of less than 10 s. When the ISFET is employed as an electrolyte FET, the current variation of 18 % was measured as the gate bias changes from zero to 0.3 V given a drain-source voltage of 0.1 V. The high current (resistance) variation ratio is attributed to the ultrathin epilayer and an unusual phenomenon of intrinsic strong electron accumulation on InN surface, which enables a chemical/biological sensor with high sensitivity and resolution and permits detection of a slight concentration variation of the electrolyte. The *p*H response measurement of 10-nm-thick InN ISFETs investigated was performed in an aqueous solution titrated with diluted NaOH and HCl. The Helmholtz potential built at the electrolyte-InN interface is governed by direct adsorption of H^+ ions at the surface metal oxides, modulating the channel current of the InN ISFETs. The channel current monotonically decreases as the *p*H value of an aqueous solution increases from 2 to 10. The sensitivity and resolution were found to be 58.3 mV per decade and 0.02 *p*H change, respectively. Besides, the detection of DNA hybridization was further performed after the InN surface was modified with MPTMS and probe DNA. A complementary target DNA solution of 100 nM led to a current decrease of approximate 6 μA, corresponding to the current variation of 0.74 %. The hybridization between negatively charged complementary DNA and the immobilized probe DNA caused the depletion of carriers at the InN surface, suppressing the channel current. The functionalized InN ISFETs are suitable for genetic analysis in clinical diagnostics without any labeling reagent. Such an InN-based sensor is appealing in the regime of chemical and biological sensing applications.

INTRODUCTION

Group-III nitrides, including GaN, AlGaN/GaN, and InN, have emerged as promising materials for next-generation chemical and biological sensors because of their biocompatibility, high-sensitivity and robust surface properties against possible damages from harsh chemical and thermal environments [1-15]. At present, a majority of GaN-based sensors utilize the configuration of high electron mobility transistor (HEMT) structure with a piezoelectric-induced two-dimensional electron gas (2DEG) to detect ions, polar liquid and even biological materials [1-9]. Recently, InN, exhibiting a phenomenon of unusually strong surface electron accumulation, is also regarded as a candidate for superior chemical detection. It has been shown that as grown, nominally undoped InN film exhibits large free surface electron accumulation (typically in excess of 1×10^{18} cm^{-3}). Intrinsic electron accumulation at InN surfaces has been confirmed by various experimental techniques [16-19]. These electron properties are believed to be related to the particularly low conduction band minimum at Γ-point and the Fermi stabilization energy (E_{FS}) deep inside the conduction band [18,19]. Since the location of Fermi level is below E_{FS}, the native defects in InN tend to become positively charged donor states, giving rise to the phenomena of high n-type background concentration and surface electron accumulation. The *ab initio* calculations of InN electronic structure have also confirmed that positively charged surface donor states exist on InN for emitting electrons into the conduction band [18,19]. The positively charged donor surface state density was found to be as high as the order of 10^{13} cm^{-2} (the largest native electron accumulation observed at III-V semiconductors) [16-19]. The high surface charge density and robust surface properties of InN are beneficial for sensor applications. Indeed, InN material has been proposed to be useful for sensing applications [10-15].

Sensors based on ion selective field effect transistor (ISFET) structure, firstly utilized as a pH sensor by Bergveld in 1970, are promising for chemical and biological applications [20]. The performance of ISFET strongly depends on the sensing material and the FET. The sensing material absorbs ions on the gate region, changing the gate potential and modulating the source-drain currents of the FET. The sensing materials with high sensitivity (close to theoretical Nernst Response) are really desired. Therefore, several materials, such as Si_3N_4 [21], Al_2O_3 [21], or Ta_2O_5 [22], have been studied and investigated to improve sensitivity and stability in pH sensing applications. On the other hand, the performance of the FET is typically characterized by resistance or current variation ratio with respect to the change of the gate potential [2, 23-26]. The FET with high current variation ratio enables to effectively transduce the change of gate potential into current variation, which improves detection resolution and is suitable to detect slight ion changes in electrolyte concentrations. Therefore, a biological sensing such as deoxyribonucleic acid (DNA) hybridization, proteins immobilization, and biomarker detection etc, usually adopts the transducer with high current variation ratio [25,26].

In this study, we have investigated the electrical responses of open-gate InN ISFETs under various gate biases (V_{GS}) in electrolyte solutions. Figure 1 shows two investigated ISFET structures where carriers flowing in the surface, interface, and bulk channels are schematically illustrated. The InN films studied here include undoped N-polar wurtzite InN ($-c$-InN) of 1 μm and 10 nm in thickness. The ISFETs based on 1-μm, undoped $-c$-InN epilayer are primarily employed to verify the small current modulation effect of the bulk channel. The 10-nm, undoped $-c$-InN epilayer is used to analyze the influences of the surface and the interface channels on the ISFET current variation ratio. The one with higher current (resistance) variation ratio enables a chemical sensor with higher sensitivity and resolution, which permits detection of a slight concentration variation of the electrolyte and was used for detecting pH variations and DNA hybridization.

EXPERIMENTAL DETAILS

The heteroepitaxial growth of high-quality wurtzite-type InN ($000\bar{1}$) films on AlN/Si_3N_4/Si(111) substrates was performed using the double-buffer layer technique [27]. The growth of InN was conducted in a molecular-beam epitaxy (MBE) system (DCA-600) equipped with a radio-frequency nitrogen plasma source. The base pressure of this MBE system was maintained at ~6×10^{-11} Torr. The nitrogen (N$_2$) plasma (instead of ammonia) has high reactivity with the elemental indium flux at the substrate surface, allowing InN deposition at low temperature. In the study, the deposited InN epilayers have 10 nm and 1 μm in thickness.

The microfabrication process of InN ISFET began with evaporation of Au/Al/Ti (50 nm/200 nm/50 nm) to ensure ohmic contact onto InN surfaces. The definition of channels in length and contact pads of the ISFETs was achieved by a lift-off process. The channel width of the ISFETs is defined by etching the conductive InN film using inductively coupled plasma (ICP) with BCl$_3$ and Cl$_2$ discharges. The RF power was set at 800W for 90s. After mesa isolation, a 2-μm-thick polyimide was patterned to define the geometry of the ISFET contact (gate) windows. The polyimide encapsulated both drains and sources of the ISFETs, preventing short circuitry via the exposure to aqueous environments. Wafers were diced and each individual ISFET was attached to the side-brazed ceramic material and bonded using aluminum wires, followed by packaging with polydimethylsiloxane (PDMS). The InN ISFET after the package is shown in figure 2.

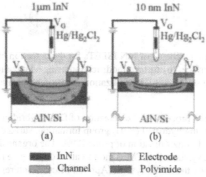

Figure 1. Schematic diagrams of two investigated ion sensitive field effect transistor (ISFET) structures, including (a)1-μm-thick −c-InN and (b)10-nm-thick −c-InN epilayers.

Figure 2. Ultrathin InN ISFET packaged using polyimide and PDMS on side-brazed ceramic materials. The inset shows the top-view photograph of InN ISFET.

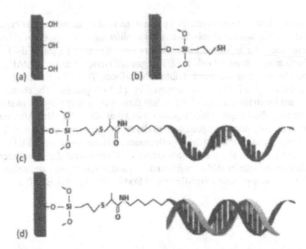

Figure 3. Functionalization scheme of InN ISFETs for the detection of DNA hybridization: (a) hydroxylation of InN surface, (b) surface modification using silanization of MPTMS, (c) immobilization of probe ss-DNA, (d) hybridization with complementary DNA.

The functionalization process for the detection of DNA hybridization is schematically shown in figure 3, involving hydroxyl (-OH) group formation, surface modification of silane cross linker MPTMS and immobilization of probe DNA. The organosilane 3-mercaptopropyltrimethoxysilane (MPTMS) modification on the gate region of 10-nm-thick InN ISFETs was carried out by the MVD system (Applied Microstructures, MVD 100). The MVD technique enhances the vapor deposition of ultrathin organic molecules by incorporating an *in-situ* surface plasma treatment and the precise delivery of precursor vapors [28]. In the beginning, the O_2 plasma was utilized to clean the bare InN surface and form hydroxyl (-OH) groups atop the gate region of InN ISFETs for 10 min. Subsequently, gas-phase silanization of MPTMS was conducted under a purge pressure of 2 Torr for 1.5 h. MPTMS molecules were covalently bonded onto the gate surface of the InN epilayer and left the terminal group, thiol (-SH) group, to further immobilize probe DNA (acrylic phosphoramidite-$(CH_2)_6$-5′-CCT AAT AAC AAT-3′). After surface modification, MPTMS-silanized ISFETs were dipped into a 10 μM probe DNA aqueous solution for 12 h. After immobilization of probe DNA, the functionalized InN ISFETs were rinsed carefully with DI water and employed to detect the hybridization with complementary ss- DNA (5′-ATT GTT ATT AGG-3′).

The current-voltage characteristics of these studied InN ISFETs were measured using an Agilent 4156B parameter analyzer with the gate voltage modulated through an Hg/Hg_2Cl_2 reference electrode (Hanna HI-5412) immersed in the electrolyte. The pH response measurement of ultrathin (~10 nm) InN ISFETs were performed in an aqueous solution titrated with diluted NaOH and HCl. The pH value was measured by a commercial pH meter (Hanna HI-111) in real time. The experimental setup for pH sensing is shown in figure 4.

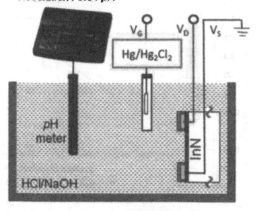

Figure 4. Illustration of experimental setup for measuring current variation ratio and pH resolution of InN ISFET

DISCUSSION

InN ISFETs under electrolyte-gate-bias

I_{DS}-V_{GS} characteristics of InN ISFETs are shown in figure 5. The drain-source currents (I_{DS}), normalized to the current at zero bias, are plotted as functions of gate bias (V_{GS}) with a fixed drain-source voltage (V_{DS}) of 0.25 V while the gates are biased through a pH 7 buffer solution. The 1-μm-thick $-c$-InN ISFET shows an insensitive case of response (<0.1 %) to the change of gate bias from –0.3V to 0.3V. For the thick epilayer, the strong electron accumulation at the InN surface reduces the field effects, which allows the field-effect affecting region only within a few nanometers from the surface (small electrostatic screening length) [19,29]. Therefore, the carriers in the bulk channel and in the interface channel cannot be efficiently induced or suppressed (i.e. the channel current can only be slightly varied). On the other hand, the 10-nm-thick $-c$-InN ISFET exhibits a current variation of 18 % (15 %) under the gate bias of 0.3 V (–0.3 V), demonstrating the current in the ultrathin channel is sensitive to the gate bias. In this case, the carriers are mainly from the surface channel compared with that of the 1-μm-thick $-c$-InN ISFET, leading to a smaller current than the 1-μm-thick $-c$-InN ISFET (0.23 mA vs. 2.52 mA at zero gate bias for V_{DS}= 0.25V). When biased by a gate voltage, the 10-nm-thick $-c$-InN ISFET has a large relative change of carrier number and we can obtain a higher current variation ratio than that of the 1-μm-thick $-c$-InN ISFET. The ultrathin InN ISFET with a high current variation ratio has the potential to detect small pH variation and improve the pH resolution.

Other III-nitride based pH sensors based on AlGaN/GaN high-electron-mobility transistor (HEMT) structures have a current variation ratio of 7.5 % at a gate bias of 0.3 V for a V_{DS} of 0.25 V, demonstrating a sensitive response to pH variation and the pH resolution is estimated to be 0.05 pH [2].

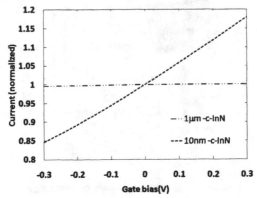

Figure 5. I_{DS}-V_{GS} characteristics of InN ISFETs immersed in a pH 7 buffer solutions within V_{GS} of 0.3 V.

Sensitivity of ultrathin InN ISFETs

The ISFET is a three-terminal device, including source, drain and gate. When ISFETs were immersed in different pH buffers, the potential change at the solid and electrolyte interface can directly obtained by adjusting the gate bias (V_{GS}) via the Hg/Hg$_2$Cl$_2$ reference electrode to compensate the ion-induced change in channel current. The ultrathin InN ISFET was immersed in pH buffers (Hanna) from pH 2 to 10, showing the variation of gate voltage has a linear relationship with the change of pH value shown in figure 6. The InN ISFET has a sensitivity of 58.3 mV/pH, which is close to the Nernst response of H$^+$ (59.16 mV/pH at 25 Φ). The gate sensitivity of InN ISFETs is similar to other III-nitride-based pH sensing devices such as GaN-capped AlGaN/GaN high electron mobility transistors (HEMTs) (56.0 mV/pH) [2] and bare AlGaN/GaN HEMTs (57.5 mV/pH) [3]. Furthermore, when keeping the gate voltage at zero bias ($V_{GS}=0$), we observed a monotonic current decrease in the range from pH 2 to10 upon exposure to pH buffer solutions (Hanna). The average current variation with respect to the pH change per decade can be found to be 17 μA with a current variation ration of 3.5 % at $V_{GS}=0$ V and $V_{DS}=0.1$ V as shown in figure 7.

As-deposited InN analyzed by x-ray photoemission spectroscopy (XPS) via the In $3d_{5/2}$ and O $1s$ core level spectra reveals there is a thin layer of metal oxides (In$_2$O$_3$) existed at the InN surface. The metal oxide is responsible for the pH response. The site-bonding model has been widely accepted to describe the pH response of metal oxide surface for several decades and can also used to explain the pH response of InN ISFETs [30-33]. When the oxidic surface is in contact with aqueous solution, amphoteric hydroxyl groups would be formed at the interface of InN and electrolyte. The amphoteric hydroxyl groups can be in deprotonized (InO$^-$), neutral

20

(InOH), or protonized (InOH$_2^+$) state, depending on the pH values of the solution as shown in figure 8. When immersed in the aqueous solution with smaller H$^+$ concentration, there are more deprotonized hydroxyl groups (InO$^-$) formed at the InN surface, changing the net surface charge and enlarging the voltage drop in the Helmholtz layer, which would decrease the channel current of InN ISFETs. Conversely, the current of InN ISFETs would increase upon exposure to the aqueous solution with higher H$^+$ concentration.

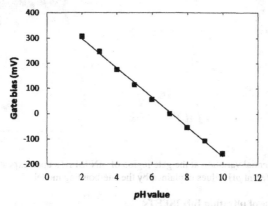

Figure 6. The gate voltage change as a function of pH for ultrathin (~10 nm) InN ISFETs. The sensitivity is estimated to be 58.3 mV/pH.

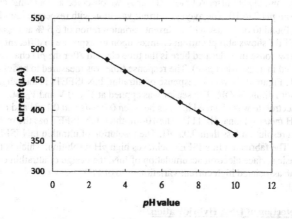

Figure 7. The current response of ultrathin InN ISFET to pH buffers (Hanna). The source-drain current shows a monotonic decrease with pH value from pH 2 to 12.

21

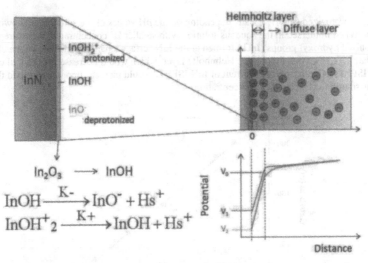

Figure 8. The potential variation at the gate region of InN ISFETs upon exposure to the solutions with different pH values explained by the site-bonding model.

<u>**Dynamic response of ultrathin InN ISFETs**</u>

Upon exposure to pH buffer solutions (Hanna), we observed a monotonic current decrease in the range from pH 2 to 10. The average current variation with respect to the pH change per decade can be found to be 17 µA with a current variation ration of 3.5 % at V_{GS}=0 V and V_{DS}=0.1 V. The InN ISFET shows abrupt current changes upon its exposure to different pH buffer solutions. The response time defined here is the time elapsed after the pH changes from 10 % to 90 % of intended increase (decrease). The response time was measured to be less than 10 s for the InN ISFET. Figure 9 shows the response of ultrathin InN ISFETs in the range from pH 7 to pH 9 titrated with dilute NaOH. The sensor was applied at V_{GS}=0 V and V_{DS}=0.25 V. The pH value of the electrolyte was changed in steps between 0.05 pH and 0.1 pH and monitored by a commercial pH meter (Hanna, HI-111). The 10-nm-thick InN ISFET has the response time less than 10 s and a resolution less than 0.05 pH. The resolution of ultrathin InN ISFETs is estimated to be 0.02 pH. The fabricated InN ISFET achieves high pH resolution, which is mostly beneficial from the intrinsic surface electron accumulation of InN, the design of ultrathin conduction channel and the associated high current variation ratio of FET.

<u>**Electronic Detection of DNA Hybridization.**</u>

Electronic measurement on the current variation of InN ISFETs provides the real-time monitoring on the hybridization behavior. The MPTMS modified InN ISFETs after the

immobilization of probe DNA (acrylic phosphoramidite-$(CH_2)_6$-5'-CCT AAT AAC AAT-3')
were used to detect the hybridization with complementary DNA (5'-ATT GTT ATT AGG-3').
The noncomplementary DNA (5'-ATT GTT ATT AGG-3') with single-base mismatch was used
as a control experiment. The functionalized InN ISFETs were encapsulated at drain and source
terminals and left the sensing gate region exposed to the aqueous solution. The encapsulation
enables to prevent water permeation and provides good electrical insulation property. DI water
was used as the solvent to prepare the DNA aqueous solutions for minimizing the effect of
counter ions screening and improving the sensitivity of charge-based detection. The dynamic
response of the functionalized InN ISFETs to noncomplementary/complementary DNA is shown
in figure 10. During the sensing process, V_{DS} was remained at 0.2 V while V_G was kept at zero
potential through the reference electrode.

The functionalized ISFETs were immersed in aqueous solution, followed by the addition of
noncomplementary or complementary DNA at 120 s. When the functionalized ISFETs was
exposed to noncomplementary DNA solution of 100 nM, there was no current change shown in
figure 10(a). On the other hand, figure 10(b) shows a drain-source current decrease of
approximate 6 μA was observed for functionalized InN ISFET upon exposure to a
complementary target DNA solution of 100 nM. The corresponding current variation is 0.74 %.
The hybridization between negatively charged complementary DNA and the immobilized probe
DNA increased the overall negative charges at gate region and caused the depletion of carriers at
the ultrathin InN surface, suppressing the channel current.

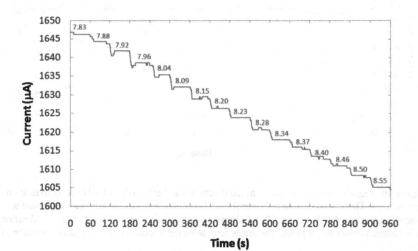

Figure 9. The transient behavior of the drain-source current to pH variation from 7 to 9. The pH
change is achieve by titration with NaOH. the pH value is measured using a commercial pH
meter (Hanna HI-111). The resolution of ultrathin InN ISFET is less than 0.05 pH.

23

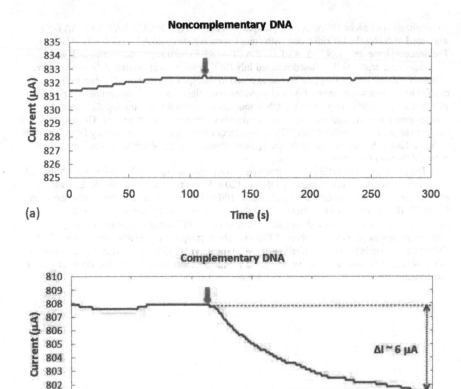

Figure 10. Real-time current response of the functionalized InN ISFETs for DNA hybridization detection. (a) There is no current response for the functionalized InN ISFETs upon exposure to noncomplementary DNA. (b) The ISFET shows a decrease of current of 6 μA after hybridization with complementary target DNA. The arrow represents the injection of DNA aqueous solution at 120 s.

CONCLUSIONS

I_{DS}-V_{GS} characteristics demonstrate that the 10-nm-thick InN ISFET has a higher current variation ratio than the 1-μm-thick $-c$-InN InN ISFETs, which is attributed to the ultrathin epilayer and an unusual phenomenon of intrinsic strong electron accumulation on InN surface. The surface channel can be effectively influenced by gate bias, thus enhancing the high current

24

variation ratio (18 % at a gate bias of 0.3 V for a V_{DS} of 0.25 V). As a pH sensor, the ultrathin ISFET shows a sensitivity of 58.3 mV/pH and a current variation ratio of 3.5 % per pH decade with a response time of less than 10 s. Besides, the functionalized InN ISFETs with probe DNA was used for label-free DNA hybridization detection. The hybridization between 12-mer probe DNA and 100 nM complementary DNA resulted in a current decrease of 6 µA, corresponding to a current variation ration of 0.74 %. On the other hand, the noncomplementary DNA with single-base mismatch can not be effectively attached to the probe DNA, resulting in an insignificant current variation of ISFETs. These results show that the ultrathin InN ISFETs are very promising for a wide variety of sensing applications.

REFERENCES

1. R. Neuberger, G. Müller, O. Ambacher, and M. Stutzmann, *Phys. Status Solidi A* **185**, 85 (2001).
2. G. Steinhoff, M. Hermann, W. J. Schaff, L. F. Eastman, M. Stutzmann, and M. Eickhoff, *Appl. Phys. Lett.* **83**, 177 (2003).
3. T. Kokawa, T. Sato, H. Hasegawa, and T. Hashizume, *J. Vac. Sci. Technol. B*, **24**, 1942 (2006).
4. R. Mehandru, B. Luo, B. S. Kang, J. Kim, F. Ren, S. Pearton, C. Pan, G. Chen, and J. Chyi, *Solid State Electron.* **48**, 351 (2004).
5. B. S. Kang, F. Ren, M. C. Kang, C. Lofton, W. Tan, S. J. Pearton, A. Dabiran, A. Osinsky, and P. P. Chow, *Appl. Phys. Lett.* **86**, 173502 (2005).
6. B. Kang, S. Pearton, J. Chen, F Ren,. J. Johnson, R. Therrien, P. Rajagopal, J. Roberts, E. Piner, and K. Linthicum, *Appl. Phys. Lett.* **89**, 122102 (2006).
7. B. S. Kang, H. T. Wang, F. Ren, and S. J. Pearton, *J. Appl. Phys.* **104**, 031101 (2008).
8. Y. Alifragis, A. Georgakilas, G. Konstantinidis, E. Iliopoulos, A. Kostopoulos, and N. A. Chaniotakis, *Appl. Phys. Lett.* **87**, 253507 (2005).
9. N. Chaniotakis, and N. Sofikiti, *Anal. Chim. Acta* **615**, 1 (2008).
10. H. Lu, W. Schaff, and L. Eastman, *J. Appl. Phys.* **96**, 3577 (2004).
11. O. Kryliouk, H. J. Park, H. T. Wang, B. S. Kang, T. J. Anderson, F. Ren, and S. J. Pearton, *J. Vac. Sci. Technol. B* **23**, 1891 (2005).
12. C.-F. Chen, C.-L. Wu, and S. Gwo, *Appl. Phys. Lett.* **89**, 252109 (2006).
13. Y.-S. Lu, C.-C. Huang, J. A. Yeh, C.-F. Chen, and S. Gwo, *Appl. Phys. Lett.* **91**, 202109 (2007).
14. Y.-S. Lu, C.-L. Ho, J. A. Yeh, H.-W. Lin, and S. Gwo, *Appl. Phys. Lett.* **92** 212102 (2008).
15. Y.-S. Lu, Y. H. Chang, Y. L. Hong, H. M. Lee, S. Gwo, and J. A. Yeh, *Appl. Phys. Lett.* **95**, 102104 (2009).
16. H. Lu, W. J. Schaff, L. F. Eastman, and C. E. Stutz, *Appl. Phys. Lett.* **82**, 1736 (2003).
17. Rickert, K. A.; Ellis, A. B.; Himpsel, F. J.; Lu, H.; Schaff, W.; Redwing, J. M.; Dwikusuma, F.; Kuech, T. F. *Appl. Phys. Lett.* **82**, 3254 (2003).
18. I. Mahboob, T. D. Veal, C. F. McConville, H. Lu, and W. J. Schaff, *Phys. Rev. Lett.* **92**, 036804 (2004).
19. L Mahboob, T. D. Veal, L. F. J. Piper, C. F. McConville, H. Lu, W. J. Schaff, J. Furthmüller, and F. Bechstedt, *Phys. Rev. B* **69**, 201307 (2004).
20. P. Bergveld, *IEEE Trans. Bio-Med. Eng.* **17**, 70 (1970).

21. T. Matsuo, M. Esashi, and H. Abe, *IEEE Trans. Electron Devices.* **26**, 1856 (1979).
22. P. Gimmel, K. D. Schierbaum, W. Gopel, H. H. Van den Vlekkert, N. F. De Rooy, *Sens. Actuators B* **1**, 345 (1990).
23. Y. Cui, Q.Wei, H. Park, and C. M. Lieber, *Science* **293**, 1289 (2001).
24. X. T. Zhou, J. Q. Hu, C. P. Li, D. D. D. Ma, C. S. Lee, and S. T. Lee, *Chem. Phys. Lett.* **369** 220 (2003).
25. Z. Li, Y. Chen, X. Li, T. Kamins, K. Nauka, and R. Williams, *Nano Lett.* **4**, 245 (2004).
26. Y. L. Bunimovich, Y. S. Shin, W.-S. Yeo, M. Amori, G. Kwong, and J. R. Heath, *J. Am. Chem. Soc.* **128**, 16323 (2006).
27. S. Gwo, C.-L. Wu, C. -H. Shen, W.-H. Chang, T. M. Hsu J.-S. Wang, and J.-T. Hsu, *Appl. Phys. Lett.* **84**, 3765 (2004). 28
28. B. Kobrin, V. Fuentes, S. Dasaradhi, R. Yi, R. Nowak, J. Chinn, R. Ashurst, C. Carraro, and R. Maboudian, *Semiconductor Equipment and Materials International* **2004**.
29. C. H. Ahn, J.-M. Triscone, and J. Mannhart, *Nature* **424**, 1015 (2003).
30. D. E. Yates, S. Levine, and T. W. Healy, J. Chem. Soc., *Faraday Trans.* **70**, 1807 (1974).
31. W. M. Siu and R. S. C. Cobbold, *IEEE Trans. Electron Devices* **26**, 1805 (1979).
32. L. Bousse, N. F. De Rooij, and P. Bergveld, *IEEE Trans. Electron Devices* **30**, 1263 (1983).
33. C. D. Fung, P. W. Cheung, and W. H. Ko, *IEEE Trans. Electron Devices* **33**, 8 (1986).

Mater. Res. Soc. Symp. Proc. Vol. 1202 © 2010 Materials Research Society   1202-I06-05

Chloride Ion Detection by InN Gated AlGaN/GaN High Electron Mobility Transistors

Byung-Hwan Chu[1], Hon-Way Lin[2], Shangjr Gwo[2], Yu-Lin Wang[3], S. J. Pearton[3], J. W. Johnson[4], P. Rajagopal[4], J. C. Roberts[4], E. L. Piner[4], and K. J. Linthicum[4], and Fan Ren[1]

[1] Department of Chemical Engineering, University of Florida, Gainesville, FL 32611
[2] Department of Physics, National Tsing-Hua University, Hsinchu, 30013 Taiwan
[3] Department of Material Science and Engineering, University of Florida, FL 32611
[4] Nitronex Corporation, Durham, NC 27703

ABSTRACT

Chloride ion concentration can be used as a biomarker for the level of pollen exposure in allergic asthma, chronic cough and airway acidification related to respiratory disease. AlGaN/GaN high electron mobility transistor (HEMT) with an InN thin film in the gate region was used for real time detection of chloride ion detection. The InN thin film provided surface sites for reversible anion coordination. The sensor exhibited significant changes in channel conductance upon exposure to various concentrations of NaCl solutions. The sensor was tested over the range of 100 nM to 100 μM NaCl solutions. The effect of cations on the chloride ion detection was also studied.

INTRODUCTION

Chlorine is a widely used chemical used in the production of paper products, antiseptics, food, insecticides, paints, petroleum products, plastic, medicine, textiles and solvents. One of its main uses includes sanitizing drinking water supplies and waste water. Chlorine readily reacts with organics to form harmful disinfection by-products that may be carcinogenic. Therefore, it is important to effectively monitor chlorine residuals, typically in the form of chloride ion concentration, to ensure public health [1, 2]. Chloride ion is also an essential mineral in our bodies. Our kidneys balance the chloride in body fluids, such as serum, blood, urine, and exhaled breath condensate (EBC). Abnormal chloride ion concentration in serum may serve as an indicator for diseases such as renal diseases, adrenalism, and pneumonia [3]. Much research has been focused on the analysis and collection of EBC to develop a noninvasive method to diagnose patients. Chloride ion concentration can be a biomarker for the level of pollen exposure in allergic asthma, chronic cough and airway acidification related to respiratory disease [4-6]. Also, the Cl⁻ concentration in EBC can be used as a reference for the other biomarkers in the EBC to estimate the dilution effect from the humidity in the ambient during the EBC collection [7].

Current analytical methods for measuring chloride ion include colorimetry, ion-selective electrode, X-ray fluorescence spectrometry, activation analysis, and ion chromatography [8-13]. However, these methods involve high expertise levels and require expensive instruments that can't be readily transported. Therefore, it would be beneficial to develop a fast and accurate sensor that has a low detection limit and low production cost.

InN is a small bandgap semiconductor material, which has potential application in solar cells and high speed electronics. InN exhibits an electron accumulation phenomenon, which is very unusual for III-V semiconductors [14-16]. The positively charged surface donor states on InN function as fixed surface sites for the reversible anion coordination [17]. Because of this property, InN has been proposed as a useful material for sensing applications. There was a report of utilizing an InN thin film based potentiometric ion selective sensor to detect Cl⁻ ions down to 1 mM [17].

AlGaN/GaN high electron mobility transistors (HEMT) have shown potential as a low cost, fast and accurate sensor for a variety of chemical and biological applications [18-27]. HEMT sensors utilize a two-dimensional electron gas (2DEG) channel with high electron sheet carrier concentration that is induced by both piezoelectric polarization and spontaneous polarization in these heterostructures [28-38]. The 2DEG channel is located at the interface between the AlGaN and GaN layers while positive countercharges are located near the surface of the HEMT. Slight changes of electrical potential at the surface of the HEMT can be detected from resulting alterations in the electron concentration in the 2DEG channel. A distinct advantage of the HEMT sensors over amperometric sensors is that it does not require a fixed reference electrode in the solution to measure the potential difference. Therefore, the amount of sample needed for analysis could be reduced. Although the time response of the HEMT sensors is similar to that of electrochemical-based sensors, a lower detection limit can be achieved due to the amplifying effect of the HEMT. Chloride ion detection has recently been demonstrated using a HEMT sensor with the gate region modified with Ag/AgCl [24]. The sensor was able detect to as low as 10^{-8}M, but the Ag/AgCl layer did not show very good durability.

In this work, chloride ion detection was demonstrated using AlGaN/GaN HEMT sensors with InN film deposited by molecular beam epitaxy (MBE) in the gate region. The sensor was exposed to a series of sodium chloride solutions with different chloride ion concentrations. The effects of positive ions and pH value of the chloride ion solution on the chloride ion concentration detection was also explored.

EXPERIMENT

A cross-section schematic of the sensor is shown in Figure 1(a). The HEMT structures consisted of an undoped GaN buffer and undoped $Al_{0.25}Ga_{0.75}N$ cap layer with 3 μm and 250Å thicknesses, respectively. The epi-layers were grown by metal-organic chemical vapor deposition on Si (111) substrates. The sheet carrier concentration was ~10^{13} cm^{-2} with a mobility of 980 cm²/V s at 300 K. Inductively coupled plasma (ICP) etching with Cl₂/Ar based discharges at - 90V dc self-bias, ICP power of 300W at 2MHz and a process pressure of 2 mTorr was used for mesa isolation. Ohmic contacts consisted of e-beam deposited Ti/Al/Pt/Au with sizes of 50×50 μm² and separated by gaps of 20 μm using standard postivie resist lift-off process. After metal deposition, the contacts were annealed at 850°C for 45 seconds under flowing N₂. Then, a final metal layer composed of Ti/Au was deposited for interconnections.

Prior to the InN deposition, a silicon nitride (SiN$_x$) film of 2500Å thickness was deposited and the dielectric window openings on the SiN$_x$ film for InN growth was etched with CF₄ plasma and buffered oxide etch (BOE).The SiN$_x$ film was deposited at 250 °C with a Unaxis 790 PECVD system using ammonia and 2% SiH₄ balanced by nitrogen as the precursors. The deposition pressure of the PECVD system was kept at 900 mtorr and rf power of 30 W was generated at 13.56 MHz. The self- bias voltage during the deposition was around 1 V, so that ion bombardment induced damages during the PECVD was is not expected to occur. A 10 nm thick InN film was deposited in the gate region of the HEMT by heteroepitaxial growth in an MBE equipped with a radio frequency nitrogen plasma source. The detailed growth process has been published elsewhere [39]. An optical microscope image of the gate region HEMT sensors with each layer labeled are shown in Figure 1(b).

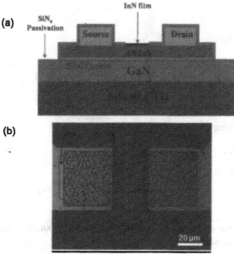

Figure 1. (a) Cross section schematic of the HEMT, (b) Microscope image of the gate region of HEMT

DISCUSSION

Figure 2 shows the results of real time detection of Cl⁻ ions by measuring the HEMT time dependent drain current with an Agilent 4156C parameter analyzer at 25°C. A constant drain bias voltage of 500 mV was applied during exposure to solutions of different chloride ion concentrations. NaCl solutions were prepared in deionized water and introduced to the sensor surface by a syringe autopipette (0.5-2µL). The pH of the solutions was not controlled with a buffer since NaCl is a salt and would not affect the pH value of the solutions. The drain current measurement begun with the HEMT surface was exposed to deionized water. Before exposure to the NaCl solutions, the sensor was moved to another water sample. Other than the small spike when at 100 seconds, no change of the drain current was detected. The spike in the current was due to mechanical disturbance of the HEMT surface and the current level went back to its original state. By sharp contrast, a rapid response of HEMT drain current was observed in less than 10 seconds when the sensors is exposed to 100nM NaCl at 200 seconds. The abrupt current change stabilized after the sodium chloride solution thoroughly diffused into water and reached a steady state. When the InN gate metal encountered chloride ion, the electrical potential of the gate was changed and resulted in the increase of the pizeo-induced charge density in the HEMT channel. A larger signal change was observed when 1 µM of NaCl solution was applied at 300 seconds. The sensor was exposed to higher Cl⁻ ion concentrations of 10 µM, 100 µM, and 1mM sequentially for a further real time test. The test was repeated with the same sensor for three times to obtain the standard deviation of source-drain current response for each concentration.

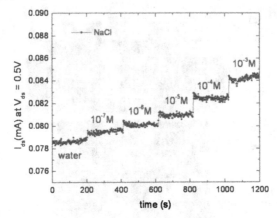

Figure2. Real time source-drain current at a constant bias of 500mV as different concentrations of Cl⁻ ions were added.

Figure 3 shows the drain current change of InN gated HEMT as a function of the Cl⁻ ion concentration solutions prepared with dissolving NaCl in DI water and diluting HCl in DI water to investigate the effect of cations on the sensor by adding HCl to adjust the pH value of the solution. The drain current changes were measured by first dipping the sensor in a glass of deionized water and then transferring it into a glass of chloride solution. The sensor was reused by washing it with deionized water and drying with nitrogen gas. Each concentration was measured three times. There was no degradation observed of the chloride ion sensor after multiple uses. As shown in Figure 3, the measured drain current of the InN gated AlGaN/GaN HEMT in both NaCl and HCl solutions were linearly proportional to the logarithm of chloride concentration, satisfying the Nernst equation. The result from the two solutions exhibited an identical slope with a slight deviation of the y-intercept near the range of the standard deviation. This indicated that the InN gated HEMT has no obvious response to cations. The current increases with increasing Cl⁻ concentration since the InN/AlGaN interface becomes more positive to balance the accumulation of negative charges on the InN surface. It is also worth to note that the pH value of the solutions did not affect the chloride ion concentration measurements. The pH value of the chloride ion solutions prepared with NaCl was kept at 7. However, the pH value of the chloride ion solutions prepared by diluting HCl depended on the concentration of H^+.

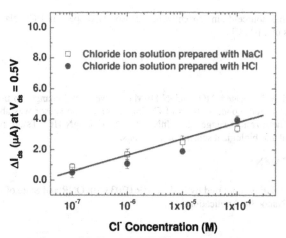

Cl⁻ Concentration (M)

Figure3. Drain current change of InN gated HEMT as a function of Cl⁻ ion concentration of solutions prepared with NaCl and HCl.

The effect of cations on chloride ion concentration sensing was further examined comparing the chloride ion solution with single charged ions, Na^+ and H^+, as well as double charged ions, Mg^{++}. As illustrated in Figure 4, a real time measurement of 1μM Cl⁻ concentration solutions, there was no drain current change when the sensor was moved from an HCl solution to the $MgCl_2$, then to the NaCl solutions. The pH of the 1 μM HCl solution was one order higher than that of the 1 μM NaCl solution and 0.5 μM $MgCl_2$ solutions. This further confirmed that the cations would not affect the Cl⁻ ion concentration measurement.

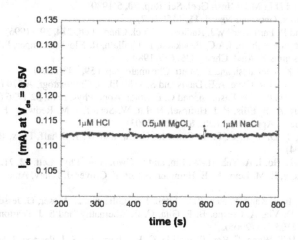

time (s)

Figure 4. Real time source-drain current when sensor is dipped in HCl, NaCl, and MgCl₂ solutions of 1 µM Cl⁻.

CONCLUSIONS

In conclusion, InN gated AlGaN/GaN HEMT showed rapid changes in the drain current when exposed to different concentrations of Cl⁻ ion solution. The sensor was able to detect as low as 100 nM. These results suggest that InN gated AlGaN/GaN HEMTs are promising for a variety of chemical and biological sensing applications.

ACKNOWLEDGMENTS

The work at UF is supported by NSF (DMR 0703340), ONR and State of Florida, Center of Excellence for Nano-Bio Applications.

REFERENCES

1. 1. J. Taylor and S. hong, J. Lab. Med., **31**, 563(2000).
2. H. Shekhar, V. Chathapuram, S. H. Hyun, S. Hong, H. J. Cho, IEEE, **1**, 67 (2003).
3. Clinical Methods; The History, Physical, and Laboratory Examinations, 3ʳᵈ ed, H. K. Walker, W. D. Hall and J. W. Husrt, Butterworths(1990).
4. A. Davidsson, M. Söderström, K.Naidu Sjöswärd, B. Schmekel, Respiration, 74, 184(2007).
5. O. Niimi, L. T. Nguyen, O. Usmani, B. Mann and K. F. Chung, Thorax, 59, 608(2004).
6. R.M. Effros, K.W. Hoagland, M. Bosbous, D. Castillo, B. Foss, M. Dunning, M. Gare, W. Lin, and F. Sun, American J. Resp. & Critical Care Med., 165, 663 (2002).
7. B. Davidsson, K. Naidu Sjöswärd, L. Lundman, B. Schmekel, Respiration, 72, 529(2005).
8. J.M. Cook and D.L. Miles, Inst. Geol. Sci. Rep. **80**, 5(1980).
9. H.N. Elsheimer, Geostand Newsl. **11**, 115(1987).
10. R. Verma and R. Parthasarthy, J. Radioanal. Nucl. Chem. Lett. **214**, 391(1996).
11. T. Graule, A. von Bohlen, J.A.C. Broekaert, E. Grallath, R. Klockenkamper, P. Tschopel and G. Telg, Fresenius Z. Anal. Chem. **335**, 637(1989).
12. S.D. Kumar, K. Venkatesh and B. Maiti, Chromatograpia **59**, 243(2004),.
13. P.A. Blackwell, M.R. Cave, A.E. Davis and S.A. Malik, J. Chromatogr. A **770** (1997), p. 93
14. H. Lu, W. J. Schaff, L. F. Eastman, and C. E. Stutz, Appl. Phys. Lett., **82**,1736 (2003).
15. K. A. Rickert, A. B. Ellis, F. J. Himpsel, H. Lu, W. Schaff, J. M. Redwing, F. Dwikusuma, and T. F. Kuech, Appl. Phys. Lett., **82**, 3254(2003).
16. I. Mahboob, T. D. Veal, C. F. McConville, H. Lu, and W. J. Schaff, Phys. Rev. Lett., **92**, 036804 (2004).
17. Y.-S. Lu, C.-L. Ho, J. A. Yeh, H.-W. Lin, and S. Gwo, Appl. Phys. Lett. **92**, 212102 (2008).
18. G. M. Chertow, E. M. Levy, K. E. Hammermeister, F. Grover, J. Daley, Amer. J. Med., **104**, 343(1998).
19. H.-T. Wang, B. S. Kang, F. Ren, R. C. Fitch, J. K. Gillespie, N. Moser, G. Jessen, T. Jenkins, R. Dettmer, D. Via, A. Crespo, B. P. Gila, C. R. Abernathy, and S. J. Pearton, Appl. Phys. Lett., **87**, 172105-1-3 (2005).
20. B. S. Kang, H.T. Wang, F. Ren, B. P. Gila, C. R. Abernathy, S. J. Pearton, J. W. Johnson, P.

Rajagopal, J. C. Roberts, E. L. Piner, and K. J. Linthicum, Appl. Phys. Lett., 91, 012110 (2007).

21. B. S. Kang, H.T. Wang, F. Ren, S. J. Pearton, T. E. Morey, D. M. Dennis, J. W. Johnson, P. Rajagopal, J. C. Roberts, E. L. Piner, and K. J. Linthicum, Appl. Phys. Lett., 91, 252103 (2007).

22. B. S. Kang, H.T. Wang, T. P. Lele, Y. Tseng, F. Ren, S. J. Pearton, J. W. Johnson, P. Rajagopal, J. C. Roberts, E. L. Piner, and K. J. Linthicum, Appl. Phys. Lett., 91, 112106 (2007).

23. H.T. Wang, B. S. Kang, F. Ren, S. J. Pearton, J. W. Johnson, P. Rajagopal, J. C. Roberts, E. L. Piner, and K. J. Linthicum, Appl. Phys. Lett., 91, 222101 (2007).

24. S. C. Hung, Y. L. Wang, B. Hicks, S. J. Pearton, D. M. Dennis, F. Ren, J. W. Johnson, P. Rajagopal, J. C. Roberts, E. L. Piner, K. J. Linthicum, and G. C. Chi, Appl. Phys. Lett., 92, 193903 (2008).

25. B. H. Chu, B. S. Kang, F. Ren, C. Y. Chang, Y. L. Wang, S. J. Pearton, A. V. Glushakov, D. M. Dennis, J. W. Johnson, P. Rajagopal, J. C. Roberts, E. L. Piner, and K. J. Linthicum, Appl. Phys. Lett., 93, 042114 (2008).

26. Yu-Lin Wang, B. H. Chu, K. H. Chen, C. Y. Chang, T. P. Lele, Y. Tseng, S. J. Pearton, J. Ramage, D. Hooten, A. Dabiran, P. P. Chow, and F. Ren, Appl. Phys. Lett., 93, 262101 (2008).

27. S. C. Hung, B. H. Chu, C. Y. Chang, C. F. Lo, K. H. Chen, Y. L. Wang, S. J. Pearton, Amir Dabiran, P. P. Chow, G. C. Chi, and F. Ren, Appl. Phys. Lett., 94, 043903 (2009).

28. A. Lupu, A. Valsesia, F. Bretagnol, P. Colpo, F. Rossi, Sensors and Actuators B, 127, 606 (2007).

29. C.A. Marquette, A. Degiuli, and L. Blum, Biosensors and Bioelectronics, 19, 433(2003)

30. A. P. Zhang, L. B. Rowland, E. B. Kaminsky, V. Tilak, J. C. Grande, J. Teetsov, A. Vertiatchikh, and L. F. Eastman, J. Electron. Mater., 32, 388(2003).

31. O. Ambacher, M. Eickhoff, G. Steinhoff, M. Hermann, L. Gorgens, V. Werss, B. Baur, M. Stutzmann, R. Neuterger, J. Schalwig, G. Muller, V. Tilak, B. Green, B. Schafft, L. F. Eastman, F. Bernadini, and V. Fiorienbini, Proceedings of the Electrochemical Society (Electrochemical Society, Pennington, NJ, 2002), p. 27.

32. R. Neuberger, G. Muller, O. Ambacher, and M. Stutzmann, Phys. Status Solidi A 185, 85 (2001).

33. J. Schalwig, G. Muller, O. Ambacher, and M. Stutżmann, Phys. Status Solidi A 185, 39 (2001).

34. G. Steinhoff, M. Hermann, W. J. Schaff, L. F. Eastman, M. Stutzmann, and M. Eickhoff, Appl. Phys. Lett., 83, 177(2003).

35. B. S. Kang, S. Kim, F. Ren, J. W. Johnson, R. Therrien, P. Rajagopal, J. Roberts, E. Piner, K. J. Linthicum, S. N. G. Chu, K. Baik, B. P. Gila, C. R. Abernathy, and S. J. Pearton, Appl. Phys. Lett., 85, 2962(2004).

36. S. J. Pearton, B. S. Kang, S. Kim, F. Ren, B. P. Gila, C. R. Abernathy, J. Lin, and S. N. G. Chu, J. Phys.: Condens. Matter 16, R961(2004).

37. G. Steinhoff, B. Baur, G. Wrobel, S. Ingebrandt, A. Offenhauser, A. Dadgar, A. Krost, M. Stutzmann, and M. Eickhoff, Appl. Phys. Lett., 86, 033901(2005).

38. B. S. Kang, F. Ren, L. Wang, C. Lofton, W. Tan, S. J. Pearton, A. Dabiran, A. Osinsky, and P. P. Chow, Appl. Phys. Lett., 87, 023508(2005).

39. S. Gwo, C.-L. Wu, C.-H. Shen, W.-H. Chang, T. M. Hsu, J.-S. Wang, and J.-T. Hsu, Appl. Phys. Lett., 84, 3765(2004).

Mater. Res. Soc. Symp. Proc. Vol. 1202 © 2010 Materials Research Society   1202-I06-02

A Novel Functionalization of AlGaN/GaN-ISFETs for DNA-Sensors

S. Linkohr[1], S. U. Schwarz[2], S. Krischok[3], P. Lorenz[3], T. Nakamura[4], V. Polyakov[1], V. Cimalla[1], C. Nebel[1] and O. Ambacher[1]

[1] Fraunhofer Institute for Applied Solid State Physics, Tullastraße 72, 79108 Freiburg, Germany
[2] Albert-Ludwigs-University of Freiburg, Department of Microsystems Engineering (IMTEK) 79108 Freiburg, Germany
[3] Institute of Micro- and Nanotechnologies, Technical University Ilmenau, P.O. Box 100565, 98684 Ilmenau, Germany
[4] AIST Tsukuba Central 1 Tsukuba, Ibaraki 305-8561, Japan

Abstract

AlGaN/GaN pH sensitive devices were functionalized and passivated for the use as selective bio-sensors. For the passivation, a multilayer of SiO_2 and SiN_x is proposed, which stabilizes the pH-sensor, is biocompatible and has no negative impact on the following bio-functionalization. The functionalization of the GaN-surface was achieved by covalent bonding of 10-amino-dec-1-ene molecules by a photochemical process. After two different surface preparations islands of TFAAD are growing on the sensor surface by exposure with UV-light. In dependence on the surface pre-treatment and the illumination wavelength the first monolayer is completed after 3 h or 7 h exposure time dependent on the pre-treatment and illumination wavelength. Further exposure results in thicker films as a consequence of cross polymerization. The bonding to the sensor surface was analyzed by X-ray photoelectron spectroscopy, while the thickness of the functionalization was determined by atomic force microscopy scratching experiments. These functionalized devices based on the pH-sensitive AlGaN/GaN ISFET will establish a new family of adaptive, selective biomolecular sensors such as selective, reusable DNA sensors.

Introduction

Group III-nitrides are chemically stable semiconductors with high internal spontaneous and piezoelectric polarization. These characteristics allow the fabrication of very sensitive and robust bio-sensors to detect ions in gases and polar liquids, monitor bio-molecules and the bioactivity of cells in solution [1–5]. The strong polarization discontinuity at the AlGaN/GaN interface oriented along the [0001] axis results in a positive polarization charge. This interface charge is compensated by a two-dimensional electron gas (2DEG) located at the GaN side near the surface, which forms the conductive channel of a high-electron-mobility transistor (HEMT). A HEMT without gate metallization realizes an ion-sensitive field effect transistor (ISFET) by direct sensing of charged particles and molecules on the exposed gate area [9–11]. AlGaN/GaN-ISFETs have a high sensitivity and are the subject of intense investigation. They have emerged as attractive candidates for pH- and ion-sensitive sensors or detectors for biochemical processes [12–14]. These pH-sensors are the basis of the AlGaN/GaN-bio-sensors. By functionalization of the open gate, selectivity to specific biomolecules can be achieved such as for DNA molecules or proteins. To bio-functionalize GaN with DNA, the open gate was functionalized with 10-amino-dec-1-ene molecules by a photochemical method [15, 16]. Therefore two different treatments of the sensor surface were used. For the first tests the GaN was terminated with hydrogen by a hydrofluoric acid dip [17] and the second tests were arranged with an oxidized GaN-surface.

In this work, we investigate the functionalization process, the dependence on the illumination wavelength and the properties of the modified sensor surface. Special attention is given to the impact of the passivation, which has to protect the inactive sensor area and the contacts but should be biocompatible and chemically stable as well in the used electrolytes. The samples were analyzed with atomic force microscopy (AFM), X-ray photoelectron spectroscopy (XPS) and the sensor characteristics were monitored by titration experiments.

Experimental Details

The epitaxial growth of AlGaN/GaN heterostructures confining a 2DEG was accomplished by metal-organic chemical vapor deposition (MOCVD). On a thin GaN nucleation layer (25 nm) a 1.8 µm thick GaN layer is deposited to obtain the structural quality required for the 2DEG in the AlGaN/GaN heterostructure. The growth is completed with 22 nm $Al_{0.22}Ga_{0.78}N$ and a cap layer of 3 nm GaN, which enhances the chemical stability of the device. The location of the conductive channel only 25 nm below the surface permits an effective current modulation by manipulation of charges or electric fields at the sensor surface. For comparison of the functionalization process, GaN layers on SiC substrates were grown by MOCVD as well.

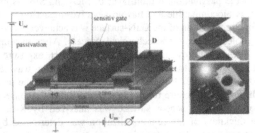

Figure 1: Schematic illustration of the sensor structure and measurement principle (left) and measurement setup (right).

pH-sensor The AlGaN/GaN-ISFETs result from leaving a bare GaN surface by omitting the gate metallization. To analyze the pH-sensor the gate was placed in an appropriate measuring solution. The definition of the potential of the solution is usually carried out by using a reference electrode (Fig. 1). This affects the surface charge by incident ions which enhance or deplete the 2DEG. At low pH, i.e. high concentration of H_3O^+, the surface acts as proton acceptor and causes the enhancement of the 2DEG. At high pH, i.e. high concentration of OH^-, it acts as proton donator and causes depletion of the 2DEG. With an achievable sensitivity (S) of 59 mV/pH these sensors are able to measure the pH-value at the Nernst limit. The sensors have to be appropriately protected with a passivation which has chemical stability in acidic and alkaline liquids. Various passivations such as SiO_2, SiN_x and a SiO_2-SiN_x-doublelayer with a thickness up to 500 nm were investigated. These layers reveal memory effects and ionic diffusion. A multilayer passivation composed of SiO_2(150 nm)-SiN_x(150 nm)-SiO_2(150 nm)-SiN_x(150 nm) shows the best chemical stability in long-term-(Fig. 2a.) and titration-measurements (Fig. 2b.). Table 1 shows the comparison of the various passivations.

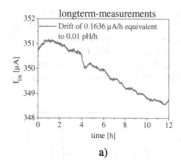
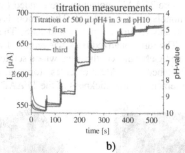

a)   b)

Figure 2: The AlGaN/GaN-ISFET with a multilayer passivation shows the best chemical stability in long-term- and titration-measurements. a) long-term-measurements revealing a drift of 0.02 pH/h. b) repetition accuracy of the titration-measurements.

Table 1: The comparison of the different investigated passivations clarifies the best chemical stability of the multilayer passivation.

Passivation	Thickness [nm]	Drift [μA/h]	Drift [pH/h]
SiN$_x$	500	25.9	1
SiO$_2$-SiN$_x$-doublelayer	500	18	1
SiO$_2$-SiN$_x$-SiO$_2$-SiN$_x$-layer	700	12.6	0.9
SiO$_2$-SiN$_x$-SiO$_2$-SiN$_x$-layer	600	0.2	0.01

Bio-sensors To bio-functionalize GaN with DNA, the open gate region with an active area of 0.04 – 3 mm² was pretreated. For the hydrogen terminated surface the sample was dipped for 5 min in 2% hydrofluoric acid. This method has three benefits: it terminates the surface with hydrogen, the sample is purged from organic contamination and the functionalization is faster. For the oxidized surface the samples were oxidized in air. The 10-amino-dec-1-ene molecules, which are used for functionalization, are passivated at one side with a trifluoroacetic acid group (CH$_2$=CH(CH$_2$)$_8$-NHCOCF$_3$) (TFAAD). To obtain covalent bonding of the olefin molecules, a photochemical method is used. The GaN-surface was illuminated with a UV lamp for different times in a nitrogen filled chamber after wetting with TFAAD. The illumination was realized in two different ways. On the one hand the samples were irradiated with the whole spectrum of the UV-lamp. On the other hand the spectrum was filtered. The filter limited the wavelength and let pass 200 to 400 nm of the UV-spectrum. After illumination, the samples were cleaned in chloroform, isopropanol and DI-water for 3 min in an ultrasonic bath. Then the samples were analyzed with AFM and XPS.

Results and discussion

By using UV-radiation hydrogen atoms were removed from the surface. The result is an electrostatic interaction with the electron rich double bonds of TFAAD, which results in the formation of a C-Ga-bond. The resulting carbocation is an unstable intermediate, and is saturated by hydrogen from a bordering Ga-atom [18]. By the binding of only one TFAAD-molecule-layer with a length of about 15 Å [19] a monolayer of about 1.5 nm thickness is formed. After different exposure times the samples were analyzed with AFM and XPS. The measurements showed that the first amines are already linked at the GaN-surface after only 30 min UV-radiation which is seen in the AFM-picture in figure 3a. By extending the exposure time, more TFAAD molecules are bonded forming islands with increasing diameter until the first monolayer is completed. These monolayers were further analyzed by AFM-scratching-experiments (figure 3b and c) and XPS. The scratching experiments were made in contact-mode. The functionalized film was removed with a force of 100 to 200 nN. Figure 3b shows a scratched GaN-surface, the area, where scratching removed the monolayer appears dark while the detached TFAAD-molecules form a bright border. These experiments allow the measurements of the thickness of the TFAAD-layer.

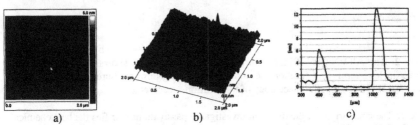

a)   b)   c)

Figure 3: AFM images of the GaN-surface a) the first amines are already linked after 30 min. After scratching: b) 3D morphology, c) line scan across the centre revealing the thickness of the TFAAD-layer on the GaN-surface.

The bonding of the TFAAD-molecules is illustrated in the XPS results in figure 4a by comparing the fluor- and the carbon-peak before (black) and after (red) the functionalization. The HF cleaned sample exhibits typical structures for GaN with some minor F and C adsorbates as illustrated in figure 4a. The XPS results indicate a successful functionalization with TFAAD-molecules. For compensation of charging effects the spectra were shifted to a reference binding energy of the Ga $2p_{3/2}$ state to 1118.6 eV. Most evident is the appearance of a new F 1s component at 689 eV binding energy, which is caused by the CF_3 group of the TFAAD [15]. In parallel the expected changes in the C1s emission, namely a strong rise at about 285.5 eV caused by the alkyl chain is observed. The carbon atoms of the TFAAD in a different chemical environment cause the additional intensity at higher binding energies.

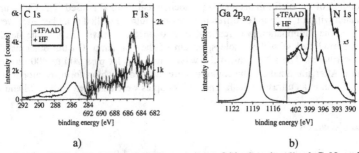

a)   b)

Figure 4: The XPS (mon. Al Kα) measurements of bio-functionalized GaN surfaces (a) comparison of the adsorbate-induced F 1s and C 1s core levels emission (b) Ga $2p_{2/3}$ and N 1s spectra. The spectra were shifted to a reference binding energy of the Ga $2p_{3/2}$ state of 1118.6 eV for compensation of charging effects.

Moreover, the functionalization procedure leads only to minor changes in GaN substrate related states, the core levels of the N 1s and the Ga 2p$_{3/2}$ are shown in figure 4b. The Ga 2p$_{3/2}$ core level exhibits no significant changes, while the N 1s spectrum a small new component at 400.5 eV is observed. This state is most likely related to the NH component of the TFAAD-molecule. Note, that the indicated binding energies are with respect to the Fermi level of the sample holder and might be affected by a slight surface charging.

Due to the fact that the illumination was realized in different ways we have to compare the different effects on the functionalization. On the one hand the samples were irradiated with the whole spectrum of the UV-lamp. The result is shown in figure 5a, it is clearly visible that the TFAAD-molecules are linked faster then with filtered illumination. Because of the fast pre-cross-polymerization caused by the visible light there are no reproducible measurements. An influence of the HF- or oxygen-treatment cannot be detected. On the other hand the spectrum was filtered. The filter let pass the UV-spectrum between 200 and 400 nm (figure 5b). There is an influence of the pre-treatment of the surface shown in figure 5b. The monolayer on the H-terminated surface is achieved faster (blue line) than on the oxidized surface (red line). The measurements of a 2 nm layer thickness by achieving the first monolayer results because of the roughness of the surface. By prolongation of the exposure, more TFAAD-layers are linked to the GaN-surface in consequence of polymerization.

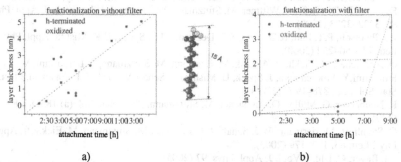

a)   b)

Figure 5: The comparison of the dependency of the different photochemical exposure and the different pre-treatments of the GaN-surface: the illumination a) without filter b) with filter.

Conclusion

The presented results demonstrate that the H-terminated and the oxidized GaN-surface can be functionalized by olefin-molecules by using a photochemical exposure. It was demonstrated that the resulting thickness of the TFAAD-layer is dependent on the time and the wavelength of the exposure. The resulting surface is the starting point for the functionalization of the pH-sensors. Such GaN/AlGaN/GaN-devices are promising bio-sensors because of the particularly good bonding stability of bio-molecules to GaN. This could be used for very robust and chemical stabile bio-sensors and DNA-sensors.

Acknowledgements

The authors would like to acknowledge the assistance of Nadine Geldmacher, Gudrun Kaufel, Mechthild Korobka, Brigitte Wieber, Arnulf Leuther and Brian Raynor. This work was supported by the Fraunhofer Attract program and the Fraunhofer "Doktorandinnenprogramm: Mehr Frauen in die angewandte Forschung". The authors also would like to thank the help of the ZMN in Ilmenau which is supported by the BMBF Optimi.

References

1. O. Ambacher, J. Phys. D: Appl. Phys. 31, 2653–2710 (1998).
2. S.J. Pearton, J.C. Zolper, R.J. Shul, F. Ren, J. Appl. Phys. 86, 1–78 (1999).
3. S.C. Jain, M. Willander, J. Narayan, R. Van Overstraeten, J. Appl. Phys. 87, 965–1006 (2000).
4. M. Stutzmann, G. Steinhoff, M. Eickhoff, O. Ambacher, C.E. Nebel, J. Schalwig, R. Neuberger, G. Müller, Diamond Relat. Mater. 11, 886–891 (2002).
5. M. Eickhoff, J. Schalwig, G. Steinhoff, O. Weidemann, L. Görgens, R. Neuberger, M. Hermann, B. Baur, G.Müller, O. Ambacher, M. Stutzmann, Phys. Stat. Sol. (c), 1908–1918 (2003).
6. O. Ambacher, J. Smart, J.R. Shealy, N.G. Weimann, K. Chu, M. Murphy, W.J. Schaff, L.F. Eastman, R. Dimitrov, L. Wittmer, M. Stutzmann, W. Rieger, J. Hilsenbeck, J. Appl. Phys. 85, 3222–3233 (1999).
7. J.P. Ibbetson, P.T. Fini, K.D. Ness, S.P. DenBaars, J.S. Speck, U.K. Mishra, Appl. Phys. Lett. 77, 250–252 (2000).
8. O. Ambacher, M. Eickhoff, A. Link, M. Hermann, M. Stutzmann, F. Bernardini, V. Fiorentini,Y. Smorchkova, J. Speck, U. Mishra,W. Schaff,V. Tilak, L.F. Eastman, Phys. Stat. Sol. (c), 1878–1907 (2003).
9. R. Neuberger, G. Müller, O. Ambacher, M. Stutzmann, Phys. Stat. Sol. (a) 185, 85–89 (2001).
10. G. Steinhoff, M. Hermann, W.J. Schaff, L.F. Eastman, M. Stutzmann, M. Eickhoff, Appl. Phys. Lett. 83, 177–179 (2003).
11. M. Bayer, C. Uhl, P.Vogl, J. Appl. Phys. 97 (2005).
12. B.S. Kang, F. Ren, L. Wang, C. Lofton, W. Tan Weihong, S.J. Pearton, A.Dabiran, A. Osinsky, P.P. Chow, Appl.Phys. Lett. 87 (2005).
13. G. Steinhoff, B. Baur, G. Wrobel, S. Ingebrandt, A. Offenhäusser, Appl.Phys. Lett. 86 (2005).
14. Y. Alifragis, A. Georgakilas, G. Konstantinidis, E. Iliopoulos, A. Kostopoulos, N.A. Chaniotakis, Appl. Phys. Lett. 87, (2005).
15. H. Kim, P.E. Colavita, K.M. Metz, B.M. Nichols, B. Sun, J. Uhlrich, X. Wang, T.F. Kuech, R.J. Hamers, LANGMUIR 22 (19), 8121-8126 (2006)
16. T. Strother, R.J. Hamers, L.M. Smith, Nucleic Acids Research, 28, 3535–3541 (2000).
17. S.W. King, J.P. Barnak, M.D. Bremser, K.M. Tracy, C. Ronning, R.F. Davis, R.J. Nemanich., J. Appl. Phys., 84, 5248–5260 (1998).
18. C.L. Hu, J.Q. Li, Y. Chen, W.F. Wang, Journal of Physical Chemistry C, 112(43), 16932–16937 (2008).
19. N. Yang, H. Uetsuka, H. Watanabe, T. Nakamura, C.E. Nebel Chemistry of Materials 19 (2007) 2852-2859

Mater. Res. Soc. Symp. Proc. Vol. 1202 © 2010 Materials Research Society   1202-I09-09

Cheng-Yi Lin[1], Yen-Sheng Lu[1], Shih-Kang Peng[2], Shangjr Gwo[3] and J. Andrew Yeh[4]
[1] Institute of Electronics Engineering, National Tsing-Hua University, Hsinchu 30013, Taiwan
[2] Department of Power Mechanical Engineering, National Tsing-Hua University, Hsinchu
30013, Taiwan
[3] Department of Physics, National Tsing-Hua University, Hsinchu 30013, Taiwan
[4] Institute of Nanoengineering and Microsystems, National Tsing-Hua University, Hsinchu
30013, Taiwan

ABSTRACT

Ultrathin (~10 nm) InN ion sensitive field effect transistors (ISFETs) are functionalized by immobilized label-free oligonucleotide probes with 3-mercaptopropyltrimethoxysilane (MPTMS) through molecular vapor deposition (MVD) technique. This layer on the InN surface serves the function of selectively detecting the hybridization of complementary deoxyribonucleic acid (DNA). Using MVD technique to perform the gas-phase silanization of MPTMS provided a time-saving and simple method to reach 68° water contact angle after 1.5 h treatment. High resolution X-ray photoelectron spectroscopy (HRXPS) was employed to analyze the surface characteristics after functionalization. Modified probes DNA were covalently bonded to MPTMS-covered gate surface of InN ISFETs. And further hybridized with complementary DNA For a 12-mer oligonucleotide probe, a significant drain-source current decrease (~ 6 µA) was observed for the hybridization with complementary DNA solution of 100 nM. In contrast, the noncomplementary DNA with single-base mismatch did not show obvious current changes. Functionalized ultrathin InN ISFETs for DNA sequence detection demonstrate the promise of biological sensing and genetic diagnosis applications.

INTRODUCTION

Deoxyribonucleic acid (DNA) sequence can naturally hybridize with its complementary sequence due to the specific base-pairing rule. This useful property renders DNA probes can be used to detect genetic disease. There have been a variety of methods adopted to detect the DNA hybridization [1-4]. However, these approaches, normally, require prelabeling steps, time-consuming optical measurements, or expensive instrumentation. To pursue label-free, rapid and accurate electrical detection, sensors based on semiconductor materials employing field effect transistor structure have been widely studied [5-7]. In particular, ultrathin InN ion sensitive field effect transistors (ISFETs) have attracted a lot of attention to be biosensors due to high sensitivity, biocompatibility and robust surface stabilities [8]. This purpose, however, requires a proper functionalization for obtaining the desired molecular recognition ability. To date, gas phase preparations have shown can provide better surface modification quality [9]. Molecular vapor deposition (MVD) is one kind of novel and effective coating technology to modify the surface due to a superior capacity of precisely controlling the process conditions [10].

In this work, we functionalize ultrathin InN ISFETs using label-free single strand DNA (ss-DNA) probes which are immobilized by 3-mercaptopropyltrimethoxysilane (MPTMS) molecules through MVD technique. High resolution X-ray photoelectron spectroscopy (HRXPS) and contact angle measurements are adopted to characterize the surface properties. Electrical measurement is performed to investigate the hybridization with complementary DNA targets.

EXPERIMENT

Materials

Adopted InN film as the sensing element was heteroepitaxially grown on a Si(111) substrate via a molecular-beam epitaxy (MBE) system. The InN/AlN film is composed of an ultrathin (10 nm) InN epilayer and a thick (190 nm) AlN buffer layer. Fabrication of InN ISFETs was started from evaporation of Au/Ti (150/80 nm) for obtaining Ohmic contact. Using lift-off process defined the channel length and contact pads of ISFETs. Mesa isolation was performed by inductively coupled plasma (ICP) to define the channel width of ISFETs. Eventually, polyimide was patterned to define the ISFET contact (gate) windows. After microfabrication process, wafer was diced and then ISFETs were bonded onto printed circuit boards using aluminum wires, followed by packing with polydimethylsiloxane (PDMS). The detailed growth process of InN film and fabrication of InN ISFETs can be found elsewhere [8, 11].

The organosilane MPTMS was purchased from Sigma-Aldrich. The adopted oligonucleotides were purchased from MDBio, Inc. The probe DNA modified at the 5' end is acrylic phosphoramidite-(CH$_2$)$_6$-5'-CCT AAT AAC AAT-3'. The 12-mer complementary DNA and noncomplementary DNA are 5'-ATT GTT ATT AGG-3' and 5'-ATT GTT ACT AGG-3', respectively. All water used was deionized (DI) water with a resistivity 18 MΩ.

Functionalization of InN ISFETs

Prior to O$_2$ plasma treatment, UV adhesives (Letonnd 5802) were used to encapsulate the fabricated InN ISFETs except for the open gate region. The O$_2$ plasma with 10 min treatment was performed to clean the samples surface and form a hydroxyl-terminated surface as well. Gas-phase silanization of MPTMS using MVD system was conducted for 1.5 h subsequently. MVD system purged MPTMS gas for continual three 0.5 h cycles. Each cycle comprised eight times purges which were set at 2 torr for each purge. MPTMS molecules were covalently bonded to the gate surface of InN ISFETs and left thiol (-SH) terminal groups to further immobilize probe DNA. MPTMS-covered ISFETs were dipped into 10 µM solution of probe DNA for 12 h. All DNA solutions were prepared by DI water. Furthermore, the experimental samples were rinsed carefully with DI water. After immobilization of probe DNA, the functionalized InN ISFETs were employed to detect the hybridization with complementary ss-DNA.

(a)   (b)   (c)

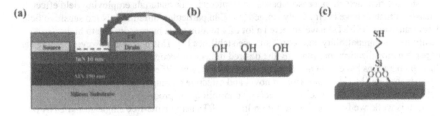

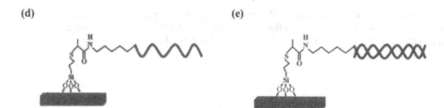

(d) **(e)**

Figure 1. Modification scheme of functionalizing ultrathin InN ISFETs for the detection of DNA hybridization: (a) Fabricated ultrathin InN ISFET; (b) Hydroxylation of InN surface; (c) modification using gase-phase MPTMS through MVD technique; (d) immobilization of probe ss-DNA; (e) hybridization with complementary ss-DNA.

<u>**Instrumentation**</u>

The gas-phase modification on the gate region of ultrathin InN ISFETs was carried out by MVD system (Applied Microstructures, MVD 100). Contact angle goniometer (Sindatek, Model 100SB) was used to measure the water contact angle at room temperature for investigating the surface properties. High resolution X-ray photoelectron spectrometer (HRXPS) was used to analyze the surface characteristics on the open-gate region of InN ISFETs with approximate 60 μm channel length. HRXPS (ULVAC-PHI, PHI Quantera SXM) was equipped with monochromatic 1486.6 eV X-ray source from the Al Kα line. The incident angle was 45° measured from the sample surface and the electron pass energy was 55 eV. Electrical measurement was conducted by a DSP Lock-in amplifier (Stanford Research System, SR830). A 200 mV Vpp at 1k Hz voltage source (V_{SD}) was applied via a function generator (Agilent, 33250A) to InN ISFET. One standard resistor of 47.7 Ω was in series with InN ISFET because of the consideration of impedance match for better measurement results. Reference electrode HgCl/Hg$_2$Cl$_2$ (Hanna, HI-5412) was kept at a zero potential during the real-time measurement.

DISCUSSION

<u>**Surface Characterization of Functionalized ISFETs**</u>

The extent of MPTMS silanization can be investigated by water contact angle. The change of water contact angle on InN surface was observed from 3°, indicating the O$_2$ plasma cleaning, to 68° after 1.5 h reaction with MPTMS vapor, which is closed to the reported saturation contact angle (~70°) of MPTMS silanization [12]. Moreover, we adopted HRXPS to confirm the surface characteristics of functionalized InN ISFETs. Two different samples were prepared by the following conditions. One was used as a control sample without any modification (Without Modification). The other sample was a complete functionalied ISFET through MPTMS modification and probe DNA immobilization (With Probe DNA + MPTMS). XPS spectrum demonstrates that samples With Probe DNA + MPTMS show the S 2p peak at 162.4 eV (Figure 2a). This result reveals the presence of sulfur (S) from MPTMS linkers on InN surface and S 2p band indicate other oxidized state of S species. In contrast, no S 2p peak is observed for the sample Without Modification. To investigate the probe DNA immobilization, the spectrum of P atoms was measured (Figure 2b) due to DNA molecules possessing phosphate functional group. The clear P 2*p* peak was only observed from sample (With Probe DNA+MPTMS), which supports the confirmation of probe DNA being immobilized by silane coupling agent successfully.

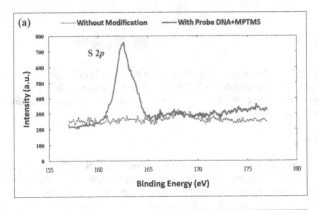

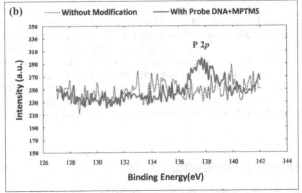

Figure 2. HRXPS spectra of (a) S 2*p* region, and (b) P 2*p* region analyzed on the gate regions of InN ISFETs based on different treatments of without modification (dotted line) and functionalization with probe DNA and MPTMS (solid line).

Electrical Detection of DNA Hybridization

Measuring the dynamic current response of ISFETs can reveal the effect of bio-conjugation such as DNA hybridization due to the charge-based sensing. To minimize the effect of counter ions screening, DI water was used to prepare the DNA solutions for obtaining higher sensitivities. During the sensing, only the gate region of InN ISFET has to be exposed to the aqueous solution. Thus, a complete encapsulation is particularly important to provide good electric insulation as well as high resistance to water penetration. Figure 3 displays the time dependence of the current response for functionalized InN ISFET on exposure to DNA solutions at approximate 100 s. For control experiment, there was no obvious current change for noncomplementary DNA solution of 100 nM (Figure 3a). On the other hand, a complementary target DNA solution of 100 nM resulted in a significant drain-source current decrease of approximate 6 μA, which current variation was 0.74% (Figure 3b). This result suggested that the attachment of negatively charged complementary DNA with the immobilized probe DNA caused the depletion of the carrier concentration in the ultrathin InN surface so that drain current decreased.

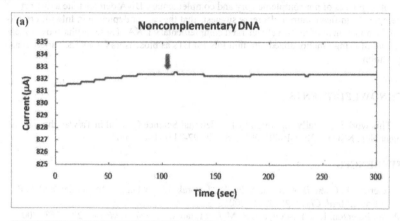

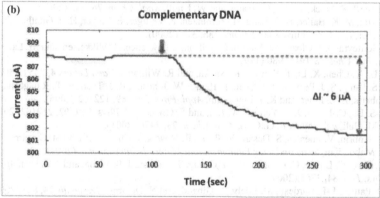

Figure 3. Real-time current responses of functionalized ultrathin InN ISFETs. (a) Response to noncomplementary DNA; (b) Response to complementary target DNA, where arrow represents the injection of DNA solution at ~100 sec to InN ISFETs in DI water.

CONCLUSIONS

Label-free and sequence-specific DNA sensors have accomplished using functionalized ultrathin InN ISFETs. We demonstrate the functionalization on InN surface by immobilizing probes DNA with gas-phase silanization of MPTMS through MVD technique. MVD technique provides a simple, time-saving and effective approach to surface modification with batch processing. Functionalized InN ISFETs with covalently immobilized ss-DNA probes could observe the hybridization with complementary DNA through electrical detection. Real-time current responses of noncomplementary and complementary DNA demonstrate a distinct difference. Significant current decrease suggests that the carrier depletion in InN film results from the attachment of negatively charged complementary DNA. These results show a great potential of using functionalized ultrathin InN ISFETs as biosensors for a wide variety sensing application.

ACKNOWLEDGMENTS

This work is partially supported by the National Science Council in Taiwan through the project Nos. NSC 98-3011-P-007-001 and NSC 97-3114-E-007-002.

REFERENCES

1. D. Gerion, F. Chen, B. Kannan, A. Fu, W. J. Parak, D. J. Chen, A. Majumdar, and A. P. Alivisatos, *Anal. Chem.* **75**, 4766 (2003).
2. A. W. Peterson, R. J. Heaton, and R. M. Georgiadis, *Nucleic Acids Res.* **29**, 5163 (2001).
3. X. Su, R. Robelek, Y. Wu, G. Wang, and W. Knoll, *Anal. Chem.* **76**, 489 (2004).
4. J. Fritz, M. K. Baller, H. P. Lang, H. Rothuizen, P. Vettiger, E. Meyer, H.-J. Güntherodt, Ch. Gerber, and J. K. Gimzewski, *Science* **288**, 316 (2000).
5. E. Souteyrand, J. Cloarec, J. Martin, C. Wilson, I. Lawrence, S. Mikkelsen, and M. Lawrence, *J. Phys. Chem. B* **101**, 2980 (1997).
6. Z. Li, Y. Chen, X. Li, T. Kamins, K. Nauka, and R. Williams, *Nano Letters* **4**, 245 (2004).
7. B. S. Kang, S. J. Pearton, J. J. Chen, F. Ren, J. W. Johnson, R. J. Therrien, P. Rajagopal, J. C. Roberts, E. L. Piner, and K. J. Linthicum, *Appl. Phys. Lett.* **89**, 122102 (2006).
8. Y.-S. Lu, C.-L. Ho, J. A. Yeh, H.-W. Lin, and S. Gwo, Appl. Phys. Lett. **92**, 212102 (2008).
9. C. M. Halliwell and A. E. G. Cass, *Anal. Chem.* **73**, 2476 (2001).
10. B. Kobrin, V. Fuentes, S. Dasaradhi, R. Yi, R. Nowak, J. Chinn, R. Ashurst, C. Carraro, and R. Maboudian, *Semiconductor Equipment and Materials International* (2004).
11. S. Gwo, C. L. Wu, C. H. Shen, W. H. Chang, T. M. Hsu, J. S. Wang, and J. T. Hsu, *Appl. Phys. Lett.* **84**, 3765 (2004).
12. B. Pattier, J.-F. Bardeau, M. Edely, A. Gibaud, and N. Delorme, *Langmuir* **24**, 821 (2008).

Mater. Res. Soc. Symp. Proc. Vol. 1202 © 2010 Materials Research Society   1202-I09-07

Formation of Antibody Microarrays on Aluminum Nitride Surfaces Using Patterned Organosilane Self-Assembled Monolayers

Chi-Shun Chiu, Hong-Mao Lee, and Shangjr Gwo
Department of Physics, National Tsing-Hua University, Hsinchu 30013, Taiwan

ABSTRACT

Surface biofunctionalization of group-III nitride semiconductors has recently attracted much interest due to their biocompatibility, nontoxicity, and long-term chemical stability under demanding physiochemical conditions for chemical and biological sensing. Among III-nitrides, aluminum nitride (AlN) and aluminum gallium nitride (AlGaN) are particularly important because they are often used as the sensing surfaces for sensors based on field-effect transistor or surface acoustic wave sensor structures. Patterned self-assembled monolayer (SAM) templates are composed of two types of organosilane molecules terminated with different functional groups (amino and methyl), which were fabricated on AlN/sapphire substrates by combining photolithography, lift-off process, and self-assembly technique. Clear imaging contrast of SAM micropatterns can be observed by field emission scanning electron microscopy (FE-SEM) operating at a low accelerating voltage in the range of 0.5–1.5 kV. In this work, the formation of green fluorescent protein (GFP) antibody microarrays was demonstrated by the specific protein binding of enhanced GFP (EGFP) labeling. The observed strong fluorescent signal from antibody functionalized regions on the SAM-patterned AlN surface indicates the retained biological activity of specific molecular recognition resulting from the antibody–EGFP interaction. The results reported here show that micropatterning of organosilane SAMs by the combination of photolithographic process and lift-off technique is a practical approach for the fabrication of reaction regions on AlN-based bioanalytical microdevices.

INTRODUCTION

Group-III nitride semiconductors (AlN, GaN, InN) are attracting much attention because of the increasing requirements of stability, miniaturization and integration in bio-microelectromechanical systems (Bio-MEMS) and micro-bioanalytical device [1]. In addition, surface biochemical functionalization of group-III nitride semiconductors has also attracted much recent interest due to their biocompatibility, nontoxicity, and long-term chemical stability for biochemical detection [2,3]. Biomolecules have sophisticated functions such as catalysis and molecular recognition. To combine these functions with nitride-based microdevices, selective immobilization of biomolecules onto the device surface is required. The design and microfabrication of functional molecular surfaces are the key steps in the development of such nitride-based microdevices. At present, organosilane self-assembled monolayers (SAMs) have been intensely used as the cross-linkers to immobilize capture or probe molecules onto group-III nitride surfaces [4-9].

Organosilane SAMs have been widely used to the creation and regulation of surface functionalities due to their excellent and diverse capability in surface functionalization [10-13]. These functionalities can play important roles in the promotion or suppression of biomolecular adhesion [14,15]. Moreover, fabrication of patterned microstructures composed of organosilane SAMs terminated with different functional groups in specific regions is a critical technique for the immobilization of biomolecules on selective regions of device surface. Several methods have been developed for the fabrication of patterned SAMs. These include photolithography, photochemical processes, ultraviolet irradiation, microcontact printing, and microchannel-flowed plasma process [16,17]. However, it is essential to develop a general process to immobilize the biomolecules on nitride-based device surfaces without losing their activity during the immobilization process. In general, surface patterning technique requires that patterned proteins maintain their biological activity. Among possible surface labeling molecules, GFP is particularly suitable for testing the surface immobilization schemes.

Besides demonstrating the surface functionalization technique for the AlN film, we also illustrate methods of direct microscopic imaging for patterned SAMs on AlN. In particular, scanning electron microscopy (SEM) has been successfully used to characterize patterned alkanethiolate SAMs on gold substrates [18-21] and organosilane SAMs on silicon [17,22]. Furthermore, SEM imaging of proteins adsorbed on gold surfaces modified with patterned SAMs has also been reported [23,24]. Previous results, however, focused on conductive or semi-conductive substrates (e.g., gold and silicon). Patterned SAMs on insulting materials are scarcely studied by SEM because of the difficulty due to the sample charging.

In this article, we report the fabrication of microstructures composed of two types of organosilane SAMs terminated with different functional groups (amino- and methyl- terminated) on AlN/sapphire films. These two-dimensional microstructures of functional SAMs were fabricated by using the combination of photolithographic process and lift-off technique. The image contrasts resulting from the different terminal functional groups of patterned SAMs were clearly observed by FE-SEM at a low voltage. Finally, we show the process for site-directed covalent immobilization of antibodies on AlN surface using the patterned silane self-assembled monolayers as cross-linkers to GFP antibodies, which function as capture molecules for the detection of enhanced GFP (EGFP) proteins.

EXPERIMENT

Aminopropyltrimethoxysilane (APTMS) and octadecyltrichlorosilane (OTS) were purchased from Aldrich. Aqueous glutaraldehyde solution (25%) was purchased from Alfa Aesar Company. Recombinant EGFP (1 mg/ml) and affinity purified rabbit anti-GFP polyclonal antibody (0.2 mg/ml) in phosphate buffered saline (PBS) were purchased from BioVision. AlN epitaxial films with a thickness of ~200 nm were grown on c-plane sapphire substrates by plasma-assisted molecular beam epitaxy (PA-MBE).

After standard wafer cleaning, positive photoresist patterns were prepared by photolithography process. Surface modification of the samples with photoresist micropatterns was performed by air plasma. Next, the AlN surface was covered with a hydrophobic OTS SAM by liquid deposition. Freshly prepared 0.1% (v/v) OTS solutions in toluene were used for the monolayer formation. AlN samples were dipped in the OTS solution for 120 s at room temperature. Cleaned samples were then amino-functionalized by an APTMS solution with the original concentration as received. The silanization proceeded in an APTMS solution for 2 h at room temperature, and then the regions previously masked by the photoresist were replaced with a hydrophilic APTMS SAM. Therefore, APTMS and OTS molecules were selectively immobilized onto the specific regions, and finally two-dimensional microstructures were formed. Figure 1 shows the process of fabricating GFP antibody microarrays on AlN surfaces.

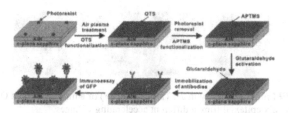

Figure 1. Schematic representation of the fabrication process of two-dimensional GFP antibody micropatterns on AlN surfaces by using patterned organosilane SAM templates.

High resolution FE-SEM was used to characterize microstructures composed of organosilane SAMs terminated with –NH$_2$ and –CH$_3$ functional groups. FE-SEM has two kinds of secondary electron (SE) detectors at different positions. These detectors include a "through the lens" SE-detector (TTL SE-detector, directly above the sample in the electron lens) and an Everhart-Thornley SE-detector (ET SE-detector, off-normal position in chamber). The APTMS/OTS patterns were observed by FE-SEM using an accelerating voltage ranging from 0.5 to 5.0 kV. All the samples were observed and imaged without a conductive coating layer.

GFP antibodies were immobilized on the surface of AlN/sapphire substrate using the process shown in figure 1. First, the APTMS/OTS silanized AlN/sapphire samples were cured in oven at 80°C for 2 h. Next, the samples were modified by incubating in 2.5% (v/v) glutaraldehyde solution in PBS for 2 h at room temperature. Finally, GFP antibodies were adsorbed onto the aldehyde-reactive modified substrates by reaction with GFP antibodies (20 µg/ml) for 4 h at room temperature. GFP polyclonal antibodies were labeled by the EGFP and allowed us to identify the immobilization of GFP antibodies onto the aldehyde-modified surface. Images were acquired using a fluorescence microscope equipped with fluorescence detection optics.

DISCUSSION

We applied low voltage FE-SEM (< 5.0 kV) to visualize micropatterns of organosilane SAM. The patterned OTS SAMs formed on AlN surfaces were observed by low voltage FE-SEM using a TTL SE-detector. Moreover, the accelerating voltages ranging from 0.5 kV to 5.0 kV was varied to investigate the contrast dependence of operating voltages for patterned organosilane SAM formed on insulating AlN surfaces (figure 2). At the voltage of 0.5–1.5 kV, there are clearly better contrasts in the SEM images. In addition, the height difference between OTS SAM and bare AlN surfaces can be distinguished (figure 2a, 2b). In contrast, the image drift and charging at the edge occurred with an increase of operating voltages. Above the voltage of ~4 kV, the SEM image contrast becomes less distinctive and allows only limited perception of the third dimension (figure 2c, 2d). It becomes faint and difficult to image at the voltage above 5.0 kV due to the effect of electron charging. It is found that OTS-covered region has a higher emission efficiency of secondary electrons than the AlN-exposed regions. Consequently, these results suggest that high image contrast due to the SAM micropatterning could be obtained at a low voltage (below 1.5 kV).

Figure 2. FE-SEM micrographs (with a TTL SE-detector) of a micropatterned OTS SAM fabricated on an AlN epitaxial film at different accelerating voltages.

A patterned SAM composed of different regions formed by the adsorption of APTMS and OTS onto an AlN surface was characterized by FE-SEM using an ET SE-detector (figure 3b). An OTS-modified pattern on an AlN surface was observed by FE-SEM operating at 0.5 kV (ET SE-detector) after the photoresist mask had been completely removed to expose the AlN regions (figure 3a). In the second step, the formation of APTMS/OTS two dimensional microscale patterns on AlN surface were observed by FE-SEM at 0.5 kV. The acquired SEM image shows a remarkable contrast between the amino-terminated APTMS regions and the methyl-terminated OTS regions, suggesting different chemical identities of surface groups on different regions. In comparison, the brighter regions correspond to the hydrophilic amino groups and the darker regions correspond to the hydrophobic methyl groups (figure 3b). The area that corresponds to APTMS appears uniformly brighter than that before exposure to the APTMS solution (i.e., AlN regions). These results indicate that FE-SEM could provide surface information to distinguish the amine- (APTMS) and methyl- (OTS) modified regions of the micropatterned SAMs with only a monolayer thickness. The relative contrasts of FE-SEM imaging are dependent on the numbers of collected secondary electrons by the detector. At low beam energies, these quantities are mainly related to the yield of secondary electrons emitting from the sample surface, detector collection efficiency (e.g., a TTL SE-detector and an ET SE-detector), surface morphology, and surface chemistry. Therefore, there are differences between micrographs recorded by TTL and ET SE-detectors for the same sample (reversed contrast between figure 2a and figure 3a) [25].

Figure 3. FE-SEM imaging (with an ET SE-detector) for micropatterns with different compositions during the immobilization process of antibody microarray.

GFP antibodies were immobilized onto micropatterned SAMs formed on AlN films. These antibody microarrays were first characterized by FE-SEM imaging (figure 3c). The image of antibody/OTS micropatterns shows a reversed contrast with that of APTMS/OTS SAMs. The image of APTMS-modified regions apparently changes from bright to dark after it had been exposed to GFP antibody solution (figures 3b and 3c). This change can be mainly attributed to the different surface properties between the GFP antibody-covered and APTMS-covered regions. These differences in surface properties include composition, surface charge, etc. The existence of APTMS SAMs on AlN films was confirmed by X-ray photoelectron spectroscopy (XPS). After surface functionalization by APTMS, AlN epitaxial films show additional N 1s characteristics originating from amine terminal groups. In figure 4, the XPS spectrum of an APTMS-functionalized AlN surface is shown to exhibit two N 1s peaks centered at 401.5 and 399.6 eV, indicating the presence of two types of amine terminal groups on the sample surface: the peak at 399.6 eV is assigned to be the uncharged amine groups and that at 401.5 eV is related to the positively charged (protonated) ones (NH_3^+).

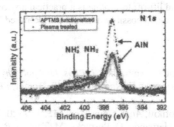

Figure 4. XPS spectra of N 1s core-level emissions for plasma-treated AlN surface and APTMS-functionalized AlN surface.

The capability to selectively biofunctionalize the AlN surface was examined by fluorescence measurement. The observed strong fluorescent signal from EGFP conjugated with antibody functionalized regions on the SAM-patterned AlN surface indicates the retained biological activity of specific molecular recognition resulting from the antibody–EGFP interaction. To evaluate the binding efficiency of GFP antibodies on aldehyde-reactive modified substrates, EGFP was used as a reporter to label immobilized GFP antibodies. GFP antibody solution (20 µg/ml) was employed for the patterning of antibody microarrays, and then different concentrations (0.1–100 µg/ml) of EGFP solution was incubated with patterned GFP antibodies for 2 h. The fluorescent signals are detectable when the EGFP concentration is above the level of 1 µg/ml. The fluorescence imaging results show that EGFP microarray (3-µm-diameter circle) was formed on AlN surface by immunoaffinity with antibody microarray (figure 5).

Figure 5. A fluorescence optical image of an EGFP microarray formed on an AlN surface resulting from the molecular recognition between EGFP (10 µg/ml solution) and GFP antibody.

CONCLUSIONS

Low voltage FE-SEM was employed here to characterize the surface chemistry of APTMS/OTS organosilane SAMs. It is shown that FE-SEM operating at low voltage can be used as a powerful tool to visualize ultrathin SAM patterns on insulating solid substrates. We demonstrate that low voltage SEM can be used as an analytical inspection tool for rapid evaluation of SAM patterns and antibody microarrays. The capability of visualizing different regions in patterned SAMs is not only useful for AlN films but also suitable for other insulating materials. Furthermore, the formation of GFP antibody microarrays was confirmed by the specific molecular recognition of EGFP labeling. The observed strong fluorescent signal from EGFP adsorbed onto the antibody functionalized regions indicates the retained biological activity of EGFP. These results show that patterned organosilane SAMs fabricated by the combination of photolithographic process and lift-off technique is a practical approach for the fabrication of reaction regions on AlN-based and possibly other III-nitride-based bioanalytical microdevices.

REFERENCES

1. V. Cimalla, J. Pezoldt, and O. Ambacher, *J. Phys. D: Appl. Phys.* **40**, 6386–6434 (2007).
2. M. Stutzmann, J. A. Garrido, M. Eickhoff, and M. S. Brandt, *Phys. Stat. Sol. A* **203**, 3424–3437 (2006).
3. N. Chaniotakis and N. Sofikiti, *Anal. Chim. Acta* **615**, 1–9 (2008).
4. B. Baur, G. Steinhoff, J. Hernando, O. Purrucke, M. Tanaka, B. Nickel, M. Stutzmann, and M. Eickhoff, *Appl. Phys. Lett.* **87**, 263901 (2005).
5. C.-F. Chen, C.-L. Wu, and S. Gwo, *Appl. Phys. Lett.* **89**, 252109 (2006).
6. R. M. Petoral, G. R. Yazdi, A. L. Spetz, R. Yakimova, and K. Uvdal, *Appl. Phys. Lett.* **90**, 223904 (2007).
7. T. Cao, A. Wang, X. Liang, H. Tang, G. W. Auner, S. O. Salley, and K. Y. S. Ng, *Colloids and Surfaces B: Biointerfaces* **63**, 176–182 (2008).
8. C.-S. Chiu, H.-M. Lee, C.-T. Kuo, and S. Gwo, *Appl. Phys. Lett.* **93**, 163106 (2008).
9. C.-S. Chiu, H.-M. Lee, and S. Gwo, *Langmuir* (in press).
10. A. Ulman, *Chem. Rev.* **96**, 1533–1554 (1996).
11. J. Sagiv, *J. Am. Chem. Soc*, **102**, 92–98 (1980).
12. D. L. Allara, A. N. Parikh, and F. Rondelez, *Langmuir* **11**, 2357–2360 (1995).
13. A. N. Parikh, M. A. Schivley, E. Koo, K. Seshadri, D. Aurentz, K. Mueller, and D. L. Allara, *J. Am. Chem. Soc.* **119**, 3135–3143 (1997).
14. A. S. Blawas and W. M. Reichert, *Biomaterials* **19**, 595–609 (1998).
15. N. Faucheux, R. Schweiss, K. Lützow, and C. Werner, Groth, T. *Biomaterials* **25**, 2721–2730 (2004).
16. R. K. Smith, P. A. Lewis, and P. S. Weiss, *Prog. Surf. Sci.* **75**, 1–68 (2004).
17. M.-H. Lin, C.-F. Chen, H.-W. Shiu, C.-H. Chen, and S. Gwo, *J. Am. Chem. Soc.* **131**, 10984–10991 (2009).
18. G. P. López, H. A. Biebuyck, and G. M. Whitesides, *Langmuir* **9**, 1513–1516 (1993).
19. E. W. Wollman, C. D. Frisbie, and M. S. Wrighton, *Langmuir* **9**, 1517–1520 (1993).
20. A. G. Bittermann, S. Jacobi, L. F. Chi, H. Fuchs, and R. Reichelt, *Langmuir* **17**, 1872–1877 (2001).
21. C. Srinivasan, T. J. Mullen, J. N. Hohman, M. E. Anderson, A. A. Dameron, A. M. Andrews, E. C. Dickey, M. W. Horn, and P. S. Weiss, *ACS Nano* **1**, 191–201 (2007).
22. L. Hong, H. Sugimura, T. Furukawa, and O. Takai, *Langmuir* **19**, 1966–1969 (2003).
23. G. P. López, H. A. Biebuyck, R. Härter, A. Kumar, and G. M. Whitesides, *J. Am. Chem. Soc.* **115**, 10774–10781 (1993).
24. N. H. Mack, R. Dong, and R. G. Nuzzo, *J. Am. Chem. Soc.* **128**, 7871–7881 (2006).
25. D. C. Joy and C. S. Joy, *Micron* **27**, 247–263 (1996).

Mater. Res. Soc. Symp. Proc. Vol. 1202 © 2010 Materials Research Society 1202-I09-10

Fast Detection of Perkinsus Marinus--A Prevalent Pathogen of Oysters and Clams From Sea Waters

Yu-Lin Wang[1], B. H. Chu[2], K. H. Chen[2], C.Y. Chang[2], T. P. Lele[2], G. Papadi[3], J. K. Coleman[3], B. J. Sheppard[3], C. F. Dungan[4], S. J. Pearton[1], J.W. Johnson[5], P. Rajagopal[5], J.C. Roberts[5], E.L. Piner[5], K.J. Linthicum[5] and F. Ren[2]

[1]Department of Materials Science and Engineering, University of Florida, Gainesville, FL 32611
[2]Department of Chemical Engineering, University of Florida, Gainesville, FL 32611,
[3]Department of Infectious Diseases & Pathology, University of Florida, Gainesville, FL 32611
[4]Cooperative Oxford Laboratory, Maryland Department of Natural Resources, Oxford, MD 21654
[5]Nitronex Corporation, Raleigh, NC 27606

ABSTRACT

Antibody-functionalized, Au-gated AlGaN/GaN high electron mobility transistors (HEMTs) were used to detect Perkinsus marinus. The antibody was anchored to the gate area through immobilized thioglycolic acid. The AlGaN/GaN HEMT were grown by a molecular beam epitaxy system (MBE) on sapphire substrates. Infected sea waters were taken from the tanks in which *Tridacna crocea* infected with P. marinus were living and dead. The AlGaN/GaN HEMT showed a rapid response of drain-source current in less than 5 seconds when the infected sea waters were added to the antibody-immobilized surface. The recyclability of the sensors with wash buffers between measurements was also explored. These results clearly demonstrate the promise of field-deployable electronic biological sensors based on AlGaN/GaN HEMTs for Perkinsus marinus detection.

INTRODUCTION

Perkinsus marinus (P. marinus), a protozoan pathogen of the oyster, is highly prevalent along the east coast of the United States. Perkinsus species (Perkinsozoa, Alveolata) are the causative agents of perkinsosis in a variety of mollusc species. Perkinsus species infections cause widespread mortality in both natural and farm-raised oyster populations, resulting in severe economic losses for the shellfish industry and detrimental effects on the environment [1-5]. Currently, the standard diagnostic method for Perkinsus species infections has been fluid thioglycollate medium (FTM) assay detection [6]. However, this method of detection requires several days. The polymerase chain reaction (PCR)-based technique is also used to diagnose Perkinsus, but it is quite expensive and time-consuming, and requires exquisite controls to assure specificity and accuracy [7]. Clearly, such methods are slow and impractical in this age of global trade that requires rapid detection of such pathogens.

AlGaN/GaN high electron mobility transistors (HEMTs) have shown promise for bio-sensing applications [8-22], since they include a high electron sheet carrier concentration channel induced by piezoelectric polarization of the strained AlGaN layer and spontaneous polarization [23-25]. There are positive counter charges at the HEMT surface layer induced by the two-dimensional electron gas (2DEG) located at the AlGaN/GaN interface. Any slight changes in the ambient can affect the surface charge of the HEMT, thus changing the electron concentration in the channel at AlGaN/GaN interface. The typical detection time for the AlGaN/Gan HEMT based sensor was in the range of 5-20 seconds.

In this paper, we report the use of antibody-functionalized Au-gated AlGaN/GaN high electron mobility transistors (HEMTs) for detecting P. marinus in water. The water samples were taken from the tanks in which *Tridacna crocea* infected with P. marinus were living and dead. The recyclability of the sensors with wash buffers between measurements was also explored.

EXPERIMENT

The HEMT structures consisted of a 3 μm thick undoped GaN buffer, 30Å thick $Al_{0.3}Ga_{0.7}N$ spacer, 220Å thick Si-doped $Al_{0.3}Ga_{0.7}N$ cap layer. The epi-layers were grown by a molecular beam epitaxy system (MBE) on sapphire substrates. Mesa isolation was performed with an Inductively Coupled Plasma (ICP) etching with Cl_2/Ar based discharges at –90 V dc self-bias, ICP power of 300 W. 10×50 $μm^2$ Ohmic contacts separated with gaps of 5 μm consisted of e-beam deposited Ti/Al/Pt/Au patterned by lift-off and annealed at 850 °C, 45 sec under flowing N_2. The source-drain current-voltage characteristics were measured at 25°C using an Agilent 4156C parameter analyzer with the gate region exposed.

Figure 1 (left) shows a schematic device cross sectional view with immobilized thioglycolic acid (TGA), and followed by anti-P. marinus antibody coating. The Au surface was functionalized with the specific bi-functional molecule, The TGA, $HSCH_2COOH$, contained both a thiol (mercaptan) and a carboxylic acid functional group. A self-assembled monolayer of TGA was anchored on the Au-coated gate area of the sensor device through strong bonding between gold and the thiol-group of the TGA. X-ray photoelectron spectroscopy and electrical measurements confirming a high surface coverage and Au-S bonding formation on the GaN surface have been previously published [21]. The device was incubated in a phosphate buffered saline (PBS) solution of 200 μg/ml anti-P. marinus rabbit antibody.

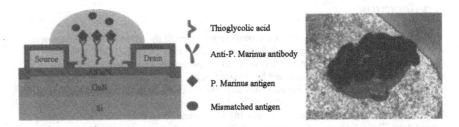

Figure 1. (left) Schematic of AlGaN/GaN HEMT sensor. The Au-coated gate area was functionalized with anti-P. marinus antibody on thioglycolic acid. (right) A picture of the clam, which may carry the perkinsus.

Figure 1 (right) shows a picture of *Tridacna crocea*, which was used as the carrier for P. marinus in this study. The *Tridacna crocea* are extremely popular ornamental reef clams imported in huge numbers into the USA from the Indo-Pacific for the aquarium trade and they are known to be vulnerable to P. marinus [26, 27]. The tanks for hosting *Tridacna crocea* were set up using bedding of crushed coral reef sold commercially as "aragonite", which added a random set of minerals to help establish a biofilm. Mimic seawater was prepared by dissolving sea salt in the water and pH value, salinity and ammonia was controlled to mimic the real seawater. The infection status of the *Tridacna crocea* in this study was verified by histopathology, FTM, and polymerase chain reaction assays.

DISCUSSION

After immobilizing the anti-P. marinus rabbit antibody on the gate area of the AlGaN/GaN HEMT sensor, the sensor surface was thoroughly rinsed off with PBS. Figure 2 shows that the drain current of the HEMT sensor was measured from 0 to 3V in drain bias voltage before and after the sensor was exposed to the water from the first tank (in which the sick and dying clams were housed). Any slight changes of the surface charges in the ambient of the HEMT affected the surface charges on the AlGaN/GaN and led to the decrease in the conductance for the device after exposure to P. marinus-infected waters.

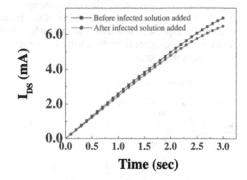

Figure 2. Drain I-V characteristics of AlGaN/GaN HEMT sensor before and after exposure to the P. marinus-infected water from tank 1

Figure 3 (left) shows a real time P. marinus detection using the source and drain current change with constant drain bias of 500 mV. No current change can be seen with the addition of buffer solution mixed showing the specificity and stability of the device. By clear contrast, the current change showed a rapid response in less than 5 seconds when 2 ul of tank 2 water was added to the surface. The abrupt drain current change due to the exposure of P. marinus in a buffer solution was stabilized after the antigen thoroughly diffused into the buffer solution. Continuous 2 ul of the tank 2 water added into a buffer solution resulted in further decreases of drain current. Obviously, tank 2 housed sick and dying calms releasing P. marinus organisms into the water, which subsequently shed surface antigens readily detected by the sensors. Then, the sensor was washed with PBS (pH 6.5) and used to detect the P. marinus again. The recycled sensor still shows very good sensitivity as previously, shown as Figure 3 (right). These results demonstrated the real-time P. marinus detection and reusability of the sensor.

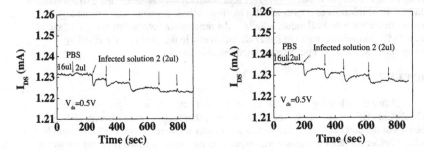

Figure 3. Real-time detection of P. marinus in an infected water from tank 2 before (left) and after recycling the sensor with PBS wash (right).

Figure 4 shows a real time P. marinus detection using the source and drain current change starting with fresh artificial seawater instead of PBS solution. No current change can be seen with the addition of diluted seawater showing the specificity and stability of the device. In contrast, the current change showed a rapid response in less than 5 seconds when 20 ul of tank 1 water was added to the surface. There was interference from the artificial sea water and the sensor was sensitive enough to use diluted tank 1 water.

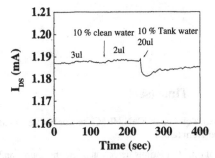

Figure 4. Drain current measured at a constant drain bias of 500 mV of an AlGaN/GaN HEMT versus time for P. marinus detection in the infected water from tank 1 and fresh artificial seawater without P. marinus.

CONCLUSIONS

In summary, we have shown that through a chemical modification method, the Au-gated region of an AlGaN/GaN HEMT structure can be functionalized for the detection of P. marinus in the infected waters directly under a variety of sampling conditions. This electronic detection of biomolecules is a significant step towards a field-deployed sensor chip, which can be integrated with a commercial available wireless transmitter to realize a real-time, fast response and high sensitivity P. marinus detector.

ACKNOWLEDGMENTS

The work at UF was partially supported by a University of Florida, College of Veterinary Medicine, Intramural Research Grant.

REFERENCES

1. J. D. Andrews, American Fisheries Society Special Publication, 18,47 (1998).
2. E.M. Burreson, and L. M. Ragone-Calvo, J. Shellfish Research 15, 17 (1996).
3. S. E. Ford, J. Shellfish Research 15, 45 (1996).
4. A. Villalba, K. S. Reece, M. C. Ordás, S. M. Casas, A. Figueras, Aquatic Living Resources 17, 411 (2004).
5. W. T. Pecher, M. R. Alavi, E. J. Schott, J. A. Fernandez-Robledo, L. Roth, S. T. Berg, G. R. Vasta, J. Parasitol., 94(2), 410 (2008).
6. S. M. Ray, Proc. Natl. shellfish. Assn. 54, 55 (1966).
7. K. S. Reece, C. F. Dungan, 2006. *Perkinsus* sp. infections of marine molluscs. *In*: AFS-FHS blue book: 2005 edition, Chapter 5.2.1-17 Fish Health Section, American Fisheries Society, Bethesda, Maryland.
8. A. P. Zhang, L. B. Rowland, E. B. Kaminsky, V. Tilak, J. C. Grande, J. Teetsov, A. Vertiatchikh and L. F. Eastman, J.Electron.Mater.32, 388(2003).
9. O. Ambacher, M. Eickhoff, G. Steinhoff, M. Hermann, L. Gorgens, V. Werss, B. Baur, M. Stutzmann, R. Neuterger, J. Schalwig, G. Muller, V. Tilak, B. Green, B. Schafft, L. F. Eastman, F. Bernadini, and V. Fiorienbini, Proc. ECS 02-14, 27(2002).
10. R. Neuberger, G. Muller, O. Ambacher and M. Stutzmann, Phys. Stat. Soli. A185, 85(2001).
11. J. Schalwig, G. Muller, O. Ambacher and M. Stutzmann, Phys. Stat. Solidi A185, 39(2001).
12. G. Steinhoff, M. Hermann, W. J. Schaff, L. F. Eastman, M. Stutzmann and M. Eickhoff, Appl. Phys. Lett., 83, 177(2003).
13. M. Eickhoff, R. Neuberger, G. Steinhoff, O. Ambacher, G. Muller, and M. Stutzmann, Phys. Stat. Solidi B 228, 519(2001).
14. B.S. Kang, H.T. Wang, F. Ren and S.J. Pearton, J.Apl.Phys.104, 031101(2008).
15. B. S. Kang, S. Kim, F. Ren, J. W. Johnson, R. Therrien, P. Rajagopal, J. Roberts, E. Piner, K. J. Linthicum, S. N .G. Chu, K. Baik, B. P. Gila, C. R. Abernathy and S. J. Pearton, Appl. Phys. Lett. 85, 2962(2004).
16. S. J. Pearton, B. S. Kang, S. Kim, F. Ren, B. P.Gila, C. R. Abernathy, J. Lin and S. N. G. Chu, J. Phys: Condensed Matter 16, R961(2004).
17. G. Steinoff, O. Purrucker, M. Tanaka, M. Stutzmann and M. Eickoff, Adv. Funct. Mater. 13,841(2003).
18. G. Steinhoff, B. Baur, G. Wrobel, S. Ingebrandt, A. Offenhauser, A. Dadgar, A. Krost, M.Stutzmann and M.Eickhoff, Appl. Phys. Lett.86, 033901(2005).
19. B. S. Kang, H. T. Wang, F. Ren, S. J. Pearton, T. E. Morey, D. M. Dennis, J. W. Johnson, P. Rajagopal, J. C. Roberts, E. L. Piner, and K. J. Linthicum, Appl. Phys. Lett., 91, 252103 (2007).

20. B. S. Kang, F. Ren, L. Wang, C. Lofton, Weihong Tan, S. J. Pearton, A. Dabiran, A. Osinsky, and P. P. Chow, Appl. Phys. Lett. 87, 023508(2005).
21. B. S. Kang, J.J. Chen, F. Ren, S. J. Pearton, J.W. Johnson, P. Rajagopal, J.C. Roberts, E.L. Piner, and K.J. Linthicum, Appl. Phys. Lett. 89, 122102 (2006).
22. M. Eickhoff , J. Schalwig, G. Steinhoff, O. Weidemann, L. Görgens, R. Neuberger, M. Hermann, B. Baur, G. Müller, O. Ambacher, and M. Stutzmann, phys. stat. sol. (c) 0, No. 6, 1908–1918 (2003).
23. R. Neuberger (a), G. Mu¨ller (a), O. Ambacher (b), and M. Stutzmann, phys. stat. sol. (a) 183, No. 2, R10–R12 (2001).
24. Parvesh Gangwani, Sujata Pandey, Subhasis Haldar, Mridula Gupta R.S. Gupta, Solid-State Electronics 51, 130–135(2007).
25. L. Shen, R. Coffie, D. Buttari, S. Heikman, A. Chakraborty, A. Chini, S. Keller, S. P. DenBaars, and U. K. Mishra, IEEE Electron Dev. Lett. 25, 7-9 (2004).
26. B. J. Sheppard, and A. C. Phillips, Dis Aquat Organ 79, 229 (2008).
27. B. J. Sheppard, and C. F. Dungan, J Zoo Wildl Med 40, 140 (2009).

Mater. Res. Soc. Symp. Proc. Vol. 1202 © 2010 Materials Research Society   1202-I06-03

Pressure Sensing With PVDF Gated AlGaN/GaN High Electron Mobility Transistor

Sheng-Chun Hung[1], Byung Hwan Chu[1], Chih Yang Chang[2], Chien-Fong Lo[1], Ke-Hung Chen[1], Yulin Wang[2], S J. Pearton[2], Amir Dabiran[3], P. P. Chow[3], G. C. Chi[4] and Fan Ren[1]

1. Department of Chemical Engineering, University of Florida, Gainesville, Florida
2. Department of Material Sci. and Eng., University of Florida, Gainesville, Florida
3. SVT Associates, Eden Prairie, Minnesota
4. Department of Physics, National Central University, Jhong-Li, Taiwan

Abstract

AlGaN/GaN high electron mobility transistors (HEMTs) with a polarized Polyvinylidene difluoride (PVDF) film coated on the gate area exhibited significant changes in channel conductance upon exposure to different ambient pressures. The PVDF thin film was deposited on the gate region with an inkjet plotter. Next, the PDVF film was polarized with an electrode located 2 mm above the PVDF film at a bias voltage of 10 kV and 70 °C. Variations in ambient pressure induced changes in the charge in the polarized PVDF, leading to a change of surface charges on the gate region of the HEMT. Changes in the gate charge were amplified through the modulation of the drain current in the HEMT. By reversing the polarity of the polarized PVDF film, the drain current dependence on the pressure could be reversed. Our results indicate that HEMTs have potential for use as pressure sensors.

Introduction

Piezoelectric materials are used widely as sensitive pressure sensors and piezoelectric gauges are typically fabricated with materials such as PZT, lithium niobate and quartz [1-4]. In 1969, Kawai found a very high piezoelectric effect in polarized polyvinylidene fluoride (PVDF) [5]. Since then, polarized PVDF has also become an important piezoelectric material due to its flexibility, low density, low mechanical impedance and easy fabrication as a ferroelectric. Because of its versatility, PVDF has many applications in low cost and disposable pressure sensors [6].

Recently, AlGaN/GaN high electron mobility transistors (HEMTs) have shown great potential for chemical and biochemical sensing applications [7-16]. This is due to their high electron sheet carrier concentration channel induced by both piezoelectric and spontaneous polarization [17, 18]. Unlike conventional semiconductor field effect transistors, there is no intentional dopant in the AlGaN/GaN HEMT structure. The electrons in the two dimensional electron gas (2DEG) channel are located at the interface between the AlGaN layer and GaN layer and there are positive counter changes at the HEMT surface layer induced by the 2DEG. Slight changes in the ambient can affect the surface charge of the HEMT, thus changing the

2DEG concentration in the channel [19-25]. Thus, nitride HEMTs may be excellent candidates for pressure sensors and other piezoelectric applications.

In this work, we investigated the effect of ambient pressure on the drain current of GaN/AlGaN HEMT sensors with a polarized PVDF thin film coated on the gate area. We quantified the sensitivity, the temporal resolution, and the limit of detection of these sensors for pressure detection. The influence of the bias voltage polarity on the drain current was also studied.

Experiment

The HEMT structures consisted of a 2μm thick undoped GaN buffer and 250 Å thick undoped $Al_{0.25}Ga_{0.75}N$ cap layer. The epilayers were grown by a Molecular Beam Epitaxy system (MBE) on sapphire substrates. Mesa isolation was performed with an Inductively Coupled Plasma (ICP) etching with Cl_2/Ar based discharges at −90 V dc self-bias, ICP power of 300 W at 2 MHz and a process pressure of 5 mTorr. E-beam deposited Ti/Al/Pt/Au ohmic metallization was patterned by standard lift-off and annealed at 850 °C, 45 sec under flowing N_2. and a contact resistivity of 5×10^{-6} Ω-cm^2 was obtained. Transmission line method using 100×100 μm^2 ohmic contacts separated with gaps of 5, 10, 20, and 40 μm was employed to determine the contact resistivity. E-beam deposited Ti/Au was employed as the interconnection metallizaton. AZ 1818 positive photoresist was used as the mask to define the gate electrode for PVDF coating and the gate dimension was 10×50 μm^2.

For coating PVDF on the gate region, a 10 wt% PVDF solution was made by dissolving PVDF powder in n,n-dimethyl acetamide (DMA). The PVDF is typically 50-60% crystalline and the beta phase PVDF is most interesting due to its piezoelectric properties. However, PVDF films synthesized from a melt or solution are predominantly α-phase. The α-phase PVDF is a non-polar molecule. The β-phase PVDF has an all-trans conformation, comprising fluorine atoms on one side and hydrogen atoms on the other side of the polymer backbone and has the highest dipole moment. The highly polar β-phase has excellent electrical properties, which include a high piezoelectric constant, a high dielectric constant, and good dielectric strength. The structures of α- and β-phase PVDF are shown in Figure 1. It has been demonstrated that incorporation of 10-15 % of Poly-methyl methacrylate (PMMA) in the PVDF/PMMA favored crystallization of the β-phase[26]. Therefore, 10% weight percent of PMMA was added in the PVDF solution; the average MW and Tg of PMMA were ~120,000 and 114 °C, respectively. The dielectric constant of the PVDF is in the range of 6.4-7.7. The un-polarized alpha phase PVDF thin film was coated on the gate region using a micro-plotter. The thickness of the coated PVDF film was about 2 μm. The PVDF (Sigma-Aldrich) had an average molecular weight (MW) and glass transition temperature (Tg) of ~534,000 and -38 °C, respectively.

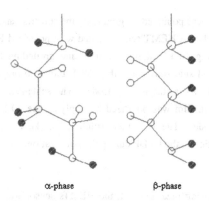

α-phase   β-phase

Figure 1. The structures of α- and β-phase PVDF.

In order to further increase the portion of β-phase in the PVDF, a high voltage was applied to the sample. A schematic of the PVDF polarization is shown in Figure 2(top). The sample was mounted on a copper chuck and immersed in a fluorinert electronic liquid F-43 (3M) to prevent arcing during the polarization. A copper wire tip with a diameter of ~ 0.2mm was kept at a distance of 1 cm above the sample to serve as a high-voltage or ground electrode. A Glassman DC power supply was used to apply 10 kV across the sample and the electrode. Figure 2(bottom) shows the schematics of HEMT sensors coated with a polarized PVDF film.

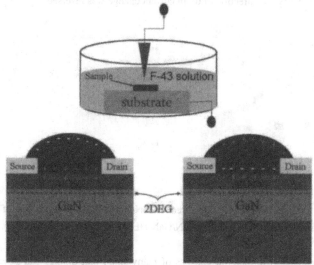

Figure 2. (top) A schematic of the PVDF polarization set up. (bottom left) A schematic of a HEMT sensor coated with polarized PVDF.

The PVDF thin film was polarized by grounding the substrate and applying 10 kV at the top electrode (bottom right). A HEMT sensor coated with polarized PVDF. The PVDF film was polarized by applying 10 kV at the substrate and grounding the top electrode. The schematic on the left hand side shows that the PVDF thin film prepared by grounding the substrate and applying 10 kV at the top electrode. The schematic on the right hand side shows that the PDVF thin film was polarized by applying the 10 kV at the substrate and grounded the top electrode. The fluor-inert solution was kept at 70 °C to enhance the carbon-carbon bond rotation in the PVDF during the polarization.

Results and Discussion

For the pressure sensing measurement, the HEMTs sensor were mounted on a carrier and put in a pressure chamber. N_2 gas was used for pressurizing the chamber and a constant drain bias voltage of 500mV was applied to the drain contact of the sensor. Figure 3 (left) shows the real time pressure detection with the polarized PVDF gated HEMT. The drain current of HEMT sensor showed a rapid decrease in less than 5 s when the ambient pressure was changed to 20 psig. A further decrease of the drain current for the HEMT sensor was observed when the chamber pressure increased to 40 psig. These abrupt drain current decreases were due to the change of charges in the PVDF film upon a shift of ambient pressure. A HEMT sensor without the PDVF coating was loaded in the pressure chamber and there was no change of drain current observed. As seen in Figure 3 (right), by reversing the PVDF film polarity, the direction of drain current change was reversed.

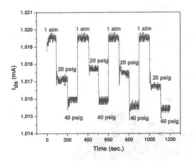 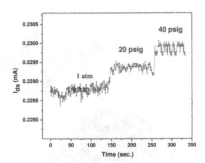

Figure 3. (left) Drain current of a PVDF gated AlGaN/GaN HEMT as a function of pressure. (right) Drain current of a PVDF gated AlGaN/GaN HEMT as a function of pressure.

The piezoelectric material consists of many small unit dipoles and possesses net positive charges at one end of the materials and negatives at the other end in the direction of the polarization. Under compressive stresses, the length of the piezoelectric sample and the length of the unit dipole in the sample are reduced. Thus the overall dipole moment per unit volume of the materials reduces accordingly. The change in dipole moment of the materials changes the charges density at the ends of the materials [27]. The sample used in Figure 2 was polarized by grounding the substrate and applying 10 kV at the top copper wire electrode. Therefore, the PVDF film near the PVDF/AlGaN interface had net positive charges and the other end of the PVDF film possessed net negative charges, as illustrated in Figure 1 (bottom left). Once the PVDF film experienced compressive force, the net charges in the PVDF film were reduced. Thus the total positive charge at the AlGaN surface was decreased and the corresponding 2DEG reduced. By reversing the polarity, the surface of the PVDF film right next to the AlGaN surface had negative charges and the top surface of PVDF film possessed positive charges. When the PVDF film was subject to the pressure, the net negative charges in the PVDF near the PVDF/AlGaN interface were reduced. Thus, the net positive charges at the AlGaN surface increased and drain current of the HEMT increased correspondingly. The pressure sensor showed a good repeatability, as illustrated in Figure 4, the drain current response of the PVDF gated HEMT sensor to pressure switching from 1 atm to 1 psig. The change of drain current for HEMT exposed between 1 atm and 1 psig was still considerably larger than the background noise. Thus, HEMTs could be used to detect small differences in pressure (1 psig).

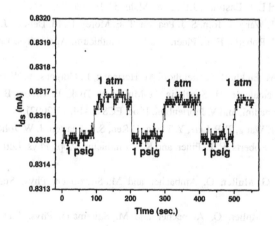

Figure 4. Drain current of a PVDF gated AlGaN/GaN HEMT as a chamber pressure.

Conclusion

In conclusion, polarized PVDF gated AlGaN/GaN HEMTs showed rapid change in the source-drain current when exposed to different pressure ambient. These results show the potential of polarized PVDF gated AlGaN/GaN HEMT for pressure-sensing applications.

Acknowledgments

The work at UF was partially supported by ONR Grant N000140710982 monitored by Igor Vodyanoy, NASA Kennedy Space Center Grant NAG 3-2930 monitored by T. Smith, by NSF DMR 0400416 and the State of Florida, Center of Excellence in Nano-Bio Sensors.

References

1. V. Mortet, R. Petersen, K. Haenen and M. D'Olieslaeger, IEEE Ultrasonics Symposium, 1456 (2005).
2. S. C. Ko, Y. C. Kim, S. S. Lee, S. H. Choi, S. R. Kim, Sens. Actuators B 103, 130 (2003).
3. R. Greaves and G. Sawyer, Phys. Technol, Vol 14, 15 (1983).
4. E. S. Kim and R. S. Muller, IEEE, IEDM 86, 8 (1986).
5. A. Odon, Measurement Science Review, 3, 3, 111, (2003).
6. A.V. Shirinov and W.K. Schomburg, Sensors & Actuators A, 142, 48(2008).
7. G. M. Chertow, E. M. Levy, K. E. Hammermeister, F. Grover, J. Daley, Amer. J. Med. 104, 343 (1998).
8. A. P. Zhang, L. B. Rowland, E. B. Kaminsky, V. Tilak, J. C. Grande, J. Teetsov, A. Vertiatchikh and L. F. Eastman, J.Electron.Mater.32 388 (2003).
9. B. S. Kang, H.T. Wang, F. Ren, S. J. Pearton, T. E. Morey, D. M. Dennis, J. W. Johnson, P. Rajagopal, J. C. Roberts, E. L. Piner, and K. J. Linthicum, Appl. Phys. Lett. 91, 252103 (2007).
10. O. Ambacher, M. Eickhoff, G. Steinhoff, M. Hermann, L. Gorgens, V. Werss, B. Baur, M. Stutzmann, R. Neuterger, J. Schalwig, G. Muller, V. Tilak, B. Green, B. Schafft, L. F. Eastman, F. Bernadini, and V. Fiorienbini, Proc. ECS 02-14, 27(2002).
11. B. S. Kang, H.T. Wang, T. P. Lele, Y. Tseng, F. Ren, S. J. Pearton, J. W. Johnson, P. Rajagopal, J. C. Roberts, E. L. Piner, and K. J. Linthicum, Appl. Phys. Lett. 91, 112106 (2007).
12. R. Neuberger, G. Muller, O. Ambacher and M. Stutzmann, Phys. Stat. Soli. A185, 85(2001).
13. J. Schalwig, G. Muller, O. Ambacher and M. Stutzmann, Phys. Stat. Solidi A185, 39(2001).
14. H.T. Wang, B. S. Kang, F. Ren, S. J. Pearton, J. W. Johnson, P. Rajagopal, J. C. Roberts, E. L. Piner, and K. J. Linthicum, Appl. Phys. Lett. 91, 222101 (2007).
15. G. Steinhoff, M. Hermann, W. J. Schaff, L. F. Eastman, M. Stutzmann and M. Eickhoff, Appl. Phys. Lett., 83, 177(2003).

16. B. S. Kang, H.T. Wang, F. Ren, B. P. Gila, C. R. Abernathy, S. J. Pearton, J. W. Johnson, P. Rajagopal, J. C. Roberts, E. L. Piner, and K. J. Linthicum, Appl. Phys. Lett. 91, 012110 (2007).

17. G. Steinoff, O. Purrucker, M. Tanaka, M. Stutzmann and M. Eickoff, Adv. Funct. Mater. 13,841(2003).

18. G. Steinhoff, B. Baur, G. Wrobel, S. Ingebrandt, A. Offenhauser, A. Dadgar, A. Krost, M.Stutzmann and M.Eickhoff, Appl. Phys. Lett.86, 033901(2005)

19. S. C. Hung, Y. L. Wang, B. Hicks, S. J. Pearton, F. Ren, J. W. Johnson, P. Rajagopal, J. C. Roberts, E. L. Piner, K. J. Linthicum and G. C. Chi, Electrochem. Solid-state Lett., 11, H241 (2008)

20. B. S. Kang, H. T. Wang, F. Ren and S. J. Pearton, J. Appl. Phys., 104, 031101 (2008)

21. B. H. Chu, B. S. Kang, F. Ren, C. Y. Chang, Y. L. Wang, S. J. Pearton, A. V. Glushakov, D. M. Dennis, J. W. Johnson, P. Rajagopal, J. C. Roberts, E. L. Piner and K. J. Linthicum, Appl. Phys. Lett., 93, 042114 (2008)

22. C. Y. Chang, B. S. Kang, H. T. Wang, F. Ren, Y. L. Wang, S. J. Pearton, D. M. Dennis, J. W. Johnson, P. Rajagopal, J. C. Roberts, E. L. Piner and K. J. Linthicum, Appl. Phys. Lett.,92, 232102 (2008)

23. S. C. Hung, Y. L. Wang, B. Hicks, S. J. Pearton, D. M. Dennis, F. Ren, J. W. Johnson, P. Rajagopal, J. C. Roberts, E. L. Piner, K. J. Linthicum and G. C. Chi, Appl. Phys. Lett., 92, 193903 (2008)

24. K. H. Chen, B. S. Kang, H. T. Wang, T. P. Lele, F. Ren, Y. L. Wang, C. Y. Chang, S. J. Pearton, D. M. Dennis, J. W. Johnson, P. Rajagopal, J. C. Roberts, E. L. Piner and K. J. Linthicum, Appl. Phys. Lett., 92, 192103 (2008)

25. B.S. Kang, H.T. Wang, F.Ren and S.J. Pearton, J. Appl. Phys. 104, 031101 (2008).

26. Rinaldo Gregorio Jr and Nadia Chaves Pereira de Souza Nociti J. Phys. D: Appl. Phys. 28 (1995) 432436.

27. W. F. Smith and J. Hashemi, Foundation of Material Science and Engineering, 4th Ed., McGraw Hill, New York, 2006.

Mater. Res. Soc. Symp. Proc. Vol. 1202 © 2010 Materials Research Society   1202-I06-06

High sensitivity of hydrogen sensing through N-polar GaN Schottky diodes

Yu-Lin Wang[1], B. H. Chu[2], C.Y. Chang[2], K. H. Chen[2], Y. Zhang[3], Q. Sun[3], J. Han[3], S. J. Pearton[1], and F. Ren[2]
[1]Department of Materials Science and Engineering, University of Florida, Gainesville, FL 32611
[2]Department of Chemical Engineering, University of Florida, Gainesville, FL 32611,
[3]Department of Electrical Engineering, Yale University, New Haven, CT 06520, USA

ABSTRACT

N-polar and Ga-polar GaN grown on c-plane sapphire by a metal-organic chemical vapor deposition (MOCVD) system were used to fabricate platinum deposited Schottky contacts for hydrogen sensing at room temperature. Wurtzite GaN is a polar material. Along the c-axis, there are N-face (N-polar) or Ga-face (Ga-polar) orientations on the GaN surface. The Ohmic contacts were formed by lift-off of e-beam deposited Ti (200 Å)/Al (1000 Å)/Ni (400 Å)/Au (1200 Å). The contacts were annealed at 850°C for 45 s under a flowing N_2 ambient. Isolation was achieved with 2000 Å plasma enhanced chemical vapor deposited SiN_x formed at 300°C. A 100 Å of Pt was deposited by e-beam evaporation to form Schottky contacts. After exposure to hydrogen, Ga-polar GaN Schottky showed 10% of current change, while the N-polar GaN Schottky contacts became fully Ohmic. The N-polar GaN Schottky diodes showed stronger and faster response to 4% hydrogen than that of Ga-polar GaN Schottky diodes. The abrupt current increase from N-polar GaN Schottky exposure to hydrogen was attributed to the high reactivity of the N-face surface termination. The surface termination dominates the sensitivity and response time of the hydrogen sensors made of GaN Schottky diodes. Current-voltage characteristics and the real-time detection of the sensor for hydrogen were investigated. These results demonstrate that the surface termination is crucial in the performance of hydrogen sensors made of GaN Schottky diodes.

INTRODUCTION

There is great interest in detection of hydrogen sensors for use in hydrogen-fuelled automobiles and with proton-exchange membrane (PEM) and solid oxide fuel cells for space craft and other long-term sensing applications. These sensors are required to detect hydrogen near room temperature with minimal power consumption and weight and with a low rate of false alarms. Due to their low intrinsic carrier concentrations, GaN- and SiC-based wide band gap semiconductor sensors can be operated at lower current levels than conventional Si-based devices and offer the capability of detection to ~600 °C [1-10]. The ability of electronic devices fabricated in these materials to function in high temperature, high power and high flux/energy radiation conditions enable performance enhancements in a wide variety of spacecraft, satellite, homeland defense, mining, automobile, nuclear power, and radar applications.

Wurtzite GaN is a polar material. Therefore, along the c-axis, there are N-face (N-polar) or Ga-face (Ga-polar) orientations on the GaN surface. The polarity determined by the GaN surface has been used to improve performance of AlGaN/GaN high electron mobility transistors (HEMTs) [11]. The Ga-polar surface is intentionally grown to make the spontaneous polarization compatible with the piezoelectric polarization to enhance the 2-dimensional electron gas in

AlGaN/GaN HEMTs [11]. Since there are different reactivities for the N-face and Ga-faces [12-14], it is interesting that how these different surfaces react with different chemicals. Our previously published result shows the improved hydrogen sensing through N-polar GaN Schottky diodes [15]. Since GaN Schottky diodes have been widely used to fabricate hydrogen sensors, it is important to investigate the surface termination effect in detail on the hydrogen sensors made of GaN Schottky diodes.

In this paper, we study the hydrogen sensing with N-face and Ga-face GaN Schottky diodes. Current-voltage characteristics and the real-time detection of the sensor for hydrogen were investigated.

EXPERIMENT

The Ga- and N-GaN layer structures were grown on c-plane sapphire substrates by a metal-organic chemical vapor deposition (MOCVD) system. Ga-face and N-face GaN were grown and fabricated as Schottky diodes. The layer structures included an initial thin undoped GaN buffer for all the two structures then followed by undoped GaN layers with different surface termination. Figure 1(top) shows the schematic of the hydrogen sensor. The undoped N-face GaN is 1.1 um and the undoped Ga-face GaN is 1.5um, respectively. These two structures have carrier concentration of 1.5×10^{18} and 1.2×10^{17} cm^{-3}, respectively, obtained by Hall measurements. The high background carrier concentrations were due to the oxygen incorporation during the growth and the details of the materials growth can be found elsewhere [16-17]. The Ohmic contacts were formed by lift-off of e-beam deposited Ti (200 Å)/Al (1000 Å)/Ni (400 Å)/Au (1200 Å). The contacts were annealed at 850°C for 45 s under a flowing N$_2$ ambient in a Heatpulse 610T system. Isolation was achieved with 2000 Å plasma enhanced chemical vapor deposited SiN$_x$ formed at 300°C. A window was opened for etching SiN$_x$ where

a 100 Å of Pt was deposited by e-beam evaporation to form Schottky contacts. Final metal of e-beam deposited Ti/Au (200 Å/1200 Å) interconnection contacts were employed on the Schottky diodes. Figure 1 (bottom) shows an optical microscope image of the completed devices. The current-voltage characteristics and real-time detection of these Schottky diodes were measured at 25°C to detect 4% hydrogen in nitrogen using an Agilent 4156C parameter analyzer with the Schottky contact exposed

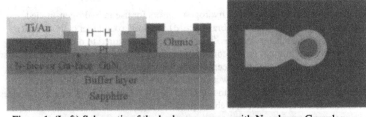

Figure 1. (Left) Schematic of the hydrogen sensor with N-polar or Ga-polar surface termination. (Right) Top view of a GaN-Schottky-diode hydrogen sensor.

DISCUSSION

Figure 2 (Left) shows the current-voltage characteristics of the two different hydrogen sensors in pure nitrogen or in 4% hydrogen for 10 minutes. The N-polar Schottky diodes become fully Ohmic when exposed to 4% hydrogen. However, the Ga-polar Schottky diodes still maintain rectifying contacts when exposed to 4% hydrogen. The current change of the Ga-polar Schottky diodes before and after exposure to hydrogen is comparable to our published results [8, 9] and the current change of Ga-polar Schottky diode was much smaller than that of the N-polar Schottky diodes. This phenomena is even more significant in the reverse bias region where the N-polar Schottky diodes have a current change from μA to mA while the Ga-polar Schottky didoes have current change in the same order of magnitude. Hydrogen sensing with Pt-deposited GaN Schottky diodes is usually explained by a mechanism in which the dipole formed on the Pt/GaN interface lowers the Schottky barrier height [18-19]. By using the thermoionic emission transport mechanism, the Schottky barrier heights were approximately evaluated as 0.4366 eV and 0.5568 eV for N-polar and Ga-polar GaN Schottky contacts, respectively, in pure nitrogen at room temperature (298 K). After these sensors were exposed to 4% hydrogen, the effective Schottky barrier heights were reduced by 3.65 meV for the Ga-polar GaN Schottky contacts. However, the change for the N-polar Schottky diode is huge because it becomes Ohmic. The only difference between these two diodes is the surface termination. Therefore, the surface potential change or the dipole forming on the surface is the most likely mechanism. High resolution electron energy loss spectroscopy (HREELS) showed a strong preference of N sites for the adsorption of hydrogen gas or atomic H [12]. The strong affinity of hydrogen with N-polar surfaces was also reported [13]. It was also shown that below 820°C, N-polar GaN has much faster reaction rate than that of Ga-polar GaN surface [14]. These experiments can explain why N-polar Schottky diodes have stronger response than Ga-polar ones.

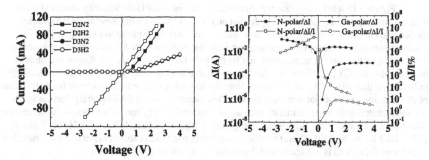

Figure 2. (Left) IV characteristics of the hydrogen sensors with N-polar or Ga-polar surface termination in pure nitrogen and 4% hydrogen gas, respectively. (Right) Current change and current change percentage of N-polar and Ga-polar GaN Schottky diodes before and after exposure to 4% hydrogen gas.

Figure 2 (Right) shows the current change and the percentage of the current change after the N-polar and Ga-polar GaN were exposed to 4% hydrogen in nitrogen. The maximum percentage change in current for N-polar termination is 10^6, which is much larger than that of Ga-polar termination, namely 10.

Figure 3 (Left) shows the real-time test of N-polar GaN Schottky diodes in pure nitrogen or in 4% hydrogen at room temperature with the applied voltage at 0.75 V. The abrupt increase of current shows when 4% hydrogen was exposed to the N-polar sensor, the current suddenly rose and quickly saturated. It shows a very strong and very fast response to hydrogen. However, when the sensor was purged with pure nitrogen again, the current did not move back to the original baseline. It is possible that some hydrogen was strongly bonded with nitrogen or trapped by defects, with the thermal energy at room temperature not being high enough to break the bonding and activate it.

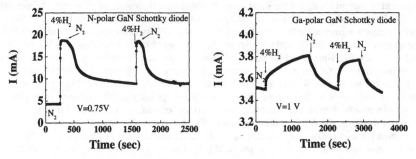

Figure 3 (Left) Real-time detection of 4% hydrogen in nitrogen using a sensor having N-face termination. (Right) Real-time detection of another sensor having Ga-face termination.

Figure 3 (Right) shows the real-time test of the Ga-polar GaN Schottky diodes at room temperature in pure nitrogen or 4% hydrogen at 1 V. Comparing with N-polar GaN Schottky diodes, the Ga-polar Schottky diodes have weaker and slower response to hydrogen. This demonstrates that the abrupt current increase from the N-polar GaN Schottky diodes was indeed determined by the surface termination, not by the bulk GaN. Therefore, the slower response to hydrogen for the Ga-polar Schottky diodes proved that the Ga-polar surface has lower reactivity to hydrogen than that of N-polar surface. Note the second response to hydrogen is faster than that of the first one for Ga-polar Schottky diodes. This indicates that in the beginning, the Ga surface is contaminated with other elements, such as oxygen, which may form Ga-O bonds (bond energy 282+/-62.7 KJ/mole) [20]. However, hydrogen may replace oxygen to form Ga-H (bond energy 282+/-20.9 KJ/mole) [20]. Therefore, the second response always shows faster response rate. For N-polar surface, N-O is not stable and thus it does not exhibit this problem.

Figure 4 shows the IV characteristics for N-polar GaN Schottky diodes, with hydrogen purging for 10 minutes initially, and then for every five subsequent minutes the IV was measured. The current change in the reversed bias region demonstrated that the effective Schottky barrier height changed with nitrogen purging.

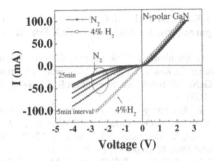

Voltage (V)

Figure 4. IV characteristics of N-polar GaN Schottky diode exposed to 4% hydrogen for 10 minutes and then purged with nitrogen and measured every five minutes at room temperature.

CONCLUSIONS

In summary, we have shown that N-polar and Ga-polar surface GaN Schottky diodes have very different hydrogen sensing reactivity. N-polar Schottky diodes have very strong and fast response to 4% hydrogen. Ga-polar surface Schottky diodes have lower and slower response to hydrogen. This difference comes from the different surface reactivity. These results demonstrate that the surface termination is crucial in the performance of hydrogen sensors made of GaN Schottky diodes.

ACKNOWLEDGMENTS

The work at UF was partially supported by ONR Grant N000140710982 monitored by Igor Vodyanoy, by NSF DMR 0400416 and the State of Florida, Center of Excellence in Nano-Bio Sensors. This work at Yale University was supported by the United States Department of Energy under Contract No. DE-FC26-07NT43227.

REFERENCES

1. A. Baranzahi, A. Lloyd Spetz and I. Lundström, Reversible hydrogen annealing of metal-oxide-silicon carbide devices at high temperatures, *Appl. Phys. Lett.* 67, 3203 (1995).
2. A. Lloyd Spetz, L. Unéus, H. Svenningstorp, P. Tobias, L.-G. Ekedahl, O. Larsson, A. Göras, S. Savage, C. Harris, P. Mårtensson, R. Wigren, P. Salomonsson, B. Häggendahl, P. Ljung, M. Mattsson and I. Lundström, SiC Based Field Effect Gas Sensors for Industrial Applications , *Phys. Status Solidi A* 185, 15 (2001).
3. B.P. Luther, S.D. Wolter and S.E. Mohney, High temperature Pt Schottky diode gas sensors on n-type GaN, *Sens. Actuat. B* 56, 164 (1999).
4. J. Schalwig, G. Muller, O. Ambacher and M. Stutzmann, Group-III-Nitride Based Gas Sensing Devices, *Phys. Status Solidi A* 185, 39 (2001).
5. J. Schalwig, G. Muller, M. Eickhoff, O. Ambacher and M. Stutzmann, Gas sensitive GaN/AlGaN-heterostructures, *Sens. Actuat. B* 87, 425 (2002).

6. J. Schalwig, G. Müller, U. Karrer, M. Eickhoff, O. Ambacher, M. Stutzmann, L. Görgens and G. Dollinger, Hydrogen response mechanism of Pt–GaN Schottky diodes, *Appl. Phys. Lett.* 80, 1222 (2002).

7. M. Eickhoff, J. Schalwig, G. Steinhoff, O. Weidmann, L. Gorgens, R. Neuberger, M. Hermann, B. Baur, G. Muller, O. Ambacher and M. Stutzmann, Electronics and sensors based on pyroelectric AlGaN/GaN heterostructures - Part B: Sensor applications, *Phys. Status Solidi C* 0, 1908 (2003).

8. J. Kim, F. Ren, B. Gila, C.R. Abernathy and S.J. Pearton, Reversible barrier height changes in hydrogen-sensitive Pd/GaN and Pt/GaN diodes, *Appl. Phys. Lett.* 82, 739 (2003).

9. J. Kim, B. Gila, G.Y. Chung, C.R. Abernathy, S.J. Pearton and F. Ren, Hydrogen-sensitive GaN Schottky diodes , *Solid-State Electron.* 47, 1069 (2003).

10. H.T. Wang, B.S. Kang, F. Ren, R.C. Fitch, J. Gillespie, N. Moser, G. Jessen, R. Dettmer, B.P. Gila, C.R. Abernathy and S.J. Pearton, Comparison of gate and drain current detection of hydrogen at room temperature with AlGaN/GaN high electron mobility transistors, *Appl. Phys. Lett.* 87, 172101-1 (2005).

11. O. Ambacher, J. Smart, J. R. Shealy, N. G. Weimann, K. Chu, M. Murphy, W. J. Schaff, L. F. Eastman, R. Dimitrov, L. Wittmer, M. Stutzmann, W. Reiger, J. Hilsenbeck, Two-dimensional electron gases induced by spontaneous and piezoelectric polarization charges in N- and Ga-face AlGaN/GaN heterostructures, J. Appl. Phys. 85, 3222 (1999).

12. U. Starke, S. Sloboshanin, F. S. Tautz, A. Seubert, J. A. Schaefer, Polarity, Morphology and Reactivity of Epitaxial GaN Films on Al2O3(0001), Phys. Status Solidi A. 177, 5 (2000).

13. J. E. Northrup, J. Neugebauer, Strong affinity of hydrogen for the GaN(000-1) surface: Implications for molecular beam epitaxy and metalorganic chemical vapor deposition, Appl. Phys. Lett. 85, 3429 (2004).

14. M. Mayumi, F. Satoh, Y. Kumagai, K. Takemoto, A. Koukitu, Influence of Polarity on Surface Reaction between GaN{0001} and Hydrogen, Phys. Status Solidi B. 228, 537 (2001).

15. Yu-Lin Wang, F. Ren, U. Zhang, Q. Sun, C. D. Yerino, T. S. Ko, Y. S. Cho, I. H. Lee, J. Han, S. J. Pearton, Appl. Phys. Lett. 94, 212108-1 (2009).

16. Q. Sun, Y. S. Cho, B. H. Kong, H. K. Cho, T. S. Ko, C. D. Yerino, I.-H. Lee, and J. Han, N-face GaN growth on c-plane sapphire by metalorganic chemical vapor deposition, J. Cryst. Growth 311, 2948 (2009).

17. Q. Sun, Y. S. Cho, I.-H. Lee, J. Han, B. H. Kong, and H. K. Cho, Nitrogen-polar GaN growth evolution on c-plane sapphire, Appl. Phys. Lett. 93, 131912-1 (2008).

18. M. Miyoshi, Y. Kuraoka, K. Asai, T. Shibata, M. Tanaka, Electrical characterization of Pt/AlGaN/GaN Schottky diodes grown using AlN template and their application to hydrogen gas sensors , J. Vac. Sci. Technol. B, 25, 1231 (2007).

19. Y. Y. Tsai, K. W. Lin, H. I. Chen, I. P. Liu, C. W. Hung, T. P. Chen, T. H. Tsai, L. Y. Chen, K. Y. Chu, W. C. Liu, Hydrogen sensing properties of a Pt-oxide-GaN Schottky diode, J. Appl. Phys. 104, 024515-1 (2008).

20. J. F. Shackelford, W. Alexander, J. S. Park, CRC Materials Science and Engineering handbook, second edition, CRC Press Inc.S-117., Boca Raton, FL (1992).

Mater. Res. Soc. Symp. Proc. Vol. 1202 © 2010 Materials Research Society   1202-I09-08

Surface Acoustic Wave Sensors Deposited on AlN Thin Films

J. L. Justice[1], O. M. Mukdadi[2], and D. Korakakis[1,3]

[1]Lane Department of Computer Science and Electrical Engineering, West Virginia University, Morgantown, WV, U.S.A.

[2]Department of Mechanical and Aerospace Engineering, West Virginia University, Morgantown, WV, U.S.A.

[3]National Energy Technology Laboratory, 3610 Collins Ferry Road, Morgantown, WV, U.S.A.

ABSTRACT

Over the past few decades, there has been considerable research and advancement in surface acoustic wave (SAW) technology. At present, SAW devices have been highly successful as frequency band pass filters for the mobile telecommunications and electronics industries. In addition to their inherent frequency selectivity, SAW devices are also highly sensitive to surface perturbations. This sensitivity, along with a relative ease of manufacture, makes SAW devices ideally suited for many sensing applications including mass, pressure, temperature, and biosensors. In the area of biosensing, surface plasmon resonance (SPR) and quartz crystal microbalances (QCM) are still in the forefront of research and development, but advancement in SAW sensors could prove to have significant advantages over these technologies. This study investigates the advantages of using aluminum nitride (AlN) as a material for SAW sensors. AlN retains its piezoelectric properties at relatively high temperatures when compared to more common piezoelectric materials such as lead zirconium titanate (PZT), lithium tantalate (LiTaO$_3$) and zinc oxide (ZnO). AlN is also a very robust material making it suitable for biosensing applications where the sensing target is selectively absorbed by an active layer on the device which may attack the piezoelectric layer. AlN thin films of different thicknesses have been deposited on Si substrates by DC reactive sputtering. Rayleigh-wave SAW devices have been fabricated by the deposition of platinum contacts and interdigital transducers (IDTs) onto AlN thin films using standard photolithographic processes. Experiments have been conducted to measure Rayleigh velocities, resonant frequencies, and insertion loss. Experimental results are compared to theoretical calculations.

INTRODUCTION

The stress-free boundary imposed by the surface of a crystal gives rise to a unique acoustic mode whose propagation is confined to the surface and is therefore known as a surface acoustic wave (SAW) [1]. SAW devices take advantage of the piezoelectric property of certain materials to convert an electrical potential into a mechanical strain. Because of their surface confinement, SAWs are conveniently generated by surface electrodes on a piezoelectric substrate. Many materials exhibit piezoelectricity including crystals (quartz), semiconductors (AlN), ceramics (PZT), and polymers (vinylidene polyfluoride). The simplest SAW device consists of a piezoelectric layer and a pair of IDTs. More complex arrangements are utilized for different SAW device applications.

SAW devices have been developed as delay lines, bandpass filters, matched filters, resonators, oscillators [2] and sensors [1]. SAW delay lines simply retain signal information for a period of time similar to a buffer in an integrated circuit. Because acoustic waves travels considerably slower than signals in a digital circuit, information is temporarily stored and becomes available after an amount of time determined by the length of the delay line. A typical

SAW delay line of 1cm can store 2 to 3 ms of information [2]. When implemented as a sensor, the SAW interacts with the device surface causing measurable differences between the input and output signals. Coupling to any medium contacting the surface strongly affects the velocity and/or amplitude of the wave [3]. Because a SAW propagating in a piezoelectric medium generates mechanical deformation and an electrical potential, both mechanical and electrical coupling between the SAW and surface contacting medium are possible [1].

The three main SAW modes that are utilized in SAW sensors are Rayleigh, Lamb, and Love wave modes. Rayleigh and Lamb wave modes exhibit strong coupling with media contacting the piezoelectric surface. Thus, both Rayleigh-wave sensors are only and Lamb-wave sensors are mostly used as gas sensors, since the waves are damped in liquids [3]. Love waves are transverse waves and exhibit very poor coupling with media at the piezoelectric surface. Therefore, SAW devices are usually designed to utilize Love waves when they operate in liquid environments and it is undesirable for the liquid to attenuate the SAW.

EXPERIMENT

AlN thin films were deposited on 2 inch, 300 μm thick, p-type, <100>, double-sided polished, Si wafers. Each wafer was cleaned in a 10 minute bath of J.D. Baker's buffered oxide etchant, rinsed for 10 minutes in de-ionized water, blown dry with nitrogen gas (N_2), and given a 15 minute dehydration bake at 110 °C before being placed into the sputtering chamber. The sputtering chamber was pumped down, using a cryopump, to a base pressure of 5.0×10^{-6} Torr before gases were allowed to flow into the chamber. A 2 inch, 99.99% pure Al target was used and was powered to 100 W using a DC magnetron. To promote quality film growth, two pre-sputters were performed. The first pre-sputter was performed to clean the Al target of any oxidation that may have occurred while exposed to atmosphere and lasted 10 minutes using Ar at a rate of 30 sccm and 60 mTorr. The second pre-sputter was performed to adequately poison the target in an effort to promote Al-N bonding. The second pre-sputter was 10 minutes using Ar and N_2 at 3 and 27 sccm respectively and 45 mTorr. To deposit the AlN thin film onto the Si substrate, a sputter was performed using the same parameters as the second pre-sputter. The length of the sputtering time was varied to produce AlN thin films of different thicknesses. The cross section of an AlN sample is shown in Figure 1.

Figure 1. SEM image of the cross section of one AlN sample. This sample was grown over a period of 30 minutes. Other film thicknesses were measured and an average growth rate for DC reactively sputtered AlN of ~ 4 nm/min was determined for this experimental setup.

Other AlN samples were sputtered with nominal thicknesses of approximately 60, 120, and 240 nm. The AlN thin film samples were kept in a clean room environment until they were ready for the photolithographic processes. AZ5214E image reversal photoresist was spun onto the AlN sample at 7000 rpm for 40 seconds. A pre-bake was performed at 90 °C for 1 minute using a hotplate. The AlN sample was allowed to cool for a few minutes and was then place into a Karl Suss MA6 mask aligner. The AlN sample was exposed to 320 nm ultraviolet light for a total emitted power of 50 mJ. A post-bake was performed at 120 °C for 2 minutes using a hotplate. The AlN sample was allowed to cool for a few minutes and then was placed in a flood exposure and was exposed to 365 nm ultraviolet for a total emitted power of 2 J. The AlN sample was developed in an undiluted solution of MIF-300 developer for 1 minute and then blown dry using nitrogen gas. The quality of the pattern was visually verified using an optical microscope and, once verified, the AlN sample was immediately placed into the sputtering chamber.

Pt was sputtered onto the prepared AlN sample using a 2 inch, 99.999% pure Pt target. Pt was chosen as the IDT and contact metal because it has been shown that Pt produces a higher d_{33} response when used on AlN thin films as compared to other metals [4]. The sputtering chamber was pumped down to 5.0×10^{-6} Torr, the target powered to 100 W, and a pre-sputter was performed for 5 minutes using Ar at 30 sccm. Pt was sputtered onto the prepared AlN sample for 2 minutes. The sample was removed from the sputtering chamber and placed in an acetone bath for 10 minutes and then into a sonic bath for 1 minute to perform liftoff. One IDT with its contact pads is shown in Figure 2.

Figure 2. Image of one IDT and its contact pads deposited on an AlN thin film. Each pad is 100 μm × 100 μm, spacing from center of pad to center of pad is 150 μm, transmission paths are 5 μm wide, IDT digits are 2 μm × 30 μm and spaced 2 μm apart.

DISCUSSION

With IDT finger width and spacing at 2 μm, the characteristic wavelength, λ_0, is 8 μm for all SAW devices tested. The normalized thickness of the AlN thin films is kh_{AlN}, where $k = 2\pi/\lambda_0$, the wave vector modulus, and h_{AlN} is the AlN film thickness. The devices used in this experiment have nominal AlN thicknesses of 60, 120, and 240 nm with $kh_{AlN} = 0.047$, 0.094, and 0.188 respectively. The Rayleigh-mode SAW velocity of AlN on Si, with small kh_{AlN} (<1), is reported in literature as 5000 to 5150 m/s [5-7]. Using this reported value, the characteristic frequency, f_0, is calculated to be in the range of 625 to 643.8 MHz. The frequency response for SAW devices deposited on three different thicknesses of AlN is presented in Figure 3.

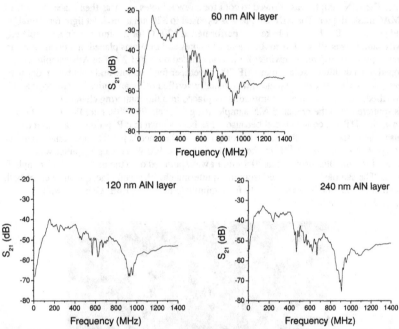

Figure 3. Measured S_{21} response of three SAW devices deposited on 60, 120, and 240 nm thick AlN layers on Si substrates with a resistivity of 10 $\Omega \cdot$cm.

For the measured results in Figure 3, a careful calibration of the network analyzer was done to eliminate certain parasitic effects, such as impedance mismatch between cables, connectors and probes. However, parasitic effects inherent to the device cannot be eliminated with a system calibration. One such parasitic is the electromagnetic feedthrough between adjacent sets of IDTs, which is greatly increased when using a conductive substrate such as Si. As a consequence, SAW devices built on conductive substrates show a poor frequency response with a low rejection rate that differs substantially from the ideal response [5]. However, each plot in Figure 3 does demonstrate a similar frequency response when the insertion loss at each of the prominent peaks is compared, as shown in Table I. The relatively low insertion loss observed from 10 to 600 MHz is due mostly to the electromagnetic feedthrough between adjacent pairs of IDTs which can be reduced by using a higher resistive Si substrate [5]. The response from 600 to 900 MHz most closely follows that of the ideal SAW filter. Considering this, the peak that appears to be the characteristic frequency for each plot in Figure 3 is the one between 723 to 739 MHz. This translates to a Rayleigh-wave SAW velocity of 5784 to 5912 m/s, which is higher than reported for SAW devices deposited on AlN/Si [5-7]. The ripples in the response can be attributed to the

Table I. SAW frequency and insertion loss of prominent peaks for 60, 120, and 240 nm AlN layers.

60 nm AlN layer		120 nm AlN layer		240 nm AlN layer	
Frequency (MHz)	Insertion Loss (dB)	Frequency (MHz)	Insertion Loss (dB)	Frequency (MHz)	Insertion Loss (dB)
132.841	22.29	155.95	39.66	141.963	32.47
411.363	28.40	406.063	42.32	405.980	35.70
520.825	39.84	507.533	45.65	507.012	42.60
No Peak in Range		588.153	48.89	581.116	49.63
668.599	49.32	662.431	50.04	643.406	49.03
723.678	47.32	739.229	48.08	734.798	46.27
810.206	50.99	No Peak in Range		No Peak in Range	
1047.370	54.84	1070.918	55.84	1007.760	53.72

electromagnetic feedthrough, triple transit reflections, acoustic processes within the IDTs, and other acoustic reflections [5-9]. The low insertion loss, at 900 MHz and higher, which deviates from the ideal response, can be attributed in part to bulk modes that travel through the substrate that are generated by the IDTs and appear at frequencies higher than the characteristic frequency of the device [8].

To increase the center frequency of the SAW devices, the resolution of the IDTs must be increased. Current image reversal photolithography processes limit the resolution size. SAW devices with IDT finger widths as small as 500 nm have been designed to be fabricated by SEM E-beam lithography. Using the acoustic velocity of AlN on Si reported as 5000 to 5150 m/s [5-7], it is believed that IDTs deposited on AlN/Si, with finger widths and spacing of 500 nm will generate SAWs with center frequencies in the 2.50 to 2.58 GHz range.

In order for SAW devices to be plausible as biosensors capable of operating in liquid environments, SAW device properties must be characterized within the presence of different liquids. Two methods for determining this are being explored. For the first method, micro-channels have been fabricated on top of existing SAW devices. A negative photolithography process and SU-8 have been used to create the micro-channels. Currently the micro-channels are either 160 μm or 230 μm wide and are located directly between two adjacent pairs of IDTs. A piece of glass is bonded to the top of the SU-8 to seal the micro-channels. One set of holes are drilled in the glass at each end of the micro-channels to insert catheters for the injection of liquid. Another set of holes is drilled in the glass to allow the probes to be placed onto the contact pads.

For the second method of testing SAW devices in liquid environments, AlN bridges have been fabricated on Si substrates [10]. The AlN bridges are fabricated by first growing separate rows of SiO_2 onto the Si substrate. A thin film of AlN is then grown on top off the Si and SiO_2 by MOVPE. The underlying SiO_2 is removed with a buffered oxide etchant to leave behind the AlN bridges. By making the thickness of the AlN bridges small enough, the SAW produced by an IDT deposited on the top of an AlN bridge can be limited to a single Lamb-wave mode. In Lamb-wave sensors, energy is not dissipated in liquids present at the piezoelectric surface and they are much more sensitive to mass loading then Rayleigh-wave sensors [3].

CONCLUSIONS

It is shown that SAW devices can be easily fabricated on AlN/Si; however, they exhibit a poor frequency response due to the conductivity of the Si substrate. A more careful consideration of the losses associated with conducting substrates is needed to improve the response of the SAW devices. All SAW devices in this experiment were deposited on 10 Ω·cm Si. Using a Si substrate with a higher resistivity will reduce the effect of the electromagnetic feedthrough between IDTs. A device with an out-of-band rejection similar to that of an ideal SAW filter (26dB) requires a substrate with a resistivity of a few kΩ·cm [5]. An improvement to the existing SAW device design shown in this experiment can reduce the distortions caused by reflections of the input and output signal due to impedance mismatches and other acoustic reflections such as triple transit reflections. The losses associated with the capacitance between IDT fingers and with the substrate most also be incorporated into a new design. E. Iborra *et al.* have proposed a new circuital model of the SAW filter to account for these capacitive losses [9] which will be applied to future iterations of higher frequency designs of SAW devices on AlN/Si.

ACKNOWLEDGEMENTS

This technical effort was performed in support of the National Energy Technology Laboratory's on-going research in high temperature flow control hardware for advanced power systems under the RDS contract DE-AC26-04NT41817. This work was also supported in part by NSF RII contract EPS 0554328 for which WV EPSCoR and WVU Research Corp matched funds.

REFERENCES

1. D.S. Ballantine Jr., R.M. White, S.J. Martin, E.T. Zellers, and H. Wohltjen, *Acoustic Wave Sensors: Theory, Design, and Physico-Chemical Applications*, 1st ed. (Academic Press, San Diego, 1997) p.70-99.
2. F.S. Hickernell, *12th Int. Con. on Micro. and Rad.*, **4**, 159 (1998).
3. T.M.A. Gronewold, *Anal. Chim. Acta.*, **603**, 119 (2007).
4. J. Harman, A. Kabulski, V.R. Pagán, P. Famouri, K.R. Kasarla, L.E. Rodak, J. Peter Hensel, D. Korakakis, *J. Vac. Sci. Technol. B*, **26(4)**, 1417 (2008)
5. E. Iborra, L. Vergara, J. Sangrador, M. Clement, A. Sanz-Hervá, and J. Olivares, *IEEE Trans. Ultrason., Ferroelect., Freq. Contr.*, **54(11)**, 2367 (2007).
6. C.W. Nam and K.C. Lee, *Proc. 5th KORUS Inter. Symp. on Sci. and Tech.*, **1**, 206 (2001).
7. K. Tsubouchi, K. Sugai, N. Mikoshiba, *Proc. IEEE Ultrason. Symp.*, **1**, 375 (1981).
8. R.C. Bray, T.L. Bagwell, R.L. Jungerman, and S.S. Elliot, *HP Application Note 5954-1569*, (1986).
9. E. Iborra, L. Vergara, J. Sangrador, M. Clement, A. Sanz-Hervá, and J. Olivares, *IEEE Trans. Ultrason., Ferroelect., Freq. Contr.*, **3**, 1880 (2004).
10. L.E. Rodak, Sridhar Kuchibhatla, and D. Korakakis, *Mat. Ltrs.*, **63**, 1571 (2009).

Surface and Interface Properties

Mater. Res. Soc. Symp. Proc. Vol. 1202 © 2010 Materials Research Society 1202-I04-01

Manipulation of Surface Charge on GaN

J. D. Ferguson[1], M. A. Foussekis[1], M. D. Ruchala[1], J. C. Moore[2], M. A. Reshchikov[1], and A. A. Baski[1]
[1]Department of Physics, Virginia Commonwealth University, Richmond, VA, U.S.A.
[2]Department of Chemistry and Physics, Longwood University, Farmville, VA, U.S.A.

ABSTRACT

We have characterized the surface charge on a variety of GaN samples using two surface potential techniques, conventional Kelvin probe and scanning Kelvin probe microscope (SKPM). Kelvin probe was primarily used to measure the change in surface potential under UV illumination, otherwise known as the surface photovoltage (SPV). Due to band bending near the semiconductor surface of about 1 eV in dark conditions, the SPV signal for n-type GaN typically reaches 0.5 to 0.6 eV upon switching on UV light. This value can slowly decrease by up to 0.3 eV during UV illumination in air ambient for 2-3 hours. We report that samples with many hours of ambient UV exposure do not show this slow decrease during SPV measurements, consistent with the UV-induced growth of a thicker surface oxide that limits charge transfer. In addition to prolonged UV exposure, the surface contact potential was also manipulated by local charge injection. In this procedure, the surface is charged using a metallized atomic force microscope tip which is scanned in contact with the sample. Subsequent SKPM measurements indicate an increase or decrease in the surface contact potential for the charged region, depending on the applied voltage polarity. Measurements of the discharge behavior in dark for these regions show a logarithmic time behavior, similar to the decay behavior during our observations of SPV transients after switching off the light. As expected, illumination of the surface increases the discharge rate and restores the charged area to its original state.

INTRODUCTION

While GaN is widely used in optoelectronic applications in the blue and UV spectral regions, an understanding of its surface-related electrical and optical properties remains limited. The presence of surface charge can be an important factor for the proper functioning of GaN devices; however, little is known about the origin of the surface charge in n- and p-type GaN. It is well established that n-type GaN shows an upward band bending of about 1 eV due to negative charge at the surface [1,2,3,4], while p-type GaN exhibits even larger downward band bending due to positive charge at the surface [3,5,6,7]. The origin of such surface charges may be due to surface states or adsorbed species on the surface. The band bending in n-type GaN can be reduced by 0.3 to 0.9 eV via the surface photovoltage (SPV) effect, where UV illumination causes photogenerated holes to reach the surface [3,8,9,10]. The SPV and presence of charge at the surface can be determined using Kelvin probe measurements. Previously, we reported preliminary results on local charging of the n-type GaN surface with negative charge [11]. In this work we present results on charging and its dynamics for n- and p-type GaN with both signs of charge.

EXPERIMENTAL DETAILS

The surface photovoltage data were acquired using a commercial Kelvin probe (McAllister) attached to an optical cryostat. Prior to illumination, the sample was maintained in dark for an extended period to minimize any residual SPV from previous light exposure. The sample was illuminated from the backside through a sapphire window using a 75-W xenon lamp passing through a monochromator and long-pass filters. All data presented here are acquired using band-to-band illumination at 365 nm with a light power density of 0.03 W/cm^2.

We also used a commercial atomic force microscope (Veeco Icon) to perform local charging of the surface in a scanned region and then subsequently measured any changes in the surface contact potential. The charging was performed by applying an external voltage (±10 V) to the sample while scanning in contact mode with a metallized tip (Ti-Pt). The resulting change in surface charge was then measured by acquiring local surface contact potential data using scanning Kelvin probe microscopy (SKPM). A standard imaging mode was used to characterize the extent of the initially charged region, whereas the discharge behavior of this region was monitored by collecting surface potential data along a scan line crossing both the charged and uncharged regions.

As summarized in Table 1.1, this study examines several c-plane GaN films grown on sapphire using molecular beam epitaxy (MBE), metal-organic chemical vapor deposition (MOCVD), and/or hydride vapor phase deposition (HVPE). All quoted free carrier densities are obtained using a four-point Hall effect apparatus.

Table 1.1: GaN samples

Sample	Doping type	Technique	Source
750	n-type (undoped, n = 2×10^{16} cm^{-3}))	MBE	VCU - H. Morkoç
1906	n-type (undoped, n = 5×10^{16} cm^{-3})	MBE on MOCVD	VCU - H. Morkoç
2015	n-type (Si-doped)	HVPE	TDI, Inc.
1321	p-type (Mg-doped)	HVPE	TDI, Inc.
2040	p-type (Mg-doped, p = 2×10^{16} cm^{-3})	MOCVD	SUNY-Albany

RESULTS and DISCUSSION

Surface photovoltage (SPV) data indicate that the surface charge on GaN can change during UV illumination [12]. Figure 1(a) shows SPV data taken in both air and vacuum for an undoped GaN sample (750). In air ambient the SPV gradually decreases during illumination, whereas it is observed to increase with time in vacuum. An increase (or decrease) in SPV corresponds to a corresponding increase (or decrease) in the surface contact potential. In air, this behavior can be explained by the photo-induced chemisorption of negatively charged surface species. Our previous experiments indicate that this species is most likely physisorbed oxygen that becomes chemisorbed when electrons are transferred from the bulk to the surface. In the case of vacuum, the increasing SPV is caused by the photo-induced desorption of these negatively charged species. We have recently observed an aging effect on samples exposed to UV illumination for many hours under air ambient conditions. This change in SPV behavior was further investigated by examining the effect of exposing pieces from a large wafer (2015) to a 100-W Hg lamp in ambient. Figure 1(b) illustrates that samples with more than 10 h UV exposure do not show as significant a change in surface charge during ambient SPV measurements. This result can be explained by the presence of a thicker oxide layer on the UV-exposed samples which inhibits the transport of charge between the bulk and surface. Preliminary experiments do indicate the presence of a thicker oxide on the UV-exposed samples, which can then be chemically etched away to restore the behavior seen for the native surface.

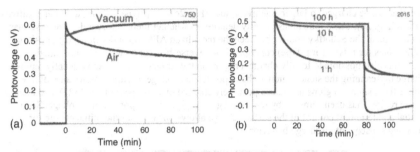

Figure 1. (a) SPV of n-type GaN (750) in both ambient and vacuum with UV illumination turned on at t = 0 s. (b) SPV of n-type GaN (2015) for three samples having the indicated exposure times to a UV lamp in ambient. The SPV illumination is turned on at t = 0 s and turned off at approximately 80 min. for all samples.

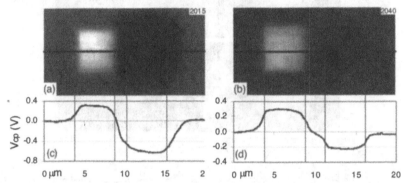

Figure 2. (a,b) SKPM surface potential images (10 μm × 20 μm) of n-type (2015) and p-type (2040) films, respectively. The left region on each image has been positively charged (–10 V on sample) while scanning in contact-mode AFM (5 μm × 5 μm square), whereas the right region has been negatively charged. (c,d) Line scans along the indicated black lines in (a) and (b), respectively, showing the relative surface contact potentials V_{cp} of the charged and uncharged regions.

In addition to SPV measurements using a Kelvin probe, we have used contact-mode AFM to perform local charging experiments and to observe the change in surface potential with SKPM. Using this technique, we have observed potential differences between charged and uncharged regions ranging from 0.1 V to over 2 V. The polarity of the applied sample bias determines whether the scanned region becomes positively or negatively charged. An applied sample voltage of –10 V causes electrons to flow from the sample-to-tip and positively charges the surface in the scanned region, whereas a voltage of +10 V negatively charges the surface. In SKPM potential images, more positively (or negatively) charged regions appear brighter (or darker). Examples of charging with either polarity on n-type and p-type samples are shown in Fig. 2. There is an asymmetry between the maximum voltages obtained for oppositely charged regions on a given sample. As indicated by the cross-sections for the SKPM images, n-type samples can be charged to larger negative values (Fig. 2c) and p-type samples to larger positive values (Fig. 2d). The applied bias polarity determines whether there is forward- or reverse-bias conduction at the Schottky contact between the metallized AFM tip and sample, e.g., positive sample bias for n-type (or p-type) corresponds to reverse-bias (or forward-bias) conduction.

It is also possible to locally charge a region on the surface and then subsequently remove charge by scanning the same region at the opposite bias voltage. Figure 3 shows an example where three square regions have been positively charged on a p-type sample (1321), and then one of the regions has been "erased" by re-scanning the region at the opposite bias voltage. Such spatial and temporal control of the surface charge allows one to tailor the initial charge for studying subsequent discharge behavior.

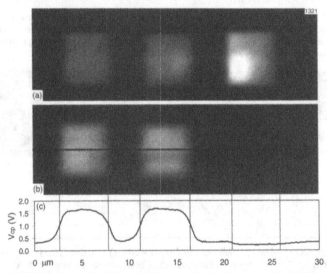

Figure 3. (a) SKPM surface potential image (10 μm × 30 μm) of a p-type film (1321), where three regions have been positively charged (−10 V on sample) from left-to-right while scanning in contact-mode AFM (5 μm × 5 μm squares). (b) SKPM image of the same area after re-scanning the square region on the right at the opposite bias voltage (+10 V). (c) Line scan along the black line in (b) showing the relative surface contact potentials Vcp of the charged and uncharged regions.

The charged regions can persist for hours if the sample is maintained in a dark environment. We monitored the discharge behavior of these regions as a function of time for a variety of samples and charging polarities. Figures 4(a) and 4(b) illustrates time-dependent discharge behaviors for n-type (2015) and p-type (2040) samples, respectively. These graphs plot the normalized difference in surface contact potential between charged (positive or negative) and uncharged regions as a function of time on a logarithmic scale. In all cases, the discharge behavior appears to be logarithmic, where negatively charged regions discharge more quickly than positively charged ones. In Fig. 5, the effect of sub-bandgap illumination by a HeNe laser is illustrated for an n-type sample (1906). As expected, the discharge is faster under illumination, but it is still on the order of minutes. Above-bandgap light discharges the surface within a few minutes or less. It is interesting to note that a similar logarithmic time dependence is observed for the decay of the SPV signal after turning off illumination, indicating a similar mechanism for charge restoration [12].

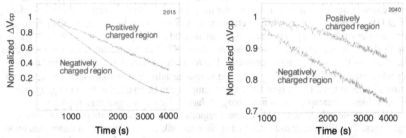

Figure 4. (a,b) Graphs of normalized differences in surface contact potential between charged and uncharged regions as a function of time on a logarithmic scale for n-type (2015) and p-type (2040) GaN, respectively. The upper (blue) lines indicate the discharge behavior for positively charged regions, while the lower (black) lines are for negatively charged regions.

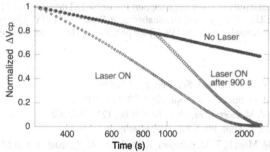

Figure 5. Graph of normalized differences in surface contact potential between charged and uncharged regions as a function of time on a logarithmic scale for n-type GaN (1906). The three lines indicate the discharge behavior in dark (labeled 'No laser'), during illumination with below-bandgap HeNe light ('Laser ON'), and a data run where the light was only turned on at ~900 s.

As seen in Fig. 4, we observe a faster decay for negatively charged regions as opposed to positively charged ones for both n- and p-type samples. This behavior indicates a faster surface-to-bulk versus bulk-to-surface electron transfer rate. Native GaN surfaces have a thin oxide layer (~ 1 nm) that would form a barrier to charge transfer [13]. If this oxide layer is sufficiently thin, however, electrons can tunnel through the oxide from the surface into the bulk. Our data show that the reverse process of tunneling from the bulk to surface is significantly slower. In the case of n-type GaN with upward band bending, electrons would encounter a thicker effective tunneling barrier, thereby decreasing the transfer rate. In the case of p-type GaN with downward band bending, the barrier for electrons would be comparable to the oxide thickness, but electrons are minority carriers and their contribution to charge transfer would be very low.

CONCLUSIONS

The surface charge on a variety of GaN samples has been manipulated either indirectly by the UV-induced growth of a surface oxide or directly by charge transfer using an AFM. Surface photovoltage measurements indicate that prolonged UV exposure under ambient conditions inhibits charge transfer. This is consistent with the growth of a thicker surface oxide layer during ambient UV illumination, which can be etched to restore the initial behavior. Another method to manipulate the surface charge is by scanning a metallized AFM tip in contact with the surface. If the applied sample bias is negative (or positive), then SKPM measurements indicate an increase (or decrease) of the surface contact potential. The resulting change in surface potential for two oppositely charged regions (± 10 V) on a given sample usually demonstrates an asymmetry as a function of sample type, i.e., n-type (or p-type) samples can be charged to larger negative (or positive) values. All samples show a discharge behavior for the charged regions that is logarithmic in time, where negatively charged regions decay at a faster rate than positively charged ones. This difference indicates that surface-to-bulk electron transfer is faster than bulk-to-surface transfer, which supports an electron tunneling mechanism through a thin surface oxide layer. Future studies will investigate the extent of thermionic processes in this system by performing temperature-dependent SKPM studies after charge transfer.

ACKNOWLEDGMENTS

The authors gratefully acknowledge support by the NSF (DMR-0804679). We also thank Dr. H. Morkoç (VCU), TDI, Inc., and Dr. Shahedipour-Sandvik (SUNY-Albany) for providing GaN samples. We acknowledge X. Ni (VCU) for performing Hall effect measurements.

REFERENCES

1   L. Kronik and Y. Shapira, Surf. Sci. Rep. **37**, 1 (1999).

2   G. Koley and M. G. Spencer, J. Appl. Phys. **90**, 337 (2001).

3   J. P. Long and V. M. Bermudez, Phys. Rev. B **66**, 121308 (2002).

4   V. M. Bermudez, J. Appl. Phys. **80**, 1190 (1996).

5   M. Eyckeler, W. Mönch, T. U. Kampen, R. Dimitrov, O. Ambacher, and M. Stutzmann, J. Vac. Sci. Technol. B **16**, 2224 (1998).

6   S. Barbet, R. Aubry, M.-A. di Forte-Poisson, J.-C. Jacquet, D. Deresmes, T. Mélin, and D. Théron, Appl. Phys. Lett. **93**, 212107 (2008).

7   M. Petravic, V. A. Coleman, K.-J. Kim, B. Kim, and G. Li, J. Vac. Sci. Technol. A **23**, 1340 (2005).

8   I. Shalish, Y. Shapira, L. Burstein , and J. Salzman, J. Appl. Phys. **89**, 390 (2001).

9   S. Sabuktagin, M. A. Reshchikov, D. K. Johnstone, and H. Morkoç, Mat. Res. Soc. Symp. Proc. **798**, Y5.39 (2004).

10 I. Shalish, L. Kronik, G. Segal, Y. Rosenwaks, Y. Shapira, U. Tisch, and J. Salzman, Phys. Rev. B **59**, 9748 (1999).

11 J.C. Moore, M.A. Reshchikov, J.E. Ortiz, J. Xie, H. Morkoç, and A.A. Baski, Proc. of SPIE **6894**, 68940B1-9 (2008).

12 M. Foussekis, A. A. Baski, and M.A. Reshchikov, Appl. Phys. Lett. **94** 162116 (2009).

13 T. Sasaki and T. Matsuoka, J. Appl. Phys., **64**, 4531 (1998).

Mater. Res. Soc. Symp. Proc. Vol. 1202 © 2010 Materials Research Society   1202-I04-03

Electronic Properties of III-Nitride Surfaces and Interfaces Studied by Scanning Photoelectron Microscopy and Spectroscopy

Cheng-Tai Kuo[1], Hong-Mao Lee[1], Chung-Lin Wu[2], Hung-Wei Shiu[1,3], Chia-Hao Chen[3], and Shangjr Gwo[1]
[1]Department of Physics, National Tsing-Hua University, Hsinchu 30013, Taiwan
[2]Department of Physics, National Cheng-Kung University, Tainan 70101, Taiwan
[3]National Synchrotron Radiation Research Center (NSRRC), Hsinchu 30076, Taiwan

ABSTRACT

We report on a method based on cross-sectional scanning photoelectron microscopy and spectroscopy (XSPEM/S) for studying electronic structure of III-nitride surfaces and interfaces on a submicrometer scale. Cross-sectional III-nitride surfaces prepared by *in situ* cleavage were investigated to eliminate the polarization effects associated with the interface charges/dipoles normal to the cleaved surface. In contrast to the as-grown polar surfaces which show strong surface band bending, the cleaved nonpolar surfaces have been found to be under the flat-band conditions. Therefore, both doping and compositional junctions can be directly visualized at the cleaved nonpolar surfaces. Additionally, we show that the "intrinsic" valence band offsets at the cleaved III-nitride heterojunctions can be unambiguously determined.

INTRODUCTION

III-nitride semiconductors such as indium nitride (InN), gallium nitride (GaN), aluminum nitride (AlN), and their alloys are pyroelectric materials, which have large spontaneous polarizations along the polar axis. The polarization effects play an important role at the III-nitride surfaces and interfaces. For example, spontaneous polarization-induced sheet charges at the surface result in the surface band bending difference between Ga- and N-polar GaN surfaces [1]. In addition, the difference in spontaneous polarization can induce a high charge density at the III-nitride heterointerface to form a two-dimensional electron (or hole) system. As a result, III-nitride heterostructures are promising materials for applications using heterojunction field-effect transistor (HFET) structures. Recently, ultrathin InN ion sensitive field-effect transistors (ISFETs) have been shown to have large sensitivity and fast response time for ion sensing in liquid [2]. Moreover, InN/AlN metal-insulator-semiconductor heterojunction field-effect transistor (MISHFET), which exhibits a higher sheet carrier density than conventional devices, can be attributed to a large spontaneous polarization difference at the InN/AlN interface [3]. However, due to the existence of large polarization differences at wurtzite-type III-nitride heterointerfaces, many reported valence band offset (VBO) values for III-nitride heterojunctions show a large discrepancy [4-7]. In this paper, we report on a method for measurements of electronic properties at the III-nitride surfaces and interfaces by photoelectron emissions from *in situ* cleaved, nonpolar cross-sectional surfaces of III-nitride multilayered structures.

EXPERIMENT

All of the III-nitride epitaxial films were grown by plasma-assisted molecule beam epitaxy (PA-MBE). Details of the growth process can be found elsewhere [8-10].The experiments were carried out at the 09A1 beamline of the National Synchrotron Radiation Research Center (NSRRC) in Hsinchu, Taiwan. Figure 1(a) shows the schematic of sample probing geometry in conventional synchrotron radiation photoelectron spectroscopy (PES). Samples were cleaned and transferred into UHV SPEM system. Conventional PES spectra were obtained with incoming energy of 130 eV and 380 eV. Figure 1(b) shows the schematic of the sample probing geometry in XSPEM/S. A combination of Fresnel zone plate and order sorting aperture was used to focus the monochromatic soft X-ray (hv =380 eV). For photon energy of 380 eV with grating 400 lines/mm, the flux is about 6×10^{12} photons/sec and the diameter of the unfocused beam is 200 μm. Samples were cleaved under UHV conditions in the SPEM chamber in order to reveal clean cross-sectional surfaces for chemical imaging and micro-area photoelectron spectroscopy (μ-PES). The sample position relative to the focused photon beam was controlled by a piezo-driven flexture stage (range: 100×100 μm^2) for raster scanning in SPEM imaging and for fine positioning in μ-PES measurements.

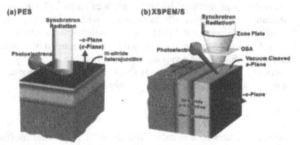

Figure 1. Sample probing geometry in (a) PES and (b) XSPEM/S for studying wurtzite III-nitride heterojunction (or *p-n* junction) grown on Si(111).

RESULTS AND DISCUSSION

Two main results will be discussed below. The first result is the direct imaging of III-nitride *p-n* junction by XSPEM/S [11]. And the second one is related to the measurement of band lineups for both polar and nonpolar InN/GaN heterojunctions [12, 13].

<u>**Direct imaging of GaN *p-n* junction**</u>

The sample structure for this experiment is *p*-GaN (1.5 μm)/*n*-GaN (1.5 μm)/AlN (25 nm)/Si$_3$N$_4$/Si(111). The *n*- and *p*-type doping was performed by using high-purity Si and Mg solid cells during the PA-MBE growth process. The Mg concentration determined by secondary ion mass spectroscopy (SIMS) is 7×10^{19} cm^{-3}. The actual *p*-type carrier concentration was confirmed by Hall effect measurement to be 1×10^{17} cm^{-3} in the van der Pauw configuration using Ni/Au Ohmic contacts on a *p*-type GaN film grown under the same conditions.

For III-nitride epilayers, the crystal orientation was confirmed to follow the following relationship, [000–1]$_{III-N}$ ∥ [111]$_{Si}$ and <–1–120>$_{III-N}$ ∥ <1–10>$_{Si}$. Therefore, by using the III-nitride epilayer samples grown on Si substrates, the nonpolar a-plane surface can be exposed by cleavage. The nonpolar, a-plane GaN p-n junction was *in-situ* cleaved under UHV conditions in order to expose a clean cleaved surface. The surface cleavage quality was confirmed by *ex situ* scanning electron microscopy (SEM) (Figure 2(a)) after the SPEM/S experiment. Figure 2(b) shows the SPEM images obtained from three selected photoelectron energy channels, corresponding to the binding energies of Ga $3d$ core levels in p-type GaN and n-type GaN, respectively, as well as Si $2p$ core level in Si substrate. Figure 2(c) shows the μ-PES spectra taken from different regions at the cross-sectional surface of GaN p-n junction. The Ga $3d$ core-level peak of n-type GaN is at 20.5 eV, while that of p-type GaN is at 18.4 eV. Therefore, the binding energy difference of Ga $3d$ core levels for two doping types of GaN can be directly determined to be 2.1 eV. This result clearly indicates that the built-in voltage at the bulk GaN p-n junction can be observed at the cleaved surface. The binding energy difference between the Ga $3d$ core level and the valence band maximum (VBM), denoted as $E_{Ga\,3d}$–E_V, is a constant value of 18.0 eV [12]. This value is independent of doping type, crystal polarity, and surface band bending (shown in Figure 3(a)). As a result, the valence band edges of p- and n-type GaN relative to the Fermi level can be determined. As shown in Figure 2(c), the surface Fermi level of p-type GaN is 0.4 eV above the VBM. This is very close to that suggested earlier for the case of flat-band p-type GaN [14]. On the other hand, the surface Fermi level of n-type GaN is 2.5 eV above the VBM. From our previous SPEM/S study on the cleaved a-plane surface of an undoped GaN layer grown on Si(111) substrate by PA-MBE, the surface Fermi level was measured to be at ~1.3 eV above the VBM [13, 15]. In comparison with the undoped GaN, we observed a large movement (~1.2 eV) of the Fermi level toward the CBM for the present Si-doped GaN case, indicating the effect of n-type doping. Discussions about Si-doped GaN can be found in the literature [11].

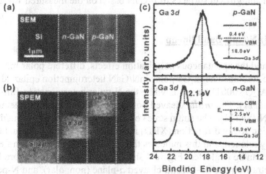

Figure 2. XSPEM/S measurement on the *in situ* cleaved a-plane GaN p-n junction cross-sectional surface (a) *Ex situ* SEM image (b) Chemical mapping images taken from selected energy channels, corresponding to binding energies of Ga $3d$ core levels in p- and n-GaN region, as well as Si $2p$ core level in Si substrate region. (c) μ-PES spectra taken in the regions of p-GaN layer and n-GaN layer. The insets show the corresponding energy-band diagrams of the cleaved GaN p-n junction surface in the p- and n- GaN region.

To compare with the results of cleaved nonpolar p-GaN surface, the surface band bending of the as-grown N-polar p-GaN surface (the same sample) was investigated by photoelectron spectroscopy and the result is shown in Figure 3 (a). The as-grown GaN:Mg surface was studied without any surface treatment and it may be nonstoichiometric and covered with surface oxides. In this case, the Ga $3d$ core level peak of as-grown GaN:Mg surface was measured to be 21.0 eV relative to the Au Fermi edge, which is very different from that of the cleaved GaN:Mg a-plane surface. Additionally, the surface Fermi-level position of the as-grown GaN:Mg layer was determined to be 3.0 eV above the linearly extrapolated VBM, in sharp contrast to 0.4 eV for the case of *in situ* cleaved a-plane GaN:Mg. Therefore, the value of surface band bending can be determined to be 2.6 eV (shown in Figure 3(b)), which is in good agreement with the literature value [14].

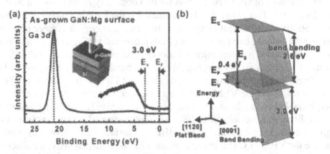

Figure 3. (a) Broad-beam photoelectron spectrum of the as-grown GaN:Mg N-polar film, which is the same sample used for XSPEM/S measurements. (b) Two-dimensional energy-band diagram of cleaved and growth surfaces of p-GaN based on the measured values of E_V-E_F in (a) and Figure2(c).

Measurement of InN/GaN band lineup

To understand the spontaneous polarization effects, different polar InN/GaN heterojunctions are studied. The In/Ga-polar InN/GaN heterojunction epitaxial film was grown on Ga-polar GaN/Al$_2$O$_3$(0001) template, while the N-polar heterojunction film was grown on N-polar AlN/Si$_3$N$_4$/Si(111). The *in situ* cleaved non-polar, a-plane InN/GaN heterojunction was grown on AlN/Si$_3$N$_4$/Si(111). The strain states of the samples were determined by *ex situ* synchrotron-radiation X-ray diffraction (XRD) measurements. Both InN/GaN heterojunctions are near completely relaxed, and this indicates the piezoelectric polarization fields at the InN/GaN interface can be neglected. Figure 4 shows the Ga $3d$ and In $4d$ core-level photoelectron spectra from In/Ga-, *in situ* cleaved a-plane (nonpolar), and N-polar InN/GaN heterojunctions. For the present case of polar InN/GaN heterojunctions, the interface dipoles would result in the change of the Ga $3d$ core-level emission energy from the buried GaN layer. As a consequence, the measured binding energies of Ga $3d$ core-level emissions can be smaller or larger, depending on the orientation of interface dipole. Thus, the binding energy difference of Ga $3d$ and In $4d$ core-levels for In/Ga-polar ($\Delta E_{CL}^{(In/Ga)}$ =2.2 eV) is larger than that for N-polar ($\Delta E_{CL}^{(N)}$ =1.7 eV), while for the cleaved nonpolar, a-plane ($\Delta E_{CL}^{(Non)}$ =2.0 eV) is just between the values of In/Ga- and N-polar cases. The reason that the measured core-level separation ($\Delta E_{CL}^{(Non)}$) between Ga $3d$ and In $4d$ for the cleaved a-plane is very different from those of polar InN/GaN heterojunctions is due to the absence of interface dipole effects for the nonpolar probing geometry.

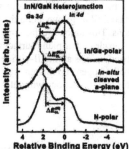

Figure 4. Ga 3*d* and In 4*d* core-level spectra of In/Ga-polar, *in situ* cleaved *a*-plane, and N-polar InN/GaN heterojunctions.

The valence band offset (VBO) can be determined according to the following equation:

$$\Delta E_V = \left(E_{In\,4d} - E_V\right)_{InN} - \left(E_{Ga\,3d} - E_V\right)_{GaN} + \left(\Delta E_{CL}\right)_{InN/GaN}. \qquad (1)$$

The first two terms in equation 1 were the binding energy differences of In 4*d* and Ga 3*d* core levels with respect to the corresponding VBM positions. The binding energy positions of valence-band maxima (VBM) (E_V) were obtained by linear extrapolation of the valence-band spectra. In our study, the $E_{CL}-E_V$ values in bulk samples should be independent of the crystal polarity. The $E_{In\,4d}-E_V$ value is ~16.8 eV both for the In polar sample and N-polar sample, while the $E_{Ga\,3d}-E_V$ is ~18.0 eV both for Ga-polar and N-polar sample. The uncertainty in the determination of $E_{CL}-E_V$, which is mainly limited by the linear extrapolation procedure of valence-band edges, is reproducible better than the instrumental resolution of 0.1 eV. As a result, the VBO values of different polar InN/GaN can be determined. By using equation 1, the VBOs are 1.0 eV, 0.8 eV, and 0.5 eV for In/Ga-, Non-, and N-polar InN/GaN heterojunctions. The asymmetry observed in VBOs measured along opposite polar directions is attributed to the inverted polarization-induced interface dipoles existing at In-polarity and N-polarity heterojunctions. Therefore, the VBO (0.8 eV) measured from the *in situ* cleaved *a*-plane InN/GaN should correspond more closely to the "intrinsic" VBO without the influence of spontaneous and piezoelectric polarization effects. The "intrinsic" conduction band offset (CBO) can also be calculated to be 2.0 eV by using the known bandgap values of 3.4 eV and 0.65 eV for GaN and InN (shown in Figure 5). The InN/GaN heterojunction lineup is found to be a type-I band alignment.

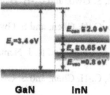

Figure 5. Schematic of the "intrinsic" band lineup of InN/GaN determined by XSPEM/S.

91

CONCLUSIONS

In summary, we have found that nonpolar, a-plane cross-sectional III-nitride surfaces prepared by *in situ* cleavage of epitaxial films grown along the polar axis on Si(111) substrate are under the surface flat-band conditions. This unique surface property allows direct imaging of GaN p-n junction by XSPEM/S on the cleaved a-plane surfaces. Moreover, under the surface flat-band conditions, the intrinsic band lineups at the III-nitrides heterojunctions can be experimentally determined. This method opens up the opportunities for studying intrinsic band lineups of other polar semiconductor heterojunctions without the influence of polarization effects

ACKNOWLEDGMENTS

This work was supported in part by a research grant (NSC 98-2112-M-007-014-MY3) from the National Science Council (NSC) of Taiwan.

REFERENCES

1. H. W. Jang, K. W. Ihm, T.-H. Kang, J.-H. Lee, and J.-L. Lee, *Phys. Status Solidi. B* **240**, 451 (2003).
2. Y.-S. Lu, C.-L. Ho, J. A. Yeh, H.-W. Lin, and S. Gwo, *Appl. Phys. Lett.* **92**, 212102 (2008).
3. Y.-S. Lin, S.-H. Koa, C.-Y. Chan, S. S. H. Hsu, H.-M. Lee, and S. Gwo, *Appl. Phys. Lett.* **90**, 142111 (2007).
4. G. Martin, S. Strite, A. Botchkarev, A. Agarwal, A. Rockett, H. Morkoç, W. R. L. Lambrecht, and B. Segall, *Appl. Phys. Lett.* **65**, 610 (1994).
5. G. Martin, A. Botchkarev, A. Rockett, and H. Morkoç, *Appl. Phys. Lett.* **68**, 2541 (1996).
6. C.-L. Wu, C.-H. Shen, and S. Gwo, *Appl. Phys. Lett.* **88**, 032105 (2006).
7. P. D. C. King, T. D. Veal, P. H. Jefferson, C. F. McConville, T. Wang, P. J. Parbrook, H. Lu, and W. J. Schaff, *Appl. Phys. Lett.* **90**, 132105 (2007).
8. C.-L. Wu, J.-C. Wang, M.-H. Chan, T. T. Chen, and S. Gwo, *Appl. Phys. Lett.* **83**, 4530 (2003).
9. S. Gwo, C.-L. Wu, C.-H. Shen, W.-H. Chang, T. M. Hsu, J.-S. Wang, and J.-T. Hsu, *Appl. Phys. Lett.* **84**, 3765 (2004).
10. C.-L. Wu, C.-H. Shen, H.-W. Lin, H.-M. Lee, and S. Gwo, *Appl. Phys. Lett.* **87**, 241916 (2005).
11. C.-T. Kuo, H.-M. Lee, H.-W. Shiu, C.-H. Chen, and S. Gwo, *Appl. Phys. Lett.* **94**, 122110 (2009).
12. C.-L. Wu, H.-M. Lee, C.-T. Kuo, S. Gwo, and C.-H. Hsu, *Appl. Phys. Lett.* **91**, 042112 (2007).
13. C.-L. Wu, H.-M. Lee, C.-T. Kuo, C.-H. Chen, and S. Gwo, *Appl. Phys. Lett.* **92**, 162106 (2008).
14. K. M. Tracy, W. J. Mecouch, R. F. Davis, and R. J. Nemanich, *J. Appl. Phys.* **94**, 3163 (2003).
15. C.-L. Wu, H.-M. Lee, C.-T. Kuo, C.-H. Chen, and S. Gwo, *Phys. Rev. Lett.* **101**, 106803 (2008).

Mater. Res. Soc. Symp. Proc. Vol. 1202 © 2010 Materials Research Society   1202-I09-03

Band Gap States in AlGaN/GaN Hetero-Interface Studied by Deep-Level Optical Spectroscopy

Yoshitaka Nakano[1,3], Keiji Nakamura[1], Yoshihiro Irokawa[2], and Masaki Takeguchi[2]
[1]Chubu University, Kasugai, Aichi 487-8501, Japan
[2]National Institute for Materials Science, Tsukuba, Ibaraki 305-0044, Japan

ABSTRACT

Planar Pt/AlGaN/GaN Schottky barrier diodes (SBDs) have been characterized by capacitance-voltage and capacitance deep-level optical spectroscopy measurements, compared to reference Pt/GaN:Si SBDs. Two specific deep levels are found to be located at ~1.70 and ~2.08 eV below the conduction band, which are clearly different from deep-level defects (E_c - 1.40, E_c - 2.64, and E_c - 2.90 eV) observed in the Pt/GaN:Si SBDs. From the diode bias dependence of the steady-state photocapacitance, these levels are believed to stem from a two-dimensional electron gas (2DEG) region at the AlGaN/GaN hetero-interface. In particular, the 1.70 eV level is likely to act as an efficient generation-recombination center of 2DEG carriers.

INTRODUCTION

Electronic devices based on AlGaN/GaN heterostrucures are of great current interest because of their capability of operating at high temperature, high power, and high frequency. In particular, the AlGaN/GaN heterostructure has a thin, high-mobility channel due to a two-dimensional electron gas (2DEG) produced at the hetero-interface. Excellent device characteristics have been reported for these high electron mobility transistors (HEMTs) [1]. However, these characteristics are not always reproducible because the device performance at high frequencies can be limited by the presence of deep-level defects in the AlGaN/GaN heterostructure [2]. That is, electrical charge trapped by the deep levels alters the 2DEG concentration in the channel and limits the switching characteristics of the devices. Therefore, it is necessary in the basic sense to investigate deep-level defects in AlGaN/GaN heterostructures. Up to date, a number of research approaches have been employed to detect and identify these defects in the AlGaN/GaN heterostructures by using various characterization techniques such as drain leakage current measurements, photoionization spectroscopy, deep-level transient spectroscopy (DLTS), and deep-level optical spectroscopy (DLOS) [3-7]. However, the electrical and physical properties of various deep-level defects existing in the AlGaN/GaN heterostructures still remain uncertain, inclusive of understanding in which region of the device structure they are located. Notably, few investigations of deep levels, particularly at the AlGaN/GaN hetero-interface, have been reported. Among these electrical techniques, capacitance-based methods have a high sensitivity of in-depth probing, utilizing variation in depletion layer width. So, they can provide direct information regarding the location of defects within the AlGaN/GaN heterostructure through the choice of the diode bias V_G. From this point of view, an AlGaN/GaN-based Schottky barrier diode (SBD) would provide a suitable device structure for detailed investigations into the hetero-interfacial electronic states. In this paper, we report on specific deep levels, related to a 2DEG region in an AlGaN/GaN SBD fabricated on sapphire substrate, using capacitance-voltage and capacitance DLOS measurements.

EXPERIMENTAL

The AlGaN/GaN heterostructure was grown on c-plane sapphire substrate by using metal-organic chemical vapor deposition (MOCVD). It consisted of a GaN buffer layer, an unintentionally doped GaN layer (~2 μm thick), and an unintentionally doped AlGaN layer (20 nm thick) with an Al mole fraction of 30 %. The AlGaN/GaN heterostructure showed typical 2DEG properties with the sheet carrier concentration of 9.98×10^{12} cm^{-2} and the mobility of 1456 cm^2/Vs as determined by room-temperature Hall-effect measurements. After growth, planar SBDs were fabricated as follows. First, Ti/Al/Pt/Au metals were sequentially deposited on the top surface by electron-beam evaporation and then were sintered at 850 °C for 30 s for ohmic contacts. Finally, Pt metal was formed with a diameter of 0.9 mm as Schottky contacts. As a reference, planar Pt/Si-doped GaN (GaN:Si, ~10 μm thick) SBDs on c-plane sapphire substrate were also prepared by using the same methodology as stated above. Here, the Si doping concentration was ~4.5×10^{16} cm^{-3} as determined by secondary ion mass spectrometry (SIMS) measurements.

The Pt/AlGaN/GaN and the Pt/GaN:Si SBDs were characterized at room temperature by means of current-voltage (I-V), capacitance-frequency (C-f), capacitance-voltage (C-V), and capacitance DLOS measurements. From I-V measurements in the dark, good rectifier characteristics of the n-type Schottky diode were confirmed for both SBD samples. C-f measurements at zero dc bias showed that the measured capacitance varied markedly with frequency for both SBD samples; the capacitance cutoff frequencies f_T in the C-f characteristics were estimated to be ~668 kHz and ~8.91 MHz for the Pt/AlGaN/GaN and the Pt/GaN:Si SBD samples, respectively. From these experimental f_T data, conventional C-V measurements were conducted in the dark at 100 kHz, in which carriers can adequately follow the high-frequency voltage modulation. Photo C-V measurements were also conducted at 100 kHz under white light illumination ($\lambda > 390$ nm) from the back side by using a 150 W halogen lamp. In both C-V measurements, diode bias V_G was applied and the capacitance was measured after a delay time of 30 s. Capacitance DLOS measurements were performed at 100 kHz by measuring photocapacitance transients as a function of incident photon energy, starting from 0.78 eV (1600 nm) up to 4.0 eV (300 nm). Details of the DLOS measurements have been reported elsewhere [8]. The DLOS signal is defined as $\Delta Css/C$. Here, C is the depletion region capacitance under the measured bias conditions in the dark, before the optical excitation, and ΔCss is the steady-state photocapacitance which is determined as a saturation value of the recorded photocapacitance transients at each wavelength.

RESULTS AND DISCUSSION

The inset of Fig. 1 shows dark C-V characteristics of the Pt/AlGaN/GaN SBD sample in addition to the C-V data of the reference Pt/GaN:Si SBD sample. As for the Pt/AlGaN/GaN SBD sample, a low depression observed in the C-V profile is due to the accumulation of carriers at the AlGaN/GaN hetero-interface. The pinch-off voltage V_{th} is -3.45 V. These C-V characteristics can be roughly classified into three regions in view of diode bias V_G, (i) 0.7 V > V_G > -1.5 V, (ii) -1.5 V > V_G > -3.45 V, and (iii) V_G < -3.45 V. In the region (i) at low |V_G|, the capacitance is almost

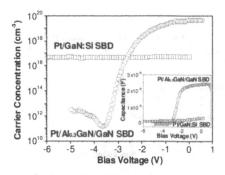

Figure 1. Carrier concentration as a function of diode bias voltage of Pt/AlGaN/GaN and Pt/GaN:Si SBDs. The inset shows dark *C-V* characteristics of Pt/AlGaN/GaN and Pt/GaN:Si SBDs.

flat due to only small changes induced by applying V_G, where the depletion layer width is calculated to be 20.7 - 22.4 nm, in reasonable agreement with the AlGaN layer thickness. So, the depletion layer probably extends to the 2DEG peak depth and the 2DEG is in an accumulation mode. Here, the 2DEG peak concentration is estimated to be ~4.8x10^{19} cm^{-3} from the slopes of $1/C^2$-V plots as shown in Fig. 1. In the region (ii), the capacitance rapidly decreases with increasing $|V_G|$ up to the $|V_{th}|$. So, in this case, the 2DEG is in a partially pinched-off mode. In the region (iii) where the 2DEG is completely pinched off, the capacitance reduces to a small constant value which reflects the combined capacitance of depleted AlGaN and GaN layers. In other word, in the pinch-off mode, the AlGaN/GaN heterostructure behaves as a bulk depletion region constituting mainly GaN. By contrast, as shown in Fig. 1, the effective carrier concentrations for the Pt/GaN:Si SBD sample are estimated to be ~4.9x10^{16} cm^{-3}, which are distributed almost uniformly over the depth of the capacitance measurements. This value is in good agreement with the Si doping concentrations as determined by the SIMS measurements.

Figure 2 shows dark and photo *C-V* characteristics of the Pt/AlGaN/GaN SBD sample. A small increase in capacitance with illumination is clearly observed, which strongly depends on the diode bias V_G; compared to the dark *C-V* characteristics, the photo *C-V* ones shift to a little more negative side in the partial pinch-off mode. As a result, the V_{th} is seen to increase from -3.45 V to -3.70 V on photo-illumination. Thus, the observed change in the *C-V* characteristics probably reflects an increase in 2DEG concentration due to deep-level photoemission [6,7]. Conversely, these experimental results suggest the presence of deep-level defects in the AlGaN/GaN heterostructure, which strongly act on the 2DEG concentration. The illumination-induced increase in capacitance ΔC is shown as a function of V_G in the inset of Fig. 2. ΔC tends to significantly increase and decrease in the partial pinch-off and the pinch-off modes, respectively, and also has a maximum value at the V_G of -2.7 V. From the integration of the ΔC peak in this figure, the increased concentration Δn_s of the 2DEG on illumination is estimated to be at least ~2.2x10^{11} cm^{-2} and is considered to be optically excited from deep-level defects to the 2DEG at the AlGaN/GaN hetero-interface.

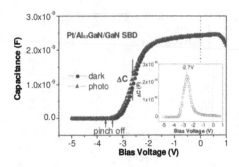

Figure 2. Dark and photo *C-V* characteristics of Pt/AlGaN/GaN SBD. The inset shows the increase in capacitance ΔC with illumination as a function of diode bias voltage.

Figure 3 shows DLOS spectra of the Pt/AlGaN/GaN SBD sample at V_G of -3.2 V, together with data of the Pt/GaN:Si SBD sample at V_G of -10 V. In both SBD samples, we can observe near-band-edge (NBE) emissions from GaN at 3.2 - 3.5 eV. For the reference Pt/GaN:Si SBD sample, three photoemission states are clearly observed with their onsets at ~1.40, ~2.64, and 2.90 eV below the conduction band, denoted as T_1, T_2, and T_3. For all the deep levels, electron emissions to the conduction band are a dominant process. These deep levels are identical to specific deep-level defects that have been commonly reported for GaN [8-11]. The detailed investigations regarding energy levels and source assignments for these levels are beyond the scope of this paper. On the other hand, DLOS spectra of the Pt/AlGaN/GaN SBD sample are significantly different from those of Pt/GaN:Si SBD. Here, the Pt/AlGaN/GaN SBD sample at V_G of -3.2V is in the partial pinch-off mode, as confirmed by the previous *C-V* measurements. In this case, three deep levels characteristic of AlGaN/GaN heterostructures, G_1, G_2 and G_3, are clearly observed. The G_2 and G_3 levels are respectively located at ~1.70 and ~2.08 eV below the conduction band, whereas the G_1 level has significant thermal contributions to the capacitance transients. Considering that the G_1 level is placed in the lower half of the band gap in *n*-type sample, electron transitions with illumination should occur from the G_1 level to the conduction band and/or from the valence band into the G_1 level. As the illumination energy is increased above the threshold energy of ~0.8 eV, the steady-state photocapacitance decreases correspondingly, indicating that the electron transition from the valence band into the G_1 level is a dominant process. So, the G_1 level behaves as a hole trap [6]. As for the G_2 level, the positive photocapacitance transients are clearly observed at photon energies above 1.70 eV, which are due to an electron transition from the G_2 level to the conduction band. Additionally, the G_2 level often appears as a "peak" in the DLOS spectra, depending on the film quality. That is, the G_2 level shows both the positive and the negative photocapacitance transients. These experimental results indicate a strong interaction of the G_2 level with both the conduction and the valence bands. Because the G_2 level is located in the half of the band gap, this deep level is potentially an efficient generation-recombination center that may impact the 2DEG carriers. The G_3 level is a deep level with an optical threshold of ~2.08 eV, which is apparently a different situation from the G_1 and G_2 levels. Considering that the G_3 level is placed in the higher half of the band gap,

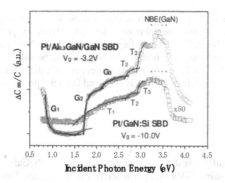

Figure 3. Room-temperature DLOS spectra of Pt/AlGaN/GaN SBD in partial pinch-off mode ($V_G = -3.2V$) and Pt/GaN:Si SBD ($V_G = -10V$).

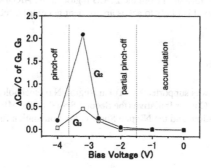

Figure 4. Steady-state photocapacitance $\Delta C_{ss}/C$ of G_2 and G_3 levels as a function of diode bias voltage.

both electrons and holes can be promoted from the G_3 level to the conduction and the valence bands, respectively. In this case, the positive photocapacitance transients observed at photon energies above ~2.08 eV are mainly due to an electron transition from the G_3 level to the conduction band. More importantly, both the G_2 and the G_3 levels show a similar strong V_G dependence of increase in steady-state photocapacitance $\Delta C_{ss}/C$ as shown in Fig. 4. The corresponding $\Delta C_{ss}/C$ of these levels is almost unchanged in the accumulation mode and significantly increases with increasing the $|V_G|$ in the partial pinch-off mode, whereas that drastically decreases in the pinch-off mode. As states above, in the pinch-off mode, the AlGaN/GaN heterostructure behaves as a bulk depletion region, comprising mainly GaN. In addition, considering that DLOS measures changes in capacitance arising from the depletion region, the photocapacitance assignable to the band gap states in AlGaN region and AlGaN/GaN hetero-interface should be strongly diminished, compared to that of GaN region in the pinch-off mode. Combining with these experimental results, the G_2 and G_3 levels probably stem from the 2DEG region at the AlGaN/GaN hetero-interface rather than the GaN region. In particular, the G_2 level, with a strong interaction with both the conduction and the valence bands, may act on the carrier transportation of 2DEG, in reasonable agreement with the experimental fact of the increase in 2DEG carriers with illumination, as observed from the previous photo C-V measurements.

CONCLUSIONS

We have successfully investigated band gap states in AlGaN/GaN heterostructure grown on sapphire substrate by using C-V and capacitance DLOS techniques. Two specific deep levels were revealed to be located at ~1.70 and ~2.08 eV below the conduction band, being clearly different from deep-level defects observed in GaN. Both deep levels showed a significant increase in their corresponding steady-state photocapacitance in partial pinch-off mode. Therefore, these levels probably stem from the 2DEG region at the AlGaN/GaN hetero-interface. In particular, the 1.70 eV level is likely to act as an efficient generation-recombination center of 2DEG carriers.

ACKNOWLEDGMENTS

A part of this work was supported by "Tokai Region Nanotechnology Manufacturing Cluster" in the Knowledge Cluster Initiative (the Second Stage) of the MEXT, the CASIO Science Promotion Foundation, and the Nippon Sheet Glass Foundation for Materials Science and Engineering, Japan.

REFERENCES

1. W. Lu, J. Yang, M. A. Khan, and I. Adesida, *IEEE Trans. Electron Devices* **48**, 581 (2001).
2. L. F. Eastman, *Phys. Status Solidi (a)* **176**, 175 (1999).
3. S. Arulkumaran, T. Egawa, H. Ishikawa, and T. Jimbo, *Appl. Phys. Lett.* **81**, 3073 (2002).
4. P. B. Klein, S. C. Binari, K. Ikossi, A. E. Wickenden, D. D. Koleske, and R. L. Henry, *Appl. Phys. Lett.* **79**, 3527 (2001).
5. A. P. Zhang, L. B. Rowland, E. B. Kaminsky, V. Tilak, J. C. Grande, J. Teetsov, A. Vertiatchikh, and L. F. Eastman, *J. Electron. Mater.* **32**, 388 (2003).
6. Z.-Q. Fang, D. C. Look, D. H. Kim, and I. Adesida, *Appl. Phys. Lett.* **87**, 182115 (2005).
7. A. Armstrong, A. Chakraborty, J. S. Speck, S. P. DenBaars, U. K. Mishra, and S. A. Ringel, *Appl. Phys. Lett.* **89**, 262116 (2006).
8. Y. Nakano, T. Morikawa, T. Ohwaki, and Y. Taga, *Appl. Phys. Lett.* **87**, 232101, (2005).
9. A. Hierro, D. Kwon, S. A. Ringel, M. Hansen, J. S. Speck, U. K. Mishra, and S. P. DenBaars, *Appl. Phys. Lett.* **76**, 3064 (2000).
10. A. Armstrong, A. R. Arehart, and S. A. Ringel, *J. Appl. Phys.* **97**, 083529 (2005).
11. P. B. Klein, J. A. Freitas, Jr., S. C. Binari, and A. E. Wickenden, *Appl. Phys. Lett.* **75**, 4016 (1999).

Mater. Res. Soc. Symp. Proc. Vol. 1202 © 2010 Materials Research Society   1202-I09-20

Experimental Observation of Sequential Tunneling Transport in GaN/AlGaN Coupled Quantum Wells Grown on a Free-Standing GaN Substrate

Faisal Sudradjat[1], Kristina Driscoll[1], Yitao Liao[1], Anirban Bhattacharyya[1], Christos Thomidis[1], Lin Zhou[2], David J. Smith[2], Theodore D. Moustakas[1], and Roberto Paiella[1]

[1]Department of Electrical and Computer Engineering and Photonics Center, Boston University, Boston, MA 02215, U.S.A.

[2]Department of Physics, Arizona State University, Tempe, AZ 85287, U.S.A.

ABSTRACT

A GaN/AlGaN multiple-quantum-well structure based on an asymmetric triple-quantum-well repeat unit was grown by molecular beam epitaxy, and its vertical electrical transport characteristics were investigated as a function of temperature. To minimize the density of dislocations and other structural defects providing leakage current paths, homoepitaxial growth on a free-standing GaN substrate was employed. The measured vertical-transport current-voltage characteristics were found to be highly nonlinear, especially at low temperatures, consistent with sequential tunneling through the ground-state subbands of weakly coupled adjacent quantum wells. Furthermore, different turn-on voltages were measured depending on the polarity of the applied bias, in accordance with the asymmetric subband structure of the sample repeat units.

INTRODUCTION

Semiconductor quantum structures based on GaN and related alloys are promising for a wide range of applications in electronics and optoelectronics. Due to their large direct bandgap, these materials are ideally suited to the development of active photonic devices operating at visible and UV wavelengths, including lasers, LEDs, photodetectors, and optical modulators. Recently, substantial research efforts have also been devoted to the study of novel device functionalities based on intersubband transitions in these materials [1-6]. The latter activities are motivated by the large conduction-band offsets of GaN/Al(Ga)N quantum wells (QWs), which allow accommodating intersubband transitions at record short wavelengths into the near-infrared spectral region, and by their large optical phonon energies, which are favorable in the context of THz intersubband optoelectronics. On the other hand, the study of tunneling transport in nitride quantum wells, which is a key feature of more complex quantum-structure devices including quantum cascade lasers [7], has so far been limited.

An important property of nitride semiconductors in this respect is their relatively large density of dislocations, especially when grown on sapphire substrates due to the associated high lattice and thermal mismatch. Dislocations can introduce parallel current paths effectively shorting out the QWs in high-resistance vertical-transport devices based on tunneling. Recently, negative differential resistance has been observed in GaN/Al(Ga)N double-barrier structures at room temperature, and attributed to resonant tunneling [8-10]. However, this behavior was found to slowly degrade after each measurement (possibly due to activated filling of trap states), and was accompanied by a hysteresis that could not be explained using simple models of interwell tunneling.

In this work we investigate vertical transport in a thick GaN/AlGaN multiple-QW structure based on a simple triple-QW repeat unit, similar in principle to a typical quantum cascade active region. To minimize the density of dislocations and other structural defects, the QWs are grown homoepitaxially on a free standing GaN substrate. Stable highly nonlinear vertical transport characteristics are observed, consistent with a picture of (incoherent) sequential tunneling. These results are significant as they indicate the feasibility of complex GaN/AlGaN QW structures with vertical transport characteristics that are not affected by defect-induced parallel current paths.

EXPERIMENT

The material used in this study was grown by rf plasma-assisted molecular beam epitaxy on a free-standing GaN substrate. A 2-μm-thick GaN film doped n-type with Si to the level of 7×10^{18} cm^{-3} was first grown on the Ga-face of the GaN substrate to produce an atomically smooth surface, since the substrate was decorated with a network of scratches due to the mechanical polishing. This layer was followed by 20 repetitions of the same three weakly coupled GaN/Al$_{0.15}$Ga$_{0.85}$N QWs, and finally by a 200-nm GaN cap layer n-doped to the level of 2×10^{18} cm^{-3}. Thus, both the bottom and the top GaN contact layers are degenerately doped, which ensures good electrical conductivity in the entire temperature range investigated [11]. The individual layers in each repeat unit, starting from the lowest Al$_{0.15}$Ga$_{0.85}$N barrier and moving up along the growth axis from the substrate, have nominal thicknesses of 12, 13, 7, 9, 9, and 15 monolayers (MLs), with 1 ML $\approx$ 2.6 Å. The 15-ML wells are n-doped to about 1.5×10^{17} cm^{-3}, while all other layers in the multiple-QW region are nominally intrinsic. The sample microstructure was investigated using a 400-keV JEM-4000EX transmission electron microscope (TEM).

To investigate the sample electrical characteristics, 400×400 μm^2 mesa-structure devices were fabricated by inductively-coupled plasma etching through the multiple-QW layer using a chlorine-based chemistry. Metal contacts consisting of Ti/Al/Ti/Au layers were then deposited by electron-beam evaporation and liftoff on top and around each mesa. These electrodes were characterized using a standard transmission line method on the same wafer, and found to be ohmic with specific contact resistance of under 1×10^{-3} ohm-cm^2, which is small enough to have negligible effect on the measured electrical characteristics. After metallization, the samples were annealed in a forming gas environment in order to passivate possible defects on the mesa sidewalls that could provide parallel current paths. The devices were then soldered on a copper block, wire-bonded, and finally mounted on the cold finger of a suitably wired continuous-flow liquid-helium cryostat.

RESULTS

Two cross-sectional TEM images of the multiple-QW region are shown in Fig. 1, where the darker and lighter regions correspond to the GaN and Al$_{0.15}$Ga$_{0.85}$N layers, respectively, and the short horizontal lines drawn on either side of each image indicate the boundaries between adjacent repeat units. As shown by these images, the sample displays high structural quality with smooth interfaces and excellent control of the structure periodicity. The individual layer thicknesses estimated from these images are also in good agreement with the nominal values, with period-to-period and in-plane variations of only about one ML.

Figure 1. Cross-sectional TEM images of different resolution showing the periodic multiple-QW structure used in this work.

Exemplary current-voltage (IV) characteristics measured from a same device using positive (i.e., top to bottom) and negative voltage pulses are shown in Figs. 2(a) and 2(b), respectively. The different curves in each plot correspond to heat sink temperatures of 20, 100, 200, and 280 K in order of decreasing resistance. All of these traces are highly nonlinear and clearly display a transition from a high-resistance state near zero applied bias to a low-resistance state as the voltage is increased. The width and repetition rate of the bias pulses are 200 ns and 20 kHz, respectively, corresponding to a duty cycle of 0.4%. It should be noted that the same results were obtained for a wide range of values of pulse width and duty cycle up to an order of magnitude higher, suggesting that the observed electrical characteristics are not affected by device heating during each scan. Furthermore, these traces were found to be completely stable and reproducible upon many repeated measurements.

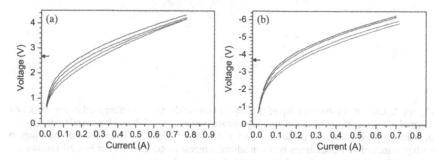

Figure 2. IV characteristics measured from a same mesa-structure device at different heat-sink temperatures, using positive (a) and negative (b) voltage pulses. The four curves in each plot correspond to temperatures of 20, 100, 200, and 280 K in order of increasing current for fixed voltage. The arrow on the vertical axis denotes the corresponding theoretical turn-on voltage.

DISCUSSION

To assist in the interpretation of the measured electrical characteristics, in Fig. 3 we plot the conduction-band lineup of two repeat units of the sample under different bias conditions, together with the squared envelope functions of the relevant subbands. These plots were calculated using a Schrödinger equation solver that includes the characteristic pyro- and piezo-electric fields of nitride heterostructures, with all relevant material parameters taken from Ref. 12. The intrinsic electric fields were computed using periodic boundary conditions, i.e., by requiring that in the absence of any external bias the voltage across each repeat unit is zero. As discussed in Ref. 13, this is appropriate to the description of thick periodic structures consisting of many repeat units (such as the material used in this work), except for the QWs immediately near the edges of the periodic stack.

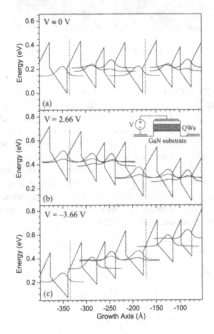

Figure 3. Conduction-band lineup of two repeat units of the QW structure used in this work and squared envelope functions of the ground-state subbands, under different bias conditions: (a) near zero applied voltage V; (b) for V = 2.66 V; (c) for V = -3.66 V. The vertical dashed lines in each plot indicate the boundaries between adjacent repeat units. The inset of Fig. 3(b) shows a schematic cross-sectional view of a processed device.

As can be inferred from Fig. 3(a), in the case of near-zero applied bias all electrons in each repeat unit reside in the ground-state subband of the widest well (the 15-ML-thick, right-most well of each triple-QW period in the figure). These subbands are energetically separated from the ground states of the adjacent wells by relatively large amounts (about 38 and 66 meV for the well immediately above and immediately below, respectively); as a result, the entire structure is in a relatively high-resistance state. As the bias is increased in either direction, the populated subbands become energetically aligned with the ground states of the neighboring wells downstream. As a result, vertical transport by sequential tunneling (either resonant or assisted by various scattering processes) becomes allowed and the resistance is lowered. Due to the asymmetric nature of the QW structure under study, the voltage drop required to achieve optimal subband alignment is different depending on whether it is applied from top to bottom or vice versa. In particular, as shown in Fig. 3(b), the ground-state subband of each 15-ML well is brought into alignment with that of its overlaying well by a voltage of about 2.66 V across the 20 repeat units. On the other hand, as shown in Fig. 3(c), a larger bias of 3.66 V is required to achieve ground-state subband alignment between each 15-ML well and its adjacent well below.

The measured IV curves are fully consistent with this picture of transport via sequential tunneling through the QW ground-state subbands. First, the expected transition from a high-resistance state near zero applied bias to a low-resistance state at higher voltage is clearly seen in the experimental traces of Fig. 2. Second, this transition occurs at significantly larger voltage when pulses of negative polarity are used as in Fig. 2(b) compared to the case of positive pulses shown in Fig. 2(a), in accordance with the simulation results of Fig. 3. For further comparison, the theoretical turn-on voltages of +2.66 and –3.66 V are also shown in Figs. 2(a) and 2(b), respectively, by the arrows on the vertical axes. Finally, the temperature dependence of the measured IV curves is also as expected in the case of tunneling transport. Specifically, as the temperature is increased the current turn-on becomes more gradual and shifts to lower values of the applied bias, due to the thermal broadening of the electronic distribution in the initially occupied subbands.

On the other hand, no signatures of negative differential resistance are observed in the data of Fig. 2, indicating that scattering-assisted tunneling plays a major role in the measured vertical transport characteristics. From a device design perspective, this suggests that selective electron injection via tunneling in these QWs will require structures with spatially separated subbands.

Finally, it should be noted that previously developed samples based on the same multiple-QW structure grown on sapphire were found to feature much weaker nonlinearities (if any) in their IV characteristics. Therefore, the use of homoepitaxy on free-standing GaN substrates appears to be critical to obtaining the high structural quality required for efficient tunneling transport.

CONCLUSIONS

In summary, we have measured the vertical transport characteristics of a periodic GaN/AlGaN multiple-QW structure grown on a free-standing GaN substrate. The measured IV characteristics are highly nonlinear with polarity-dependent turn-on voltage, in agreement with a picture of sequential tunneling through the ground-state subbands of weakly coupled QWs. These results are promising for the future development of advanced III-nitride quantum-structure devices whose operation relies on tunneling transport, such as quantum cascade lasers.

ACKNOWLEDGMENTS

This work was supported by the National Science Foundation under grant ECS-0824116. We acknowledge use of facilities in the John M. Cowley Center for High Resolution Electron Microscopy at Arizona State University.

REFERENCES

1. C. Gmachl, H. M. Ng, S.- N. G. Chu, and A. Y. Cho, Appl. Phys. Lett. **77**, 3722 (2000).
2. N. Iizuka, K. Kaneko, and N. Suzuki, IEEE J. Quantum Electron. **42**, 765 (2006).
3. Y. Li, A. Bhattacharyya, C. Thomidis, T. D. Moustakas, and R. Paiella, Opt. Express **15**, 5860 (2007).
4. L. Nevou, M. Tchernycheva, F. H. Julien, F. Guillot, and E. Monroy, Appl. Phys. Lett. **90**, 121106 (2007).
5. A. Vardi, G. Bahir, F. Guillot, C. Bougerol, E. Monroy, S. E. Schacham, M. Tchernycheva, and F. H. Julien, Appl. Phys. Lett. **92**, 011112 (2008).
6. K. Driscoll, Y. Liao, A. Bhattacharyya, L. Zhou, D. J. Smith, T. D. Moustakas, and R. Paiella, Appl. Phys. Lett. **94**, 081120 (2009).
7. C. Sirtori, F. Capasso, J. Faist, A. L. Hutchinson, D. L. Sivco, and A. Y. Cho, IEEE J. Quantum Electron. **34**, 1722 (1998).
8. A. Kikuchi, R. Bannai, K. Kishino, C.-M. Lee, and J.-I. Chyi, Appl. Phys. Lett. **81**, 1729 (2002).
9. A. E. Belyaev, O. Makarovsky, D. J. Walker, L. Eaves, C. T. Foxon, S. V. Novikov, L. X. Zhao, R. I. Dykeman, S. V. Danylyuk, S. A. Vitusevich, M. J. Kappers, J. S. Barnard, and C. J. Humphreys, Physica E **21**, 752 (2004).
10. S. Golka, C. Pflügl, W. Schrenk, G. Strasser, C. Skierbiszewski, M. Siekacz, I. Grzegory, and S. Porowski, Appl. Phys. Lett. **88**, 172106 (2006).
11. R. J. Molnar, T. Lei, and T. D. Moustakas, Appl. Phys. Lett. **62**, 72 (1993).
12. I. Vurgaftman and J. R. Meyer, J. Appl. Phys. **94**, 3675 (2003).
13. S. Gunna, F. Bertazzi, R. Paiella, and E. Bellotti, in *Nitride Semiconductor Devices: Principles and Simulations*, edited by J. Piprek (Wiley, 2007), chapter 6.

Electronic Devices

Mater. Res. Soc. Symp. Proc. Vol. 1202 © 2010 Materials Research Society   1202-I04-08

Cubic AlGaN/GaN Hetero-field effect transitors with normally on and normally off operation

D.J. As[1], E. Tschumak[1], F. Niebelschütz[2], W. Jatal[2], J. Pezoldt[2], R. Granzner[3], F. Schwierz[3], and K. Lischka[1]

[1] Universität Paderborn, Department Physik, Warburger Strasse 100, 33098 Paderborn, Germany
[2] FG Nanotechnologie, Institut für Mikro- und Nanotechnologien, TU Ilmenau, Postfach 100565, 98684 Ilmenau, Germany
[3] FG Festkörperelektronik, TU Ilmenau, Postfach 100565, 98684 Ilmenau, Germany

ABSTRACT

Non-polar *cubic AlGaN/GaN HFETs* were grown by plasma assisted MBE on 3C-SiC substrates. Both normally-on and normally-off HFETs were fabricated using contact lithography. Our devices have a gate length of 2 µm, a gate width of 25 µm, and source-to-drain spacing of 8 µm. For the source and drain contacts the $Al_{0.36}Ga_{0.64}N$ top layer was removed by reactive ion etching (RIE) with $SiCl_4$ and Ti/Al/Ni/Au ohmic contacts were thermally evaporated. The gate metal was Pd/Ni/Au. At room temperature the DC-characteristics clearly demonstrate enhancement and depletion mode operation with threshold voltages of +0.7 V and –8.0 V, respectively. A transconductance of about 5 mS/mm was measured at a drain source voltage of 10 V for our cubic AlGaN/GaN HFETs, which is comparable to that observed in non-polar a-plane devices. From capacity voltage measurements a 2D carrier concentration of about 7×10^{12} cm^{-2} is estimated. The influence of source and drain contact resistance, leakage current through the gate contact and parallel conductivity in the underlaying GaN buffer are discussed.

INTRODUCTION

AlGaN/GaN heterojunction field-effect transistors (HFETs) are presently of outstanding interest for electronic devices, in particular, for high-power and high-frequency amplifiers. This is motivated by the potential commercial and defense applications, e.g., in the area of communication systems, radar, base stations, high-temperature electronics and high-power solid-state switching [1, 2]. Currently, state of the art AlGaN/GaN HFETs are fabricated on c-plane surface material of the stabile wurzite crystal structure with inherent spontaneous and piezoelectric polarization fields which produce extraordinary high sheet carrier concentration at the heterointerface. Therefore, usually these devices are of the *normally-on* type. However, for power and consumer applications, *normally-off* operation is required to simplify the design of driving circuits and for the safety of the products. A direct way to fabricate HFETs without undesirable parasitic polarization effects is the use of *cubic* group III-nitrides.

If the cubic group-III nitrides are grown in (001) direction spontaneous and piezoelectric polarization effects can be avoided at the interfaces and surfaces and the density of the two-dimensional electron gas (2-DEG) in cubic $Al_xGa_{1-x}N$/GaN heterostructures is independent on the thickness and Al mole fraction of the $Al_xGa_{1-x}N$ barrier layer and can be controlled by doping with silicon. The combination of these effects may be used to realize cubic AlGaN/GaN HEMTs with both *normally on* and *normally off* operation as it is strongly required for logic devices [3].

In addition, the electronic structure of the cubic GaN (001) surface is different to that of the c-plane in hexagonal GaN and therefore may alters the electronic properties of the Schottky diodes and ohmic contacts [4].

In this work, non-polar *cubic AlGaN/GaN HFETs* were grown by plasma assisted MBE on 3C-SiC substrates. Both normally-on and normally-off HFETs were fabricated using contact lithography and normally on and normally off operation is demonstrated.

EXPERIMENT

For the epitaxy of cubic AlGaN/GaN hetero structures, freestanding Ar^+ implanted 3C-SiC was used. Previous to the Ar^+ implantation, the carrier concentration in the 3C-SiC substrate of $n=2\times10^{18}$ cm^{-3} was measured by Hall effect. A three energy implantation with Ar ions at doses of 6×10^{14} cm^{-2} at 160 keV, 2.4×10^{14} cm^{-2} at 80 keV and 1.2×10^{14} cm^{-2} at 40 keV was used to form a damage layer near the surface. We showed that this damage acts as insulation layer [5]. Reflection high energy electron diffraction (RHEED) was used to monitor the crystalline nature of the sample surface. Streaky RHEED patterns were observed for both the surface of the Ar^+ implanted 3C-SiC and for the surface of a 600 nm thick cubic GaN (c-GaN) grown on this substrate revealing a two dimensional surface condition.

Cubic $Al_xGa_{1-x}N/GaN$ hetero structures were grown in a Riber 32 system by plasma-assisted molecular beam epitaxy. Prior to growing process, the substrate was chemically etched by organic solvents and buffered oxide etching (BOE). In order to minimize hexagonal inclusions in our layers and to obtain an optimum interface roughness, coverage of one monolayer Ga was established during growth [6]. The substrate temperature was 720°C and the growth rate was 115 nm/h.

Two different cubic Al_xGaN_{1-x}/GaN heterostructures (Sample A and Sample B) with similar crystalline properties were investigated. The full width at half maximum (FWHM) of the cubic GaN (002) rocking curve was 25 arcmin. In both samples, the cubic $Al_xGa_{1-x}N$ was pseudomorphically strained on the cubic GaN, measured by reciprocal space mapping [7]. The RMS roughness of the surface measured by AFM in a 5×5 μm^2 scan was 5 nm.

RESULTS AND DISCUSSION

HFET with normally-off characteristics

Sample A consists of 600 nm unintentionally doped (UID) cubic GaN followed by 3 nm UID cubic $Al_{0.25}Ga_{0.75}N$ spacer layer, 2 nm cubic $Al_{0.25}Ga_{0.75}N$:Si and 15 nm UID cubic $Al_{0.25}Ga_{0.75}N$. The carrier concentration of the Si doped AlGaN layer is $n=4.5\times10^{18}$ cm^{-3}. A 5 nm thick heavily silicon doped cubic GaN:Si cap with a carrier concentration of $n=6\times10^{19}$ cm^{-3} was grown on top of the sample. For ohmic source and drain contacts, Ti/Al/Ni/Au (15 nm/50 nm/15 nm/50 nm) was thermally evaporated on GaN:Si and annealed at 850°C for 30 s in nitrogen environment. Then, mesa insulation was performed with SiCl$_4$ RIE down to the substrate. For the gate contact, the GaN:Si cap layer was removed using RIE with SiCl$_4$ down to the substrate. The gate was fabricated by evaporation of Pd/Ni/Au (15 nm/15 nm/50 nm) and a subsequent annealing processs at 400°C for 10 min. The device had a gate length of 2 μm, a gate

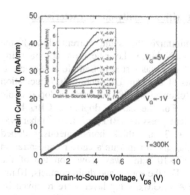

Figure 1. Drain current I_{DS} vs. drain source voltage V_{DS} of HFET A. The inset shows the same measurement curves corrected for the drain current at V_{GS}=-1 V.

width of 25 µm and a source-to-drain spacing of 8 µm. 250 nm of SiO_2 was deposited around the device to insulate the contact pads electrically.

The room temperature DC drain current-voltage curves with gate-to-source voltages from -1 V to + 5V of HFET Sample A are displayed in Fig. 1. The threshold voltage of this device is +0.7 V measured at V_{DS} =10 V by extrapolation of the transconductance curve (not shown here). This indicates a normally-off device characteristics, however, the drain-to-source current at V_G = 0 V is relatively large due to the high conductivity of c-GaN buffer layer. The inset shows the measurement data adjusted by the shunt current. A maximum drain-to-source current of 6.5 mA/mm was observed when a gate voltage of +5 V was applied.

In Table I the experimental data of our cubic normally off HFET is compared with hexagonal state of the art nonpolar a-plane HFETs and c-plan HFETs [8]. The transconductance g_m of about 3 mS/mm of our cubic device compares well with that of the non-polar a-plane devices. For our cubic device, however we clearly measured a positive threshold voltage of + 0.7 V, whereas for the a-plan devices only "nearly positive" V_{th} of - 0.5 V are observed. In addition, cubic HFET show no dependence on the orientation of the gate as observed in a-plan HFET (compare columns 3 and 4 in Table I). Compared to c-plane HFET, which only shows *normally on* behavior, the transconductance is nearly a factor of 30 lower. We attribute this fact to the still inferior material quality of cubic or a-plane GaN compared to c-plan GaN and to the fact that the dislocation density, which reduces the carrier mobility and therefore g_m, is at least one order of magnitude higher in the non-polar nitrides.

Table I: Summary of device performance of cubic, a-plane with gate in [1-100] and [0001] direction and c-plane AlGaN/GaN HFETs [8].

	cubic HFET	a-plane HFET gate in [1-100]	a-plane HFET gate in [0001]	c-plane HFET
$I_{DS\ max}$ (mA/mm)	6.5	19.5	13.5	423
$g_{m\ max}$ (mS/mm)	3	6.7	3.6	112
V_{th}(V)	+ 0.7	- 0.5	- 0.5	- 4.0
L_G (µm)	2	1	1	1

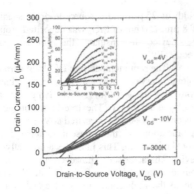

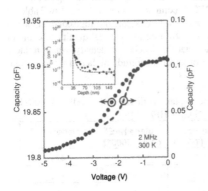

Figure 2. *Static output characteristics of the HFET B. The inset shows the same measurement curves corrected for the drain current at $V_{GS}=-10$ V.*

HFET with normally-on characteristics

To minimize the shunt current through the cubic GaN buffer layer, carbon doping using CBr$_4$ [9-11] was tested in Sample B. Sample B consists of a 60 nm UID cubic GaN nucleation layer followed by 580 nm carbon doped cubic GaN:C. For the carbon doping a CBr$_4$ beam equivalent pressure (BEP) of 1×10^{-6} mbar was used. A 34 nm thick homogeneously doped Al$_{0.36}$Ga$_{0.64}$N:Si cap with a carrier concentration of $n=1.5\times10^{18}$ cm^{-3} was grown on top of the sample. For the source and drain contacts, 10 nm of the Al$_{0.36}$Ga$_{0.64}$N top layer were removed by reactive ion etching (RIE) with SiCl$_4$. After that Ti/Al/Ni/Au (15 nm/50 nm/15 nm/50 nm) was thermally evaporated and annealed at 850°C for 30 s in N$_2$ environment. Then the Pd/Ni/Au gate contact was fabricated as described for HFET A. The device geometry was the same as for HFET Sample A.

The room temperature output characteristics of HFET Sample B are depicted in Fig. 2. The gate-to-source voltage was varied between -10 V and +4 V. Apart from the shunt current through the 3C-SiC substrate and GaN:C buffer layer (red curve), a clear field effect with normally-on characteristics is measured in this sample. The inset shows the same measurement data adjusted by the shunt current. The measurements of the source and drain contact resistance showed a slight non ohmic behaviour which limited the absolute current through the device. Therefore, the source-to-drain current difference between $V_G=-10$V and $V_G=+4$ V was 80 μA/mm only. At high positive gate voltages, an additional gate leakage is observed at low source-to-drain voltages. So, the drain current at +4 V gate voltage is reduced by the gate leakage.

CV measurements of the HFET Sample B device were performed at 2 MHz to detect the electron channel at the AlGaN/GaN interface. For this purpose, the gate was biased and the source and drain were connected in parallel and grounded. Fig. 3 shows the measured room temperature CV profile of HFET Sample B. The typical shape was observed where the

Figure 3. *CV characteristics of HFET B measured on the gate contact at 2 MHz confirming the presence of an electron channel at the AlGaN/GaN interface. The inset shows a carrier density profile N_{CV}. The red dashed lines are CV curves calculated using a Poisson-Schrödinger model.*

110

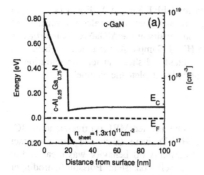

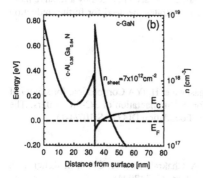

Figure 4. *Calculated conduction band edge and the electron concentration versus distance from surface at V_{GS}=0V for (a) Sample A and (b) Sample B.*

capacitance was found to be roughly constant when the electron channel was present, falling to smaller values once the electron channel had been depleted. The left-hand scale is the measured capacity which has been corrected for the parasitic parallel capacity of the contact pads (C_p=19.8 pF). The resulting gate capacity is plotted at the right hand-scale. The red dashed curve depicts calculated CV data using the self consistent Poisson-Schrödinger model [12] with a donor concentration of n=1.5×10^{18} cm^{-3} in cubic Al$_{0.36}$Ga$_{0.64}$N and n=1×10^{17} cm^{-3} in cubic GaN.

.The inset of Fig. 3 shows the apparent carrier density N_{CV} in HFET Sample B calculated from the CV characteristics using the following equations [14]:

$$N_{CV} = -\frac{C^3}{q\varepsilon\varepsilon_0 A^2}\frac{dV}{dC} \qquad (1)$$

$$z_{CV} = \frac{\varepsilon\varepsilon_0 A}{C} \qquad (2)$$

where z_{CV} is equal to the distance from the surface and A is the contact area. The resulting profile shows a carrier agglomeration at the AlGaN/GaN interface building an electron channel. The red dashed curve is the calculated carrier density using the self-consistent Poisson-Schrödinger model with a donor concentration of n_D=1.5×10^{18} cm^{-3} in cubic Al$_{0.36}$Ga$_{0.64}$N and n_D=1×10^{17} cm^{-3} in cubic GaN.

Comparison of Normally on and Normally off HFET

We measured a much larger transconductance in Sample A than in Sample B. We believe that this is due to lower source and drain contact resistance in Sample A due to highly doping of GaN:Si cap layer with n=6×10^{19} cm^{-3}. In Sample B, the same metal contacts were evaporated on Al$_{0.36}$Ga$_{0.64}$N:Si with n=1.5×10^{18} cm^{-3}. As a results the source-to-drain current and therewith the transconductance of Sample B is limited by the contact resistance. The transconductance-to-shunt ratio is 0.22 in Sample A and 0.61 in Sample B measured at V_{DS}=10 V. Obviously the carbon doping of the cubic GaN buffer layer induces a reduction of the buffer leakage, but the optimum carbon concentration for insulating cubic GaN has still to be found.

To clarify the normally-off behavior of HFET Sample A and the normally-on behavior of HFET Sample B, band structures and electron density profiles were calculated for V_G=0V using a self-consistent Poisson-Schrödinger model. For the gate contact, a Schottky barrier of 0.8 eV was assumed [4]. The simulation diagrams are shown for HFET Sample A (Fig. 4 (a)) and for HFET Sample B (Fig. 4 (b)). According to electrical measurements, the electron channel of Sample A is nearly depleted. To achieve a higher electron density at the AlGaN/GaN interface, a positive gate voltage has to be applied. In contrast to HFET Sample A, the electron channel is degenerate and conductive in HFET Sample B with a calculated sheet carrier concentration of $n_{sheet}=7\times10^{12}$ cm^{-2}. A negative gate voltage has to be applied to deplete the channel.

CONCLUSIONS

Non-polar *cubic AlGaN/GaN HFETs* were grown by plasma assisted MBE on free-standing 3C-SiC substrates. Both normally-on and normally-off HFETs were fabricated using contact lithography. A clear field effect with normally-on and normally-off behavior was measured at room temperature and verified by calculations using a self consistent Poisson-Schrödinger model. The electron channel at the cubic AlGaN/GaN interface was also detected by capacitance-voltage measurements. However the field effect was accompanied by a relative large shunt current through the GaN buffer which clearly demonstrates the need for further reduction of the buffer conductivity.

ACKNOWLEDGMENTS

The authors want to thank Dr. M. Abe and Dr. H. Nagasawa at HOYA Corporation, for supply of free-standing 3C-SiC substrates and Dr. J. Lindner for Ar$^+$ implantation of 3C-SiC wafers. The project was financial supported by the German Science Foundation DFG (AS 107/4-1).

REFERENCES

1. L. Shen, R. Coffie, D. Bultari, S.J. Heikmann, A. Chakraborthy, A. Chini, S. Keller, S.P. DenBaars, and U.K. Mishra, *IEEE Electron. Dev. Lett.* **25**, 7 (2004).
2. S. Rajan, P. Waltereit, C. Poblenz, S.J. Heikmann, D.S.Green, S.P. DenBaars, and U.K. Mishra, *IEEE Electron. Dev. Lett.* **25**, 247 (2004).
3. S. Haffouz, H. Tang, J.A. Bardwell, E.M. Hsu, J.B. Webb, and S. Rolfe, *Solid State Electron.* **49**, 802 (2005).
4. D.J. As, E. Tschumak, I. Laubenstein, R.M. Kemper, K. Lischka, *MRS. Symp. Proc. Vol.* **1108**, A01-02 (2009).
5. E. Tschumak, M.P.F. de Godoy, D.J. As, and K. Lischka, *Microelectronics Journal* **40**, 367 (2009).
6. J. Schörmann, S. Potthast, D.J. As, and K. Lischka, *Appl. Phys. Lett.* **90**, 041918 (2007).
7. E. Tschumak, J.K.N. Lindner, M. Bürger, K. Lischka, H. Nagasawa, M. Abe, and D.J. As, *phys. stat. sol. (c)* (2009) (to be published).
8. M. Kuroda, H. Ishida, T. Ueda, and T. Tanaka, J. Appl. Phys. **102**, 093703 (2007).
9. D.J. As, E. Tschumak, H. Pöttgen, O. Kasdorf, J.W. Gerlach, H. Karl, and K. Lischka, J. Crys. Growth **311**, 2039 (2009).
10. C. Poblenz, P. Waltereit, S. Rajan, S. Heikmann, U. K. Mishra, and J. S. Speck, J. Vac. Sci. Technol. **B 22(3)**, 1145 (2004).
11. D.S. Green, U.K. Mishra, and J.S. Speck, J. Appl. Phys. **95**, 8456 (2004).
12. I.H. Tan, G. Snider, and E.L. Hu, J. Appl. Phys. **68**, 4071 (1990).

Mater. Res. Soc. Symp. Proc. Vol. 1202 © 2010 Materials Research Society   1202-I09-05

Normally-Off GaN MOSFETs on Silicon Substrates With High-Temperature Operation

Hiroshi Kambayashi, Yuki Niiyama, Takehiko Nomura, Masayuki Iwami, Yoshihiro Satoh, and Sadahiro Kato
Furukawa Electric Co., Ltd., Yokohama, 220-0073, Japan

ABSTRACT

We have demonstrated enhancement-mode n-channel gallium nitride (GaN) MOSFETs on Si (111) substrates with high-temperature operation up to 300 °C. The GaN MOSFETs have good normally-off operation with the threshold voltages of +2.7 V. The MOSFET exhibits good output characteristics from room temperature to 300 °C. The leakage current at 300°C is less than 100 pA/mm at the drain-to-source voltage of 0.1 V. The on-state resistance of MOSFET at 300°C is about 1.5 times as high as that at room temperature. These results indicate that GaN MOSFET is suitable for high-temperature operation compared with AlGaN/GaN HFET.

INTRODUCTION

GaN has excellent properties such as high critical electric field, high electron mobility and good thermal conductivity compared with Si and SiC [1]. Therefore, GaN-based electronic devices are very attractive for power switching applications under high-temperature conditions [2-5]. A low-loss and a high-temperature operation enable us to eliminate cooling systems.

For GaN-based transistors, AlGaN/GaN heterojunction field effect transistors (HFETs) have been intensively reported. The operation of these devices was essentially normally-on mode. However, for power switching applications, normally-off operation is required for fail safe operation and noise margin. Recently, normally-off AlGaN/GaN HFETs have been fabricated by applying several techniques [6-8]. However, these threshold voltages are below +1 V. GaN MOSFETs are also expected for normally-off devices, and several groups have recently reported on their work [9- 19]. For example, Matocha et al. reported 200°C operation [9]. We reported 250°C operation and output current at 2.2 A [11, 12]. Otake et al. and Kodama et al. demonstrated GaN MOSFETs with trench gate structures [13-15]. These MOSFETs above were fabricated on sapphire or GaN substrates. Fabrication of GaN MOSFET on the Si substrate has the advantage on the cost side compared with these substrates and there are some reports of GaN MOSFETs on Si substrates [16-19]. However, high-temperature operation of GaN MOSFET fabricated on Si substrate has not been reported so far.

In this paper, a high-temperature operation of GaN MOSFETs on Si substrates is reported. The MOSFETs we fabricated exhibit good performance at a high-temperature operation up to 300°C.

EXPERIMENT

Figure 1 shows the schematic cross-sectional view of the fabricated GaN MOSFET. A GaN-based buffer layer and a 1.5 µm-thick GaN epilayer with Mg acceptor doping at a concentration of 1×10^{17} cm^{-3} was grown on a Si (111) substrate using metal-organic chemical vapor deposition (MOCVD). Then n$^+$ source and drain regions were selectively implanted with a silicon dose of 3×10^{15} cm^{-2} with a maximum energy of 160 KeV to achieve a junction depth of

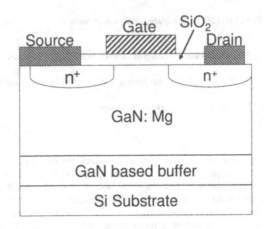

Figure 1.   A schematic cross-section of an n-channel GaN MOSFET.

300 nm. After the deposition of 500-nm-thick SiO₂, annealing at 1200°C for 10 sec in an Ar ambient was applied to activate the implanted silicon. The sheet resistance of n⁺ regions was 125 Ω/sq. After removing the SiO₂, 60-nm-thick SiO₂ was deposited by plasma-enhanced chemical vapor deposition (PE-CVD) as a gate oxide, and then annealed at 800°C for 30 min in a N₂ ambient. This annealing is useful to decrease the SiO₂/GaN interface-state density near the GaN conduction band [20]. The source and drain regions were defined by photo lithography. After opening the SiO₂, Ti and Al were deposited by sputter equipment, and then lifted off. Then, the source and drain regions were annealed at 600°C for 10 min in a N₂ ambient. The contact resistance of the regions was $7.5 \times 10^{-7} \ \Omega \cdot cm^2$. The gate electrode of Ti and Au was also defined by lift-off. The gate width and the gate length of the fabricated GaN MOSFETs were 4 µm. of 1.1 mm, respectively.

DISCUSSION

Figure 2 shows transfer I-V characteristics of the GaN MOSFET with a gate width of 1.1 mm at room temperature. The drain-to-source voltage was 0.1 V. As shown in Fig. 2, we have realized good normally-off operation on GaN MOSFETs with the threshold voltages of +2.7 V. The log-plotted transfer I-V characteristics from room temperature to 300°C are shown in Figure 3. The leakage currents at below the pinch-off voltage from 100°C to 300°C were about two orders of magnitude higher than that at room temperature. However, the leakage current at 300°C was less than 100 pA/mm. The threshold voltage slightly shifted to minus with increasing temperature. Although, we have no conclusive evidence of the origin of this threshold voltage shift. We suppose that the acceptor density of GaN epilayer decreased with increasing temperature, and the threshold voltage decreased. The maximum field-effect mobility evaluated from transfer I-V slightly decreased and the subthreshold slope slightly increased as the

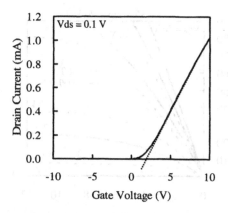

Figure 2. A typical room temperature transfer I-V characteristics of the GaN MOSFET.

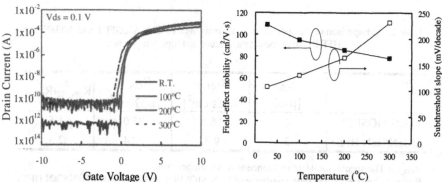

Figure 3. Transfer I-V characteristics of the GaN MOSFET from room temperature to 300°C (log plot).

Figure 4. Temperature dependency of maximum field-effect mobility and subthreshold slope.

temperature was higher as shown in Figure 4.

Figure 5 shows the output I-V characteristics of the GaN MOSFET at room temperature and 300 °C, respectively. Note that the maximum current of the measurement setup was controlled to 100 mA. Good output characteristics were observed. The specific on-state resistances at room temperature and 300 °C were 16.5 mΩ-cm^2 and 25.0 mΩ-cm^2 at the gate-to-

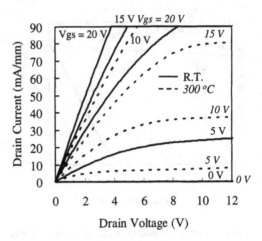

Figure 5. Output I-V characteristics of the GaN MOSFET at room temperature and 300°C.

Table 1. Comparison of the specific on-state resistance of GaN MOSFET and AlGaN/GaN HFET [20] at room temperature and high temperature.

	L_{ch} [μm]	R_{ON_RT} [mΩ-cm^2]	R_{ON_HT} [mΩ-cm^2]	R_{ON_HT}/R_{ON_RT}
GaN MOSFET	4	16.5	25.0	1.5
AlGaN/GaN HFET	2	9.8	22.4	2.3

R_{ON_RT}: The specific on-state resistance at room temperature.
R_{ON_HT}: The specific on-state resistance of GaN MOSFET at 300°C and AlGaN/GaN HFET at 225°C.
The specific on-state resistance of AlGaN/GaN HFET is calculated from the results of reference 20.

source voltage of 20 V, respectively. We compared the temperature dependency between GaN MOSFET and AlGaN/GaN HFET [20]. Table 1 shows the specific on-state resistance of GaN MOSFET and AlGaN/GaN HFET at room temperature and high temperature. The on-state resistance of GaN MOSFET at 300°C was about 1.5 times as high as that at room temperature. However, in the case of AlGaN/GaN HFET, the on-state resistance at 227°C (500 K) increased about 2.3 times as high as that at room temperature. The result indicates that the degradation ratio of the inversion careers of GaN MOSFET with increasing temperature was lower than the ratio of two dimensional electron gas of AlGaN/GaN HFET. Furthermore, Si-based FETs can not operate in such high temperature. This result indicates that GaN MOSFETs are suitable for high-temperature operation.

CONCLUSIONS

We have demonstrated enhancement-mode n-channel GaN MOSFETs on Si substrates. Good normally-off operation with threshold voltage of +2.7 V is observed. The GaN MOSFET operated up to 300°C and show little degradation of the field-effect mobility and the subthreshold slope over the temperature change. The on-state resistances are 16.5 mΩ-cm^2 at room temperature and 25.0 mΩ-cm^2 at 300°C, respectively. The characteristics of on-state resistance suggest that GaN MOSFET is suitable for high-temperature operation compared with AlGaN/GaN HFET. These results we have demonstrated indicate that GaN MOSFETs fabricated on Si substrate are promising for high efficiency power devices.

REFERENCES

1. T.P. Chow, and R. Tyagi, *IEEE Trans. Electron Devices*, Vol. 41, pp. 1481-1483, 1994.
2. O. Akutas, Z.F. Fan, S.N. Mohammad, A.E. Botchkarev, and H. Morkoç, *Appl. Phys. Lett.*, Vol. 69, pp. 3872-3874, 1996.
3. W. Yang, J. Lu, M. Asifkhan, and I. Adesida, *IEEE Trans. Electron Devices*, Vol. 48, pp. 581-585, 2001.
4. S. Yoshida, and J. Suzuki, *Jpn J. Appl. Phys. Lett.*, Vol. 37, pp. 482-484, 1998.
5. S. Yoshida, and J. Suzuki, *Jpn J. Appl. Phys. Lett.*, Vol. 38, pp. 851-853, 1999.
6. W. Saito, Y. Tanaka, M. Kuraguchi, K. Tsuda, and I. Ohmura, *IEEE Trans. Electron Devices*, Vol.53, pp. 356-362, 2006.
7. Y. Umemoto, M. Hikita, H. Ueno, H. Matsuo, H. Ishida, M. Yanagihara, T. Ueda, T. Tanaka, and D. Ueda, *IEDM Technical Digest*, 2006.
8. D. Song, J. Liu, Z. Cheng, W. C. W. Tang, K. M. Lau, and K. J. Chen, *IEEE Electron Device Lett.*, Vol. 28, no. 3, pp. 189-191, 2007.
9. K. Matocha, T. P. Chow, and R. J. Gutmann, *IEEE Trans. Electron Devices*, Vol. 52, no. 1 pp. 6-10, 2005.
10. W. Huang, T. Khan, and T. P. Chow, *IEEE Electron Device Lett.*, Vol.27, no. 10, pp. 796-798, 2006.
11. T. Nomura, H. Kambayashi, Y. Niiyama, S. Otomo, and S. Yoshida, *Solid-State Electronics* 52 pp. 150–155, 2008.
12. Y. Niiyama, H. Kambayashi, S. Ootomo, T. Nomura, S. Yoshida, and T. P. Chow, *Jpn J. Appl. Phys. Lett.*, Vol. 47, No. 9, pp. 7128-7130, 2008.
13. H. Otake, S. Egami, H. Ohta, Y. Nanashi, and H. Takasu, *Jpn J. Appl. Phys. Lett.*, Vol. 46, No. 25, pp. L599-L601, 2007.
14. H. Otake, K. Chikamatsu, A. Yamaguchi,T. Fjishima, and H. Ohta, *Appl. Phys. Expr.*, 1, pp. 011105-1-011105-3, 2008.
15. M. Kodama, M. Sugimoto1, E. Hayashi, N. Soejima, O. Ishiguro, M. Kanechika, K. Itoh, H. Ueda, T. Uesugi, and T. Kachi, *Appl. Phys. Expr.*, 1, pp. 021104-1-021104-3, 2008.
16. H.-B. Lee, H.-I. Cho, H.-S. An, Y.-H. Bae, M.-B. Lee, J.-H.Lee, and S.-H. hahn, *IEEE Electron Device Lett.*, Vol.27, no. 2, pp. 81-83, 2006.

17. S. Jang, F. Ren, S. J. Pearton, B. P. Gila, M. Hlad, C. R. Abernathy, H. Yang, C. J. Pan, J. I. Chyi, P. Bove, H. Lahreche, and J. Thuret, *J. Electron. Materials,* Vol.35, No. 4, pp. 685-690, 2006.
18. H. Kambayashi, Y. Niiyama, S. Ootomo, T. Nomura, M. Iwami, Y. Satoh, S. Kato, and S. Yoshida, *IEEE Electron Device Lett.*, Vol.28, no. 12, pp.1077-1079, 2007.
19. T. Oka, and T. Nozawa, *IEEE Electron Device Lett.*, Vol.29, no. 7, pp.668-670, 2008.
20. H. Kambayashi, J. Li, N. Ikeda, and S. Yoshida, *Mater. Res. Soc. 2006 Fall Symp. Proc. Vol. 955E,* I15-20, 2006.

Mater. Res. Soc. Symp. Proc. Vol. 1202 © 2010 Materials Research Society　　　1202-I09-02

Yongkun Sin, Erica DeIonno, Brendan Foran, and Nathan Presser
Electronics and Photonics Laboratory
The Aerospace Corporation
El Segundo, CA 90245

ABSTRACT

High electron mobility transistors (HEMTs) based on AlGaN-GaN hetero-structures are promising for high power, high speed, and high temperature operation. Especially, AlGaN-GaN HEMTs grown on semi-insulating (SI) SiC substrates are the most promising for both military and commercial applications. High performance characteristics from these devices are possible in part due to the presence of high two-dimensional electron gas charge sheet density maintaining a high Hall mobility at the AlGaN barrier-GaN buffer hetero-interface and in part due to high thermal conductivity of the SiC substrates. However, long-term reliability of these devices still remains a major concern because of the large number of traps and defects present both in the bulk as well as at the surface leading to undesirable characteristics including current collapse. We report on the study of traps and defects in two MOCVD-grown structures: $Al_{0.27}Ga_{0.73}N$ HEMTs on SI SiC substrates and $Al_{0.27}Ga_{0.73}N$ Schottky diodes on conducting SiC substrates. Our HEMT structures consisting of undoped AlGaN barrier and GaN buffer layers grown on an AlN nucleation layer show a charge sheet density of $\sim 10^{13}/cm^2$ and a Hall mobility of $\sim 1500 cm^2/V \cdot sec$. Deep level transient spectroscopy (DLTS) was employed to study traps in AlGaN Schottky diodes and HEMTs fabricated with different Schottky contacts consisting of Pt-Au and Ni-Au. Focused ion beam was employed to prepare both cross-sectional and plan view TEM samples for defect analysis using a high resolution TEM.

INTRODUCTION

AlGaN-GaN HEMT technology is an ideal technology for RF and microwave power amplifiers and for high voltage switches due to the high breakdown fields and excellent electron transport characteristics in GaN [1-3]. Electron drift velocity is lower in GaN under low electric fields compared to that in GaAs, but peak and saturation electron velocities are higher in GaN under high electric fields compared to those in GaAs [1]. Although significant progress has been made in performance characteristics of AlGaN-GaN HEMTs grown on SiC substrates over the last decade, long-term reliability of these devices under high electric field operation has not been reported yet. Rather, poor reliability often observed from AlGaN HEMTs is not fully understood at this time, but the presence of point defects and dislocations including threading dislocations due to a huge lattice mismatch between SiC substrates and GaN-based materials as well as traps in AlGaN barrier and GaN buffer layers is believed to play a critical role in electrical degradation of the devices [4-6]. In our efforts to understand the root causes responsible for poor reliability of these devices, we employed DLTS to investigate traps in AlGaN Schottky diodes and HEMTs as well as TEM to investigate extended defects in AlGaN epi wafers used to fabricate the devices.

EXPERIMENTAL METHODS AND RESULTS

Tables 1 and 2 show the profiles of MOCVD-grown AlGaN epi wafers used to fabricate Schottky diodes and HEMTs for the present study. In particular, electrical characterization of the HEMT wafer includes an electron mobility of over $1500 cm^2/V \cdot s$, channel charge density of over $10^{13} cm^{-2}$, and sheet resistance of less than $350 \Omega/\mu$, all measured at $10\mu A$ and room temperature. The composition and thickness of the AlGaN barrier layer are the most critical parameters because these parameters determine carrier confinement and piezoelectric polarization induced sheet carrier density. We chose an undoped 26nm thick $Al_{0.27}Ga_{0.73}N$ barrier layer for the present study because enhanced spontaneous and piezoelectric polarizations due to tensile strain in the $Al_{0.27}Ga_{0.73}N$ layer induce enough channel electron density at the AlGaN-GaN interface [7].

Table 1. AlGaN Schottky diode epi profile.

Layer	Material	Thickness	Al Composition	Doping
Barrier	AlGaN	~1μm	0.27	Si: $5.5 \times 10^{17}/cm^3$
Nucleation	AlN	100nm	1	Undoped
Substrate	SiC			Conducting

Table 2. AlGaN-GaN HEMT epi profile.

Layer	Material	Thickness	Al Composition	Doping
Barrier	AlGaN	26nm	0.27	Undoped
Buffer	GaN	2650nm	0	Undoped
Nucleation	AlN	100nm	1	Undoped
Substrate	SiC			SI

The following processing steps were taken to fabricate AlGaN Schottky diodes. Ti(200Å)-Al(1500Å)-Ni(500Å)-Au(800Å) was first e-beam evaporated on the epi wafer to form Ohmic contacts. After lift-off of the extraneous metal, the sample was annealed in a rapid thermal annealing (RTA) station at 900°C for 30sec under N_2 ambient. Then, either Pt(300Å)-Au(1200Å) or Ni(300Å)-Au(1200Å) was e-beam evaporated to form Schottky contacts followed by a lift-off of the extraneous metal. No alloying was performed after this step. The same processing steps were taken to fabricate AlGaN HEMTs. We chose round HEMT structures for the present study because these structures do not require a mesa etching process for electrical isolation. The Ti-Al-Ni-Au was e-beam evaporated to form Ohmic contacts for source and drain and the Pt-Au was e-beam evaporated to form Schottky contacts for the gate. Completed AlGaN HEMT devices had four different gate lengths. The minimum gate length was 1.5μm and the separation between source and drain was 5.5μm.

I-V characteristics of AlGaN Schottky diodes were measured at different temperatures ranging from 101 to 446K. Figure 1 (a) shows variable temperature I-V curves measured from a Pt-AlGaN Schottky diode (Sample A) and Figure 1 (b) shows temperature dependences of ideality factors and Schottky barrier heights measured at seven different temperatures from the same device. The same measurements were performed on the Ni-AlGaN Schottky diode (Sample B), but the results from Sample B are not shown here. Figure 1 (b) shows a monotonic increase of ideality factors as well as a monotonic decrease of Schottky barrier heights as the temperature increased from 162 to 436K. This trend clearly indicates that the current transport in AlGaN Schottky diodes cannot be explained by thermionic emission alone. In addition, I-V-T measurements were not able to detect current contributions from either defect-related trap states or from conduction along dislocations based on the Frenkel-Poole emission model.

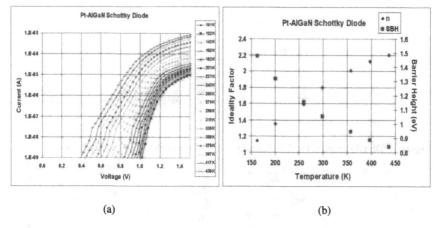

(a)   (b)

Figure 1. Variable temperature I-V curves (a) and temperature dependence of ideality factors and Schottky barrier heights (b) measured from the Pt-AlGaN Schottky diode.

In recent years, a few groups reported traps in AlGaN-GaN hetero-structures grown by different techniques on SiC substrates including MOCVD and MBE. Among various techniques used to investigate traps in these material systems, we chose a capacitance digital DLTS technique mainly because this technique provides precise information on traps by measuring changes in capacitance due to the presence of traps in the depletion region of metal-semiconductor junctions. Individual diode chips were sawed from completed wafers and mounted on carriers for the DLTS measurements. Figures 2 and 3 show a DLTS spectra (a) and the corresponding Arrhenius plot (b) from Sample A and Sample B, respectively. Pt (or Ni)-AlGaN junction areas were ~2000μm^2 and typical junction capacitances measured at zero bias were around 10pF. DLTS measurements were made on a number of Schottky diodes and the DLTS spectra shown in Figures 2 and 3 are representative spectra. Table 3 summarizes energy level (E_c-E_T), density (N_T), and capture cross section (σ) of the two traps that we identified from both samples. Since no GaN buffer layer was grown for our Schottky diode structures, both traps originate from the AlGaN barrier layer. Further investigation is necessary to fully understand the root causes of these traps, but it appears that our level 1 traps with E_a=0.415-0.464 eV and our level 2 traps with E_a=0.787-0.758 eV correspond to E2 and E3 traps, respectively, reported by Reddy et al. [8]. Densities of both traps are 2-7×10^{16} cm^{-3}. Our level 1 traps are due to point defects related to Group III-vacancy complexes or N anti-sites, whereas our level 2 traps are due to point defects related to formation of extended defects including dislocations [9, 10].

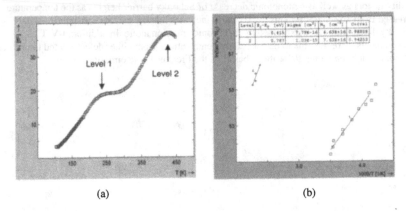

(a)   (b)

Figure 2. DLTS spectra (a) and Arrhenius plot (b) of the Pt-AlGaN Schottky diode.

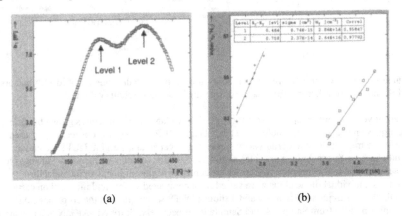

(a)   (b)

Figure 3. DLTS spectra (a) and Arrhenius plot (b) of the Ni-AlGaN Schottky diode.

Table 3. Summary of DLTS measurements.

Trap Level		E_c-E_T (eV)	N_T (cm^{-3})	σ (cm^2)
Sample A	Level 1	0.415	4.63×10^{16}	7.79×10^{-16}
	Level 2	0.787	7.63×10^{16}	1.05×10^{-15}
Sample B	Level 1	0.464	2.86×10^{16}	8.74×10^{-15}
	Level 2	0.758	2.64×10^{16}	2.37×10^{-14}

122

We employed TEM to investigate extended defects in the AlGaN Schottky diode epi wafer shown in Table 1. TEM samples were prepared using focused ion beam. Figure 4 shows cross sectional TEM images taken at two different magnifications of 80K× (a) and 500K× (b).

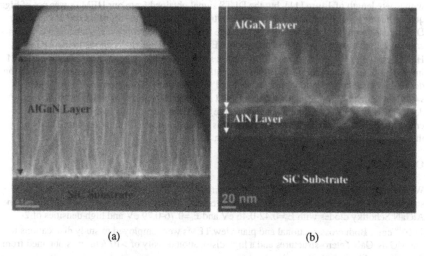

(a)　　　　　　　　　　　　　　(b)

Figure 4. Cross sectional TEM images of the AlGaN Schottky diode epi wafer.

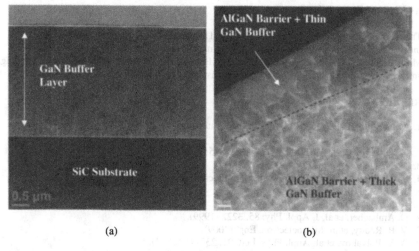

(a)　　　　　　　　　　　　　　(b)

Figure 5. Cross sectional (a) and plan view (b) TEM images of the AlGaN HEMT epi wafer.

A number of threading dislocations, clearly visible in Figure 4, are contributing to our level 2 traps. Note that two groups reported the presence of only E2 traps in aged AlGaN HEMTs [5, 6]. One group reported DLTS spectra measured from fat AlGaN HEMTs grown on SiC substrates with a gate length of 14μm [11], but the DLTS signal obtained from our HEMTs was not of high quality due to small junction capacitances resulting from short gate lengths in our devices. Our DLTS measurements on HEMTs with long gate lengths will be reported elsewhere. We employed both cross sectional and plan view TEM to investigate dislocations in the AlGaN HEMT epi wafer. Figure 5 shows a 25K× cross sectional TEM image (a) and a plan view TEM image (b), where a number of threading dislocations were also visible. To estimate a dislocation density, we prepared additional plan view TEM samples using Ar ion milling from the same wafer covering a ~3μm×3μm area and obtained a dislocation density of >10^8/cm^2 from the device active layer (TEM image not shown here). The presence of this many defects in the critical AlGaN-GaN device active areas remains a major concern for long-term reliability of these devices under high electric field operation.

CONCLUSIONS

We fabricated Schottky diodes and HEMTs using MOCVD-grown AlGaN-GaN hetero-structures on SiC substrates. Our DLTS measurements identified two different types of traps in AlGaN Schottky diodes with E_a=0.42-0.46 eV and E_a=0.76-0.79 eV and high densities of 2-7×10^{16} cm^{-3}. Both cross-sectional and plan view TEMs were employed to study dislocations in the AlGaN-GaN hetero-structures and a high dislocation density of >10^8/cm^2 was obtained from the device active layer.

ACKNOWLEDGMENTS

The work described in this paper was performed as part of The Aerospace Corporation's Independent Research and Development Program. The authors gratefully acknowledge Mr. Miles Brodie for TEM sample preparation, Ms. Barbara Hill for wafer sawing and wire bonding, and Dr. Steven C. Moss for technical discussions as well as support and encouragement.

REFERENCES

1. M. S. Shur, et al, Proceedings of 4th IEEE International Caracas Conference on Devices, Circuits and Systems, Aruba, D051 (2002).
2. T. Kikkawa, et al, IEEE MTT-S Digest, 1815 (2002).
3. M. Nagahara, et al, IEDM (2002).
4. S. C. Binari, et al, IEEE Transactions on Electron Devices, 48, 465 (2001).
5. J. Joh, et al, IEDM (2006).
6. A. Sozza, et al, IEDM (2005).
7. O. Ambacher, et al, J. Appl. Phy. 85, 3222 (1999).
8. V. R. Reddy, et al, Microelectron. Eng. (2009).
9. A. Y. Polyakov, et al, Appl. Phys. Lett. 91, 232116 (2007).
10. K. Sugawara, et al, Appl. Phys. Lett. 94, 152106 (2009).
11. M Faqir, et al, Microelectronics Reliability 47 (2007).

Energy Conversion

Mater. Res. Soc. Symp. Proc. Vol. 1202 © 2010 Materials Research Society   1202-I07-03

Improvement of Photoelectrochemical Reaction for Hydrogen Generation From Water Using N-face GaN

Katsushi Fujii[1], Keiichi Sato[1], Takashi Kato[1], Tsutomu Minegishi[1,2], Takafumi Yao[1]
[1]Center for Interdisciplinary Research, Tohoku University,
Aramaki Aza Aoba 6-3, Aoba-ku, Sendai, Miyagi 980-8578, JAPAN
[2]Department of Chemical System Engineering, School of Engineering, University of Tokyo,
7-3-1 Hongo, Bunkyo-ku, Tokyo, 113-8656, JAPAN

ABSTRACT

Photoelectrochemical properties of Ga- and N-face GaN grown by hydride vapor phase epitaxy (HVPE) were investigated. The properties were also compared with Ga-face GaN grown by metal-organic vapor phase epitaxy (MOVPE). The flatband potentials were in order of Ga-face GaN grown by MOVPE < N-face GaN < Ga-face GaN. The highest photocurrent density at zero bias was obtained from the N-face GaN. The photocurrent density was over 3 times larger than that of Ga-face GaN.

INTRODUCTION

Photoelectrochemical water splitting is one of expecting renewable energy techniques to generate hydrogen from light and water [1]. The energy conversion efficiency improvement is one of the keys to realize the hydrogen production using this technique. Photoelectrochemical water splitting used a photo-absorbed semiconductor as a working electrode to generate electron-hole pairs, which are utilized to split water. Therefore, increasing of light absorption by band gap shrinkage is important to improve the energy conversion. A few semiconductors have been reported to split water by the absorption of visible light [2]. The reduction of generated carrier loss also improves the efficiency. Co-catalyst on the working electrode is typically used to reduce the carrier loss.

Nitride semiconductors are good candidates for the photo-illuminated working electrodes considered from the band edge energy and the chemical stability in solutions [3,4]. Band gap shrinkage for nitride semiconductor using InGaN increases the light absorption [5]. Co-catalyst for GaN also improves the photoelectrochemical water splitting [6]. In addition, semiconductor properties like carrier lifetime and surface orientation affect the light to carrier conversion efficiency [4,7]. Thus, the property tuning is also important to improve the photoelectrochemical efficiency. In this report, we evaluated the surface polarity of Ga- and N-face GaN in order to reduce the carrier loss and to compare with conventional Ga-face GaN grown by MOVPE.

EXPERIMENT

We used 1.5 mm thick freestanding n-type GaN grown by hydrogen vapor phase epitaxy (HVPE) for the evaluation of Ga- and N-face GaN. The carrier concentration (C.C.) and mobility were 1.8×10^{18} cm^{-3} and 280 cm^2/Vs, respectively. The contact electrodes were the backsides of the photoelectrochemical reaction faces. The Ga-face GaN samples grown by metal-organic vapor phase epitaxy (MOVPE) (n-type, C.C. = 2.2 ~ 10.0 × 10^{18} cm^{-3}, 4.5 ~ 4.2 µm in thickness) were also used as references. The contact electrodes of the references were placed at the front surface edges because the insulating sapphire substrates were used. The electrolytes were 1.0 mol/L HCl (pH 0.2) and 1.0 mol/L NaOH (pH 13.6) aqueous solutions. In addition to these electrolytes, the electrolytes used for the references were 0.5 mol/L Na_2SO_4 (pH 6.1) and 0.5 mol/L Na_2SO_4 in addition of 0.01 mol/L NaOH (pH 12.0). The counterelectrode was Pt wire. Potentiostat (Solartron SI1280B) and Ag/AgCl/NaCl reference electrode (0.212 V vs standard electrode) were used when those were needed. The working electrode was connected to the counterelectrode directly for the current measured at zero bias. The schematic diagrams of the GaN working electrodes and experimental set up in the case of a potentiostat used are shown in Fig. 1.

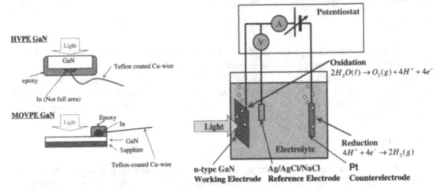

Figure 1. Schematic diagrams of the working electrodes of GaN grown by HVPE and MOVPE, and photoelectrochemical experimental set up in the case of a potentiostat used.

RESULTS AND DISCUSSION
Mott-Schottky Plot of HVPE GaN

The Mott-Schottky plots of the Ga- and N-face GaN grown by HVPE in 1.0 mol/L HCl (pH 0.2) and NaOH (pH 13.6) electrolytes are shown in Fig. 2(a). The flatband potentials, which are obtained at the points of $1/C^2 \to 0$ from Fig. 2(a), are summarized in Table I. The band edge potentials calculated from the flatband potentials as functions of pH are shown in Fig. 2(b). The band edge potentials of the Ga-face GaN grown by MOVPE are also plotted in Fig. 2(b). The band edge potentials of Ga-face GaN were about 0.2 V positive to the N-face GaN. Interestingly, even the band edge potentials of N-face GaN were about 0.3 V positive to those of the Ga-face GaN grown by MOVPE especially for the case of acidic electrolyte.

The flatband potential is generally defined by surface carrier traps, which are changed by the surface material, crystal structure, and contacted electrolyte. Thus, the flatband potentials with different surface orientation are usually different [7]. For this case, the surface orientations of the Ga-face GaN grown by HVPE and the GaN grown by MOVPE are the same. The flatband difference would be explained to be the difference of the amounts and kinds of the impurities.

The flatband potentials between the Ga- and N-face GaN were also different. This difference would be explained by the difference of the surface structures.

The gap between the conduction band edge potential and the electrode potential of hydrogen generation is a kind of the indicator of the photoelectrochemical hydrogen generation ability from water. The gap between the valence band edge potential and oxygen generation electrode potential is the indicator of the oxygen generation ability in the same manner. For this case, conduction band edge potential is important for the water splitting because the potential is close to the electrode potential of hydrogen generation, which is known as -0.057 [pH] V vs standard electrode. From this consideration, N-face GaN is expected to have better ability to split water than that of Ga-face GaN at zero bias because the gap for N-face GaN is larger than that of Ga-face GaN.

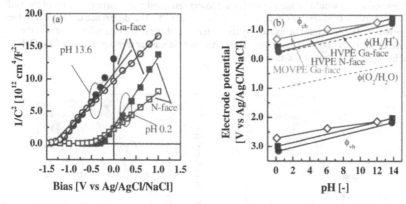

Figure. 2 (a) Mott-Schottky plot of Ga- and N-face GaN grown by HVPE. The electrolytes are 1.0 mol/L HCl (pH 0.2) and 1.0 mpl/L NaOH (pH 13.6). (b) Flatband potentials of Ga- and N-face GaN grown by HVPE as functions of pH. The flatband potentials of Ga-face GaN grown by MOVPE, the electrode potentials of hydrogen and oxygen generation are also shown.

Table I Flatband potentials of Ga- and N-face GaN grown by HVPE and Ga-face GaN grown by MOVPE.

pH	Ga-face GaN (HVPE) [V vs Ag/AgCl/NaCl]	N-face GaN (HVPE) [V vs Ag/AgCl/NaCl]	Ga-face GaN (MOVPE) [V vs Ag/AgCl/NaCl]
0.2	-0.24	-0.41	-0.61
6.1	-	-	-0.95
12.0	-	-	-1.18
13.6	-1.21	-1.38	-

Cyclic voltammetry

The cyclic voltammetries without light illumination showed Shottky diode-like relationships for all samples, i.e., negative currents were observed at the biases below the flatband potentials. The results indicate the interface between GaN and the electorolytes are good shottky contacts.

The results of cyclic voltammetries under light illuminations are shown in Fig. 3. The results of the Ga-face GaN grown by MOVPE are also shown in the case of 1.0 mol/L NaOH electrolyte used in Fig. 3(b). The order of the turn on voltage versus reference electrode was Ga-face MOVPE GaN < N-face GaN < Ga-face GaN as expected from the order of the flatband potentials. The absolute values shifted since the pH was different, but the orders in HCl and NaOH electrolytes were coincident. The order of the saturated photocurrent density at the positive biased region was Ga-face GaN > Ga-face GaN grown by MOVPE ≥ N-face GaN. As shown in Fig. 1, the contact electrode of the GaN grown by MOVPE was placed at the edge of the photo-illuminated face. The resistance of the thin current path in the sample is expected to be higher than that of the GaN grown by HVPE, which contact electrode is placed at the backside. This high resistance would be the reason for the lower saturated photocurrent density of the GaN grown by MOVPE than that of Ga-face GaN grown by HVPE. The difference of Ga- and N-face GaN grown by HVPE would be caused by the difference of the carrier transfer process at the interface of the GaN and the electrolyte. The barrier for the carrier transfer at the N-face GaN would be higher than that of Ga-face GaN.

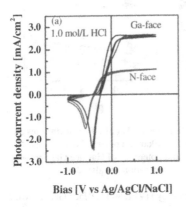 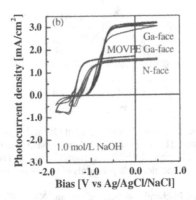

Figure. 3 Cyclic voltammetry of Ga- and N-face GaN grown by HVPE in (a) 1.0 mol/L HCl (pH 0.2) and (b) 1.0 mol/L NaOH (pH 13.6). That of Ga-face GaN grown by MOVPE is also shown in (b).

Static current – bias charactristics

The results of the static current – bias characteristics without light illumination showed Shottky diode-like relationships similar to the case of cyclic voltammetry.

The results of static current – bias characteristics under light illuminations are shown in Fig. 4. The results of the Ga-face GaN grown by MOVPE are also shown in the case of 1.0 mol/L NaOH used in Fig. 4(b). The turn on voltage for the photocurrent densities were vely close compared with those of cyclic voltammetry because the biases were measured versus counterelectrodes. The order of the turn on voltage for photocurrent density without reference electrode is N-face GaN < Ga-face GaN ≤ Ga-face GaN grown by MOVPE. This order was changed from that obtained from cyclic voltammetry. The positive shift of Ga-face GaN grown by MOVPE can be also explained by the voltage loss of the high series resistance of the sample.

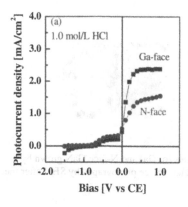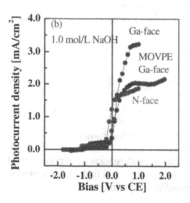

Figure. 4 Static current – bias characteristics of Ga- and N-face GaN grown by HVPE in (a) 1.0 mol/L HCl (pH 0.2) and (b) 1.0 mol/L NaOH (pH 13.6). That of Ga-face GaN grown by MOVPE is also shown in (b).

Time dependence of photocurrents

Photocurrent density at zero bias of N-face GaN was over 3 times higher than that of Ga-face GaN as expected from the flatband potentials. The density was almost stable over 600 min as shown in Fig. 5(a).

The surfaces after the time dependence observation were also shown in Fig. 5(b). The surfaces were smooth at the beginning of the experiments, but those became rough after the treatment. Orange peel-like structures were observed at the surface of N-face GaN, and pit-like shapes are observed at the surface of the Ga-face GaN. This suggests that surface photocorrosion are included to the photocurrent density. The ratio of the photocorrosion is, however, probably small because the surface is less roughened than that after the photoelectrochemical etching.

CONCLUSIONS

Photoelectrochemical properties of Ga- and N-face GaN grown by HVPE were observed. The properties are also compared with that of Ga-face GaN grown by MOVPE. The flatband potential and turn on voltage as functions of bias were coincident and in order of Ga-face GaN grown by MOVPE < N-face GaN < Ga-face GaN. The highest saturated photocurrent density was observed when the Ga-face GaN was used. The photocurrent density of N-face GaN was,

131

however, over 3 times higher than that of Ga-face GaN at zero bias. The time dependences of the photocurrent density were almost constant for both Ga- and N-face GaN. Negative flatband potential of N-face GaN would help the highest photocurrent density without bias.

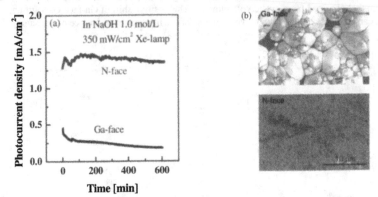

Figure. 5 (a) Time dependences of photocurrent densities for Ga- and N-face GaN grown by HVPE in 1.0 mol/L NaOH (pH 13.6) electrolytes. (b) The surface photographs by SEM after the time dependence experiments.

ACKNOWLEDGMENTS

The authors thank Professor T. Itoh for his technical assistance.

REFERENCES

1. for example, A.J. Nozik and R. Memming, *J. Phys. Chem.* **100**, 13061 (1996).
2. C. A. Grimes, O. K. Varghese and S. Ranjan, *Light, Water, Hydrogen*, (Springer, New York, 2008), pp. 399 – 400, pp. 462 – 464.
3. S. S. Kocha, M. W. Peterson, D. J. Arent, J. M. Redwing, M. A. Tischler and J. A. Turner, *J. Electrochem. Soc.*, **142**, L238 (1995).
4. M. Ono, K. Fujii, T. Ito, Y. Iwaki, A. Hirako, T. Yao and K. Ohkawa, *J. Chem. Phys.*, **126**, 054708 (2007).
5. K. Fujii, M. Ono, T. Ito, Y. Iwaki, A. Hirako and K. Ohkawa, *J. Electrochem. Soc.*, **154**, B175 (2007).
6. N. Arai, N. Saito, H. Nishiyama, Y. Inoue, K. Domen and K. Sato, *Chem. Lett.*, **35**, 796 (2006).
7. K. Fujii, Y. Iwaki, H. Masui, T. J. Baker, M. Iza, H. Sato, J. Kaeding, T. Yao, J. S. Speck, S. P. DenBaars, S. Nakamura and K. Ohkawa, *Jpn. J. Appl. Phys.*, **46**, 6573 (2007).

Controlled Light-Matter Interactions

Mater. Res. Soc. Symp. Proc. Vol. 1202 © 2010 Materials Research Society   1202-I12-03

GaN Photonic-Crystal Surface-Emitting Laser

Susumu Yoshimoto[1], Hideki Matsubara[1], Kyosuke Sakai[2], and Susumu Noda[1, 3]
[1]Department of Electronic Science and Engineering, Kyoto University, Kyoto 615-8510, Japan.
[2]Pioneering Research Unit for Next Generation, Kyoto University, Kyoto 615-8245, Japan.
[3]Photonics and Electronics Science and Engineering Center, Kyoto University, Kyoto 615-8510, Japan.

ABSTRACT

We introduce the lasing principle and important characteristics of photonic-crystal surface-emitting laser (PC-SELs). Specifically, we demonstrate two-dimensional coherent lasing oscillation with GaN PC-SELs, using a unique crystal growth technique called "air hole retained overgrowth" (AROG). Above the threshold, we obtained a two-dimensionally distributed near-field pattern, and a distinctive far-field pattern with a divergence angle less than 1°. We also investigate a suitable sample structure for the reduction of the threshold current, where the PC structure is moved from an n-cladding layer to a p-cladding one. This is an important step towards the realization of novel light sources that can be integrated two dimensionally for a variety of new scientific and engineering applications in the blue to ultraviolet wavelengths.

INTRODUCTION

There has been growing interest in photonic-crystal surface-emitting lasers (PC-SELs) [1–3] (Fig. 1). The lasing principle is based on the band-edge effect in a 2D PC, where the group velocity of light becomes zero and a 2D cavity mode is formed. The output power is coupled to the vertical direction by the PC, which gives rise to the surface-emitting function. PC-SELs have the following features: first, perfect, single longitudinal, and lateral mode oscillation can be achieved even when the lasing area becomes very large. Secondly, the polarization mode [2] and

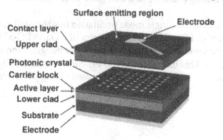

Figure 1. Schematic structure of a PC-SEL.

beam pattern [3] can be controlled by appropriate design of the unit cell and/or lattice phase in the 2D PC. For example, unique beam patterns, including doughnut shapes with radial or tangential polarizations have been successfully generated, which lead to the realization of super-high-resolution light sources that can be focused to a spot smaller than their wavelength [4, 5]. Recently, GaN PC-SELs have been realized in the blue-violet wavelengths [6] using a unique method called "air hole retained overgrowth" (AROG). The device successfully oscillated with a current injection at room temperature. A high-power laser beam in the blue to ultraviolet regime would greatly contribute to fields such as biomedical research and information technology. In this study, we overview the general characteristics of the PC-SEL, including the lasing principle and the unique beam patterns, and present the recent progress of GaN PC-SEL.

LASING PRINCIPLE AND UNIQUE BEAM PATTERNS

Figure 2 shows the lasing principle in (1) a triangular lattice photonic crystal, where one finds the six fold equivalent directions. When (2) the wavelength of light propagating in a specific Γ-X direction is equal to a lattice period a, (3) the lattice points diffract it backward as well as in the other in-plane directions. (4) These light waves, propagating in the five directions, are diffracted back again in the original direction. The (5) multiple diffractions couple the light waves, forming 2D resonant modes in the whole cavity area (band-edge effect). Because the resonant mode is determined by the periodic perturbation of the refractive index, 2D PC-SELs can operate in a perfectly single longitudinal and transverse mode, even for a large cavity area.

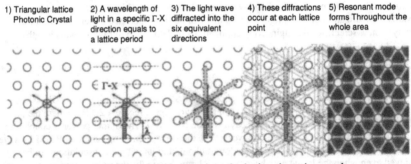

| 1) Triangular lattice Photonic Crystal | 2) A wavelength of light in a specific Γ-X direction equals to a lattice period | 3) The light wave diffracted into the six equivalent directions | 4) These diffractions occur at each lattice point | 5) Resonant mode forms Throughout the whole area |

Figure 2. Lasing principle in a triangular lattice photonic crystal.

In the infrared regime, a successful single mode oscillation has already been achieved. Figure 3 shows the near field pattern (NFP) and the lasing spectra of a GaAs PC-SEL device working in the infrared regime. All the spectra show the same peak wavelength, even for a large lasing area over a $150 \times 150 \ \mu m^2$ region. This indicates that 2D PC-SELs are inherently suited for high-power single-mode oscillation. In fact, we achieved pulsed output power up to 1W in the infrared regime.

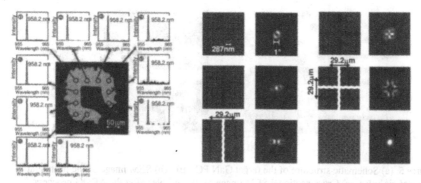

Figure 3. Near field pattern and the lasing spectra in the infra-red regime.

Figure 4. A range of beam patterns for a various PC structure. (left) SEM image. (right) Far field pattern.

The output beam is emitted in the direction normal to the crystal plane, using the crystal as a diffraction grating. Thus, the device works as a surface emitting laser. The output beam pattern, i.e., the transverse mode profile, can be determined by the photonic crystal. Figure 4 shows a scanning electron microscope (SEM) image of the fabricated crystal (left side) and the output beam pattern (right side) for a range of PC structures. (a) A PC-SEL with a square lattice of circular air holes produces a doughnut shaped beam, and (b)-(e) introducing various types of phase shift produces a range of multiple doughnut shaped beams. Moreover, (f) when the air hole shape is changed to a triangle, the beam pattern shows a single lobed profile. Therefore, we can design a beam shape by tailoring the PC structure. Among these beam patterns, a doughnut shaped beam can be applied to a range of novel applications such as optical tweezers, where we can trap opaque particles, as well as transparent particles. In addition, a doughnut beam with radial polarization is expected to produce a focal spot smaller than the wavelength of light. Thus, reproducing this type of beam in a blue-violet regime can support a range of technological innovations, such as a higher density optical memory or a super resolution microscope.

DEVICE FABRICATION AND CHARACTERISTICS

Figure 5(a) shows the schematic structure of the target GaN PC-SEL. To construct a 2D GaN/air periodic structure of the PC, we developed the new AROG method. The method is based on the unique characteristics of GaN growth [6], which proceeds much faster in the lateral direction than vertical growth on the (0001) crystal plane. First, a triangular lattice of air holes (Fig. 5(b)) with a period of 186 nm, a diameter of 85 nm, and a depth of 100 nm was formed on an epitaxial GaN/AlGaN layer situated above a GaN substrate in an n-cladding layer, and SiO_2 was then deposited at the bottom of each air hole. This SiO_2 prohibits the crystal growth process inside the air holes. In the next step, the GaN layer was overgrown using a low pressure MOVPE. Figure 5(c) shows a cross-sectional SEM image of the fabricated device. The periodic arrangement of air holes was well defined inside the GaN epitaxial layer. The GaN overgrowth appeared to proceed laterally, capping the top of the air holes, while the SiO_2 deposited at the bottom of the air holes and blocked growth.

137

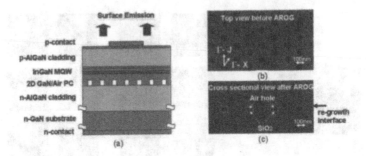

Figure 5. (a) Schematic structure of the target GaN PC-SEL. (b) SEM image of the triangular lattice of air holes. (c) Cross-sectional SEM image of the air holes after the AROG proccess.

Figure 6 shows the lasing characteristics of the fabricated sample. In the L-I curve, (a) a clear threshold at 6.7 A is apparent. Consequently, (b) the spectra show a drastic change around the threshold. The spectrum below the threshold shows a broad profile (at 6.5 A), whereas, above the threshold (at 6.9 A and 7.4 A), we observed sharp lasing peaks with the full width at half maximum of ~0.15 nm, which was equal to the resolution limit of the measurement system. The peak wavelength was 406.5 nm. The near field pattern (c) showed a large lasing area distributing two dimensionally around the $100 \times 100 \ \mu m^2$ electrode, and correspondingly the far field pattern has a divergence angle less than 1°. This indicates that over a large area coherent oscillation was successfully obtained.

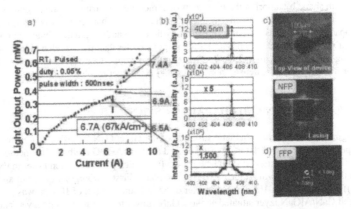

Figure 6. Lasing characteristics of the GaN PC-SEL. (a) L-I curve shows clear threshold. (b) Spectra below and above the threshold. (c) Top view of the device and the near field pattern above threshold. (d) Far field pattern of the laser beam.

DISCUSSION

In the above mentioned structure, the PC layer was formed in the n-cladding layer before the active layer (Fig. 7 (a)), where we faced the following trade-off. It was important to make the cavity effect more pronounced by decreasing the distance between the active and PC layer, and decrease the threshold. However, to keep the quality of the active layer, it was necessary to grow a buffer layer between the two layers to avoid degradation in the crystal quality due to the AROG process. To circumvent this trade-off, we investigated a new structure, where a PC layer is formed in the p-cladding layer, as shown in Fig. 7 (b). In a trial sample, we formed the PC layer 60 nm above the active layer using SiO_2 to prohibit the crystal regrowth inside the air holes.

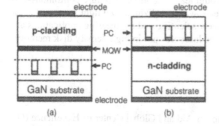
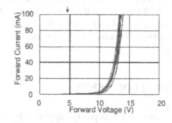

Figure 7. Schematic cross sectional structure of the GaN PC-SEL. (a) Original device. (b) New device has PC layer in the p-cladding layer.

Figure 8. I-V curve for the new device.

However, the V-I characteristics show high resistance, e.g., 13.6V at 100mA (Fig. 8), which is possibly due to the Si (n-type dopant) compensating for the p-dopant. Thus, in the second trial sample, we avoid using SiO_2 in the AROG process. As a result, although the air hole shape was slightly rounded (Fig. 9), we achieved a great decrease in the resistance (4.6 V at 100mA), as shown in Fig. 10.

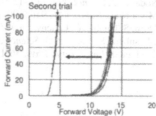

Figure 9. Cross-sectional SEM image of the air holes after the AROG process without the SiO_2.

Figure 10. I-V curve for the second trial device.

Under the pulsed condition (width: 500 nm, repetition rate: 1 kHz), we obtained the lasing oscillation just above 1A, which is almost one fifth of the previous sample. Subsequently, we measured the threshold current densities as a function of a pulse duty at room temperature and at

−45°C. We found that the lasing oscillation could be achieved with a pulse duty of 83% at −45°C and with 70% at room temperature. These promising results indicate that room temperature continuous operation of GaN PC-SELs is imminent.

CONCLUSIONS

We presented the recent progress in GaN PC-SEL research after a brief overview of the lasing principle and general characteristics of PC-SELs. We demonstrated the coherent oscillation of GaN PC-SEL at room temperature under pulsed conditions. A promising step for threshold reduction was also presented, which should lead to room-temperature continuous-wave operation. We believe that this is an important step towards realizing novel light sources that can be integrated two-dimensionally for a variety of new scientific and engineering applications at blue to ultraviolet wavelengths, including post-blue lasers for use in high-density disk memory systems.

ACKNOWLEDGMENTS

This work was partly supported by a Grant-in-Aid and Global Center of Excellence (G-COE) program and Consortium for Photon Science and Technology (C-PhoST), the Ministry of Education, Culture, Sports, Science and Technology of Japan, and also by the Core Research for Evolutional Science and Technology of the Japan Science and Technology Agency.

REFERENCES

1. M. Imada, et al, *Appl. Phys. Lett.* 75, 316 (1999).
2. S. Noda, et al, *Science* 293, 1123 (2001).
3. E. Miyai et al., *Nature* 441, 946 (2006).
4. R. Dorn, S. Quabis, G. Leuchs, *Phys. Rev. Lett.* **91**, 233901 (2003).
5. H. Matsubara, S. Noda, et al, *Science* **319**, 445 (2008)
6. A. Usui, et al, *Jpn. J. Appl. Phys.* 36, L899 (1997).

Mater. Res. Soc. Symp. Proc. Vol. 1202 © 2010 Materials Research Society   1202-I11-02

THz Emission From InN

Hyeyoung. Ahn,[1] Yi-Jou Ye,[1] Yu-Liang Hong,[2] and Shangjr Gwo[2]
[1] Department of Photonics and Institute of Electro-Optical Engineering,
National Chiao Tung University, Hsinchu 30010, Taiwan R.O.C.
[2] Department of Physics,
National Tsing Hua University, Hsinchu 30013, Taiwan R.O.C.

ABSTRACT

We report the terahertz (THz) emission from the wurzite indium nitride (InN) films grown by molecular beam epitaxy (MBE). More than two orders of magnitude of THz power enhancement has been achieved from the InN film grown along the a-axis and magnesium (Mg) doped InN with a critical carrier concentration. The primary radiation mechanism of the a-plane InN film is found to be due to the acceleration of photoexcited carriers under the polarization-induced in-plane electric field perpendicular to the a-axis. Apparent azimuthal angle dependences of THz wave amplitude and the second harmonic generation are observed from a-plane InN. In the Mg-doped films, Mg as the acceptors compensate the native donors in the InN films and large band bending over a wider space-charge region causes the enhancement of THz emission power compared to the undoped InN.

INTRODUCTION

Due to its narrow bandgap and high electrical mobility, indium nitride (InN) has actively been studied for the fundamental material researches and the fruitful prospects in applications as high-frequency electronic devices, near-infrared optoelectronics, and high-efficiency solar cells. With a unique advantage of large energy difference between the conduction band minimum and the next local minimum, InN also inspires potential applications in the THz range [1-5]. Meanwhile, the performance of short-wavelength optoelectronic devices realized by growing III-nitrides along the c-axis is typically limited by the polarization-induced internal electric fields. Piezoelectric and spontaneous polarizations are responsible for the polarization-induced electric field and in particular, the strain-dependent piezoelectric polarization along the c-axis of the wurzite crystals increases with the lattice mismatch in the nitride layers. For the layers grown along a- or m-axis direction (so called, nonpolar), on the other hand, polarization-induced electric field perpendicular to the layer interface can be minimized and the efficiency of the devices can be increased. Despite there are abundant results reported for other group-III nitrides, researches on InN grown along the a-axis (a-plane InN) are rare mainly due to the technical difficulty in growing high crystalline quality a-plane InN films.

Due to its high electron affinity, undoped InN is unintentionally doped n-type. However, in order to realize the optoelectronic devices, it is essential to have the ability to fabricate both n- and p-type InN. p-type doping of InN has been one of the main challenges of intense research efforts. In the electrolyte-based capacitance-voltage (ECV) measurement, Jones et $al.$ showed that the surface Fermi level pinning of InN:Mg causes an extremely large band bending compared to that of undoped InN. And they proposed that the InN:Mg film is consisted of a p-type bulk region separated from a thin surface electron accumulation layer by a wide depletion region [6]. It opens the possibility of obtaining p-type bulk conductivity in InN. Meanwhile, photoexcited p-type InAs film can emit at least one order of magnitude larger THz radiation than n-type InAs and the same is expected from p-type InN. Therefore, it is essential to fabricate p-type InN for strong THz emission. In general, THz radiation from semiconductors is very complex since multiple emission mechanisms compete to dominate and it mainly depends on the dynamics of photoexcited and free carriers. In order to clarify the THz emission mechanism from InN, a comprehensive study on carrier dynamics and the carrier transport mechanism is required. Among several THz emission mechanisms, the surge current-induced transient electric field is known to play an important role in the THz emission from the narrow bandgap semiconductors, such as InAs and InN. Previously, we have reported the THz power enhancement from low-dimensional InN nanostructures compared to the InN film [5]. The carrier dynamics and the fundamental optical properties of InN nanostructures were characterized by using THz time-domain spectroscopy [7,8].

Here, we summarize even more significant enhancement of THz emission from the a-plane InN surface excited at a moderate pump fluence and the carrier concentration-dependent THz emission from Mg-doped InN. The azimuthal angle dependence of THz emission and second-harmonic generation from a-plane InN shows that the THz emission from a-plane InN has different origin from that from c-plane InN. The azimuthal-angle-independent response may be due to the accelerated photocarriers in the in-plane electric field of the a-plane InN film, while angle-dependent radiation might be due to nonlinear optical processes. Typically, THz emission from the undoped InN is much weaker than that from InAs due to large screening from high intrinsic carrier density. Doping InN with Mg is expected to reduce the carrier density and enhance the radiation intensity. Recently, the enhanced radiation from Mg-doped InN compared to that from Si-doped n-type InN has been reported [9]. We also found the enhancement of THz emission from Mg-doped c-InN films with reduced carrier concentrations (n). As the carrier concentration decreases by Mg doping, THz emission increases due to the combined effects of carrier compensation and the reduced screening of the photo-Dember field. Meanwhile, for InN:Mg samples with the electron concentration below a critical value ($n_c \sim 1 \times 10^{18}$ cm^{-3}), radiation amplitude is reduced and the polarity of THz waveform becomes negative. Negative polarity of THz waveform is the same that of p-type GaAs, and it shows the dominant contribution of drift current induced by the strong surface band bending within the space-charge layer.

EXPERIMENTAL DETAILS

The a-plane InN epitaxial film (~1.2 μm) was grown by plasma-assisted molecular beam epitaxy (PA-MBE) on r-plane $\{1\bar{1}02\}$ sapphire wafer, while the $-c$-plane InN epitaxial film (~2.5 μm) was grown on Si(111) using a double-buffer layer technique. The back side of r-plane sapphire wafer was coated with a Ti layer for efficient and uniform heating during the PA-MBE growth and this metal layer was removed before performing optical measurement. The growth direction of the InN film was determined using a 2θ–ω x-ray diffraction scan. Although it is much smaller than that of c-plane InN, near-infrared photoluminescence was detected from both a-plane and Mg-doped InN films at room temperature. The carrier concentrations of 7.0×10^{18} and electron mobilities (μ) of 298 cm^2/Vs were determined by room-temperature Hall effect measurements for a-plane InN films.

Seven Mg-doped InN films with different carrier concentrations were grown by PA-MBE on Si(111). The film thicknesses of InN:Mg samples are in the range of 1–1.5 μm. Mg doping was performed with a high-purity Mg(6N) Knudsen cell and the Mg doping level was controlled by regulating the cell temperature between 180 and 270°C. The electron concentrations and electron mobilities were determined by room-temperature Hall effect measurements. Even for highly doped InN:Mg films, Hall measurement only shows the electron concentration, which is due to the strong n-type surface layer in InN. The mobility of InN:Mg films is reduced as the carrier density decreases whereas for InN:Mg with $n < n_c$, the mobility begins to increase with more reduction of carrier concentration. The same abnormal behavior has been observed previously by infrared reflectance measurement [10] and was explained to be the formation of p-type InN:Mg only within the limited range of carrier concentration. The electron concentrations of seven InN:Mg films and the undoped InN film are summarized in Table I.

Table. I. The electron concentrations of Mg-doped InN films used in the text.

Sample	Concentration (cm^{-3})
A	4.7×10^{17}
B	7.7×10^{17}
C	9.9×10^{17}
D	1.1×10^{18}
E	1.2×10^{18}
F	1.6×10^{18}
G	2.1×10^{18}
undoped	3.1×10^{18}

Time-domain terahertz emission measurements were performed using a Ti:sapphire laser system, which delivers ~50 fs optical pulses at a center wavelength of 800 nm with a repetition rate of 1 kHz. For this experiment, the pump laser beam is collimated on the samples with a spot size of ~ 2 mm at the incident angle of 70°. The terahertz pulses were detected by free-space electro-optic sampling in a 2-mm-thick ZnTe crystal as a function of delay time with respect to the optical pump pulse. The azimuthal angle dependence of THz emission and second harmonic generation was measured as the sample is rotated along the axis normal to the surface. The experimental details can be found elsewhere [11].

DISCUSSION

Despite its large interband transition energy (~2.8 eV), THz emission from InN is typically much smaller than that from InAs. That is attributed to at least one order of magnitude smaller electron mobility and higher carrier concentration of InN films. Therefore, growth of thick and high crystalline quality InN films with much reduced free electron concentrations and high electron mobilities is an essential requirement for higher THz emission from InN. Meanwhile, THz emission from InN is known to be attributed to the competition between several emission processes, such as the surge current due to the surface depletion field, photo-Dember field, and other nonlinear processes. While at very low pump fluence, THz radiation is attributed to the drift current induced by the internal surface electric field [12], for highly excited InN, stronger THz emission is due to the photo-Dember field.

THz radiation generated within a material of high refractive index N has an extraction problem: at the semiconductor surface, all rays outside an emission cone of half-angle $\alpha = \sin^{-1}(1/N)$ to the surface normal suffer total internal reflection and cannot escape from the device. The fraction of power extracted depends on the orientation of the dipole axis relative to the surface. Several methods exist to increase the extraction efficiency, such as applying an external magnetic field [13], using a coupling prism [14], or fabrication of low-dimensional nanostructures [5]. For example, under a magnetic field the Lorentz force—the force on a point charge due to electromagnetic fields—can rotate the orientation of the dipole to coincide with the emission cone. By applying strong magnetic field, about two orders of magnitude of THz power enhancement has been achieved for InAs. In the meantime, THz dipoles parallel to the surface electric field also can be achieved by growing InN along a-axis and then the same amount of power enhancement can be expected from it.

THz emission from a-plane InN

Figure 1(a) shows the THz waveforms of p-polarized THz field from the a-plane InN film, compared to that from the c-plane InN film. Under the same pump fluence at ~0.24 mJ/cm^2, the amplitude of THz radiation component from a-plane InN is about ten times larger than that from c-plane InN. In Fig. 1(b), the peak values of THz emission from a- and c-plane InN increase linearly with the increase of pump fluence up to 1 mJ/cm^2 and no saturation of the emission is observed.

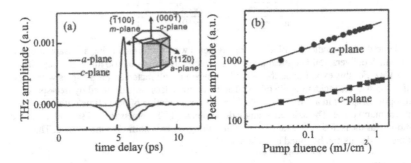

Figure 1. (a) The amplitude of the terahertz emission from a-plane InN film [black solid] compared to that from c-plane InN film [red solid]. The excitation fluence is 0.24 mJ/cm^2. (b) The amplitude of the p-polarized emission field vs. pump fluence. The solid line is a linear fit to the data.

Figure 2(a) shows the azimuthal angle dependent p-polarized THz fields from a-plane InN film excited by s- and p-polarized pump with the fluence of ~0.32 mJ/cm^2. Both curves in Fig. 2(a) show a common feature of a large angle-independent field superposed with a weak, angle-dependent field. While the angular dependence of P_{pump}-P_{THz} show a four-fold angular symmetry, S_{pump}-P_{THz} response has a two-fold, but asymmetric dependence. The polarity of

P_{pump}-P_{THz} is positive whereas that of S_{pump}-P_{THz} is negative. For a dipole radiating normal to the semiconductor surface, there is no s-polarized emission. Indeed, P_{pump}-S_{THz} and S_{pump}-S_{THz} signals were negligibly small. The angular dependence of the THz signal from c-plane InN is different from the THz emission from a-plane InN, which exhibits no significant angular dependence, as is shown in Fig. 2(b).

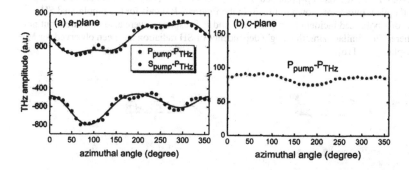

Figure 2. (a) Peak amplitudes of the p-polarized THz fields vs. azimuthal angle rotation of a-plane excited by the p- and s-polarized optical pump at the fluence of 0.32 mJ/cm^2 and (b) the same of c-plane InN excited by the p-polarized optical pump. The solid lines in Fig. 2(a) are obtained from the calculation based on eq. (2) in the text.

Figure 2(a) shows that the main radiation is due to the angle-independent radiation process and the angle-dependent radiation contributes less than 20% of the total amplitude. Usually, azimuthal angle-dependent terahertz emission is attributed to nonlinear optical mechanisms, such as the resonance-enhanced optical rectification. However, the nonlinear effects play the significant role typically under very high excitation fluence so that it may not be so important for the current experiment. The linear relation of THz radiation to the pump fluence in Fig. 1(b) shows that our pump fluence (0.2-0.35 mJ/cm^2) is well below the regime of photocarrier saturation. Slightly larger linear slope of the a-plane InN compared to that of the c-plane InN in Fig. 1(b) might be due to the slow increase of nonlinear effects with the increase of pump fluence.

The contribution of electron accumulation field to THz emission can be negligible for both c- and a-plane InN due to the narrow thickness of the electron accumulation layer. Strong radiation from a-plane InN with much smaller mobility than c-plane InN reveals that the contribution of the photo-Dember effect to the radiation can be even smaller than that for c-plane InN. Therefore, the drastic power enhancement observed for a-plane InN cannot be explained by either the electron accumulation or the photo-Dember fields. By the way, the c-plane wurtzite InN has the alternating stacking sequence of In and N atoms along the wurtzite c-axis direction so that the surface layers of c-plane InN have either an In- or a N-terminated polar surfaces. Then the electric field generated by these In-N bilayers directs perpendicular to the surface. The out-of-surface radiation from the vertically aligned dipoles can be significantly limited by the geometrical reason mentioned above. Meanwhile, a-plane InN has the same number of In and N atoms in a plane and these In-N pairs form in-plane electric field parallel to the surface. The photocarriers can then be efficiently coupled to the in-plane electric field such that a more favorable part of emission can escape the emission cone.

The second-harmonic generation from the surface is often employed to identify the nonlinear properties of semiconductors. In order to describe the abnormal azimuthal angle dependence of THz radiation from a-plane InN, the azimuthal angle-dependent SH radiation was also measured. The results of p-polarized SH radiations from a-plane InN excited by s- and p-polarized optical pump are shown in Fig. 3. In the figure, S-P stands for the p-polarized SH response of s-polarized incident laser pulse. S-S and P-S responses were also measured, but not shown here. The similar azimuthal angle dependence of SH radiation has been observed for Mg doped a-plane InN [16].

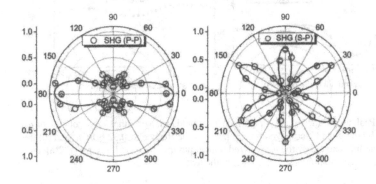

Figure 3. The azimuthal angle-dependent p-polarized SH signals from a-plane InN when the incident fundamental laser is p- (a) and s-polarized (b), respectively.

In the surface region, the inversion symmetry of the bulk is broken and we expect a large contribution to the nonlinear polarization. We present this term by the second-order polarization $P^{(2)}$ located just inside the medium. This nonlinear polarization can be related to the electric field at the fundamental frequency by

$$P_i^{(2)}(\Omega) = \chi_{ijk}^{(2)eff}(\Omega;-\omega,\omega)E_j(-\omega)E_k(\omega) \tag{1}$$

where the structural symmetry of the surface is reflected in the form of the nonlinear susceptibility tensor $\chi_{ijk}^{(2)eff}$ which has the bulk $\left(\chi_{ijk}^{(2)bulk}\right)$ and surface $\left(\chi_{ijk}^{(2)surface}\right)$ contributions. For the surface contribution, we take only the first-order term in the expansion and then the effective nonlinear susceptibility can be written as $\chi_{ijk}^{(2)eff} = \chi_{ijk}^{(2)bulk} + 3\chi_{ijkl}^{(3)surface}E_z^{surface}$. In order to examine the azimuthal angle rotation dependence of the radiation, the susceptibility tensors are calculated in the beam coordinates. The generated second-order radiation field for s- or p- polarized harmonic light due to the bulk anisotropic source as a function of the azimuthal angle become

146

$$E_{s,p}^{\text{bulk}} = A_{s,p}\Omega L_{\text{eff}}\left[\,a_0^{s,p(\text{bulk})} + b_m^{s,p(\text{bulk})}(\cos\phi)^m + c_m^{s,p(\text{bulk})}(\sin\phi)^m\,\right] \tag{2}$$

the subscript notation indicates s- or p-polarized bulk electric field. The coefficients a_0, b_m, and c_m's are related with the coordinate projection coefficients, the transmission coefficients, and the effective phase-matching length [17]. The solid lines in Fig. 2 and 3 are obtained from the best fitting to the experimental data using equation (2). The best fitting to SH radiation was obtained with the dominant bulk contribution whereas that to THz radiation was obtained by the combined contribution of bulk and surface susceptibility. The details of the analysis will be reported elsewhere.

The effect of Mg-doping on THz emission from InN

It is known that THz emission from small bandgap semiconductors such as InAs is due to the photo-Dember effect whereas in wide bandgap semiconductors such as GaAs surface field acceleration is responsible for THz emission. The distinction between the photo-Dember effect and surface field acceleration can be made by monitoring the polarity of THz wave. In wide bandgap semiconductors, Fermi level-pinning occurs in such a way to bend the bands to form a surface built-in field in the depletion region. The direction of the band bending-induced electric field depends on the dopant types so that the corresponding THz waves show the opposite polarity. However, InN has the surface Fermi level locates above the conduction band minimum and the band bending for both n- and p-type InN is downward, which induces THz wave with negative polarity for n- and p-type InN. Meanwhile, the polarity of THz wave due to the photo-Dember effect is always positive since it is due to the drift motion of electrons and holes with respect to the surface normal. THz emission due to both mechanisms depends on carrier density and its amplitude and polarity from InN are determined by the competition between these emission mechanisms depending on the carrier concentration and the excitation energy.

In order to investigate the effect of Mg doping to THz emission, we measured the time-domain THz emission from a series of InN:Mg samples with different n and compared it with that from an undoped InN film. Figure 4 shows the THz waveforms of the selected InN:Mg samples and the undoped InN excited at the pump fluence of ~ 200 μJ/cm^2.

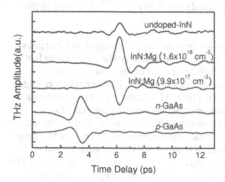

Figure 4. The time-domain THz waveforms of the undoped InN and Mg-doped InN films, sample C and F. Waveforms of sample F and the undoped InN have positive polarity whereas that of sample C shows the negative polarity. The THz waveforms of n- and p-type GaAs are plotted for comparison. The amplitude of each sample does not correspond to the absolute value.

The waveforms of InN:Mg sample with $n = 1.6 \times 10^{18}$ cm^{-3} (sample F) and the undoped InN ($n = 3.1 \times 10^{18}$ cm^{-3}) have the positive polarity while that of the samples C with $n = 9.9 \times 10^{17}$ cm^{-3} exhibit a negative polarity. THz emission from c-InN under the current excitation level is known to be dominated by the photo-Dember effect and the corresponding THz polarity should be positive for both n- and p-type InN. In order to verify the negative polarity of THz wave, THz waveforms from n- and p-type GaAs are shown at the bottom panel in Fig. 4. It shows that negative polarity of the sample C is parallel to that of p-type GaAs wafer and thus the photo-Dember effect can no longer describe THz emission from the sample C. Additionally, no significant azimuthal angle dependence of terahertz emission was measured for the InN:Mg samples (the example is shown in Fig. 5) so that we can rule out the contribution of the optical rectification effect for THz emission. Then it leaves the surface acceleration effect due to the downward surface band bending to be the only possible explanation for the negative THz polarity.

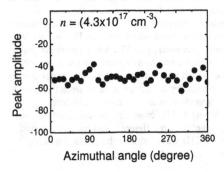

Figure 5. Peak terahertz amplitudes of InN:Mg sample ($n = 4.3 \times 10^{18}$ cm^{-3}) measured as the sample is rotated about the surface normal.

A recent research showed that THz emission from InN:Si and undoped InN increases with mobility and the enhanced THz emission was observed from InN:Mg compared to InN:Si [9]. And this enhancement was described as the high compensation between acceptors (n_A) native donors (n_D) which decreases the carrier concentration. However, from even more highly doped InN:Mg, an abnormal carrier dependence of THz field was observed. Figure 6 shows the peak THz amplitude as a function of electron concentration measured by the Hall effect measurement. It shows that the magnitude of THz radiation increases as the carrier concentration decreases from that of undoped InN film and the maximum emission occurs for InN:Mg samples with $n = 1.2–1.5 \times 10^{18}$ cm^{-3} and the THz polarities of these samples are positive. However, for InN:Mg samples with lower carrier concentrations than $\sim 1 \times 10^{18}$ cm^{-3} (n_c) the amplitude of THz radiation becomes smaller and has a negative polarity. We measured the optical pump fluence dependence of THz radiation for InN:Mg samples (not shown) and found that THz emission increases linearly with the pump fluence for our InN:Mg samples and the saturation of the emission is not observed.

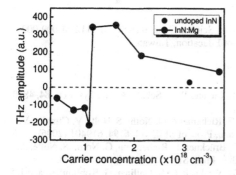

Figure 6. The peak amplitude of THz emission field vs. carrier concentration of InN:Mg films. The red data point corresponds to the THz amplitude of the undoped InN.

The Hall effect measurement shows that the electron concentration decreases as the Mg doping level increases. It indicates that doped Mg acceptors compensate the native donors in the unintentionally-doped n-type InN films and reduce the electron concentration in the InN:Mg film. According to the surface field model, the highest THz emission and the largest depletion width occur for a highly compensated semiconductors with $n_A \approx n_D$ [9,18]. Therefore, the negative THz polarity observed for InN:Mg samples with $n < n_c$ indicates that the carriers in these sample are highly compensated and the large downward band bending occurs across the space-charge layer whose depth extends to cover the penetration depth ($d \sim 133$ nm [19]) of the pump pulse (800 nm). And THz emission from these films is mainly due to the surface field acceleration effect. On the contrary, for InN:Mg samples with $n > n_c$, the space-charge layer where band bending occurs might be much shorter than d due to the partial carrier compensation and the photo-Dember effect dominates THz emission in the bulk region of InN:Mg. In this bulk region, the reduction of screening of the photo-Dember field can be more significant as n decreases further from that of undoped InN. The maximum positive polarityTHz emission is, therefore, obtained before the sharp increase of depletion depth occurs near n_c.

CONCLUSIONS

In summary, we found that the a-plane InN film radiates about two orders of magnitude more intense THz waves compared to the c-plane InN film. THz radiation from a-plane InN has small, but distinctive azimuthal angle dependence and it may be due to the nonlinear optical processes. The effective geometrical coupling between the in-plane electric field and the out of plane THz radiation is proposed to be responsible for the drastic enhancement. The role of Mg as the acceptors in Mg-doped InN were elucidated by measuring THz emission from InN:Mg samples with a wide variation of carrier concentration. The competitive play of the photo-

Dember effect and surface field acceleration occurs InN:Mg depending on the carrier concentration which varies with the carrier compensation within the surface space-charge layer.

ACKNOWLEDGMENTS

This work was supported by the National Science Council (NSC 98-2112-M-009-009-MY3) and the ATU program of the Ministry of Education, Taiwan.

REFERENCES

1. R. Acsazubi, I. Wilke, K. Denniston, H. L. Lu, and W. J. Schaff, Appl. Phys. Lett. **84**, 4810 (2004).
2. B. Pradarutti, G. Matthäus, C. Brückner, S. Riehemann, G. Notni, S. Nolti, V. Cimalla, V. Lebedev, O. Ambacher, and A. Tünnermann, Proc. of SPIE vol. **6194**, 619401 (2006).
3. V. Cimalla, B. Pradarutti, G. Matthäus, C. Brückner, S. Riehemann, G. Notni, S. Nolte, A. Tünnermann,V. Lebedev, and O. Ambacher, Phys. Stat. Sol. B **244**, 1829 (2007).
4. G. D. Chern, E. D. Readinger, H. Shen, M. Wraback, C. S. Gallinat, G. Koblmuller, and J. S. Speck, Appl. Phys. Lett. **89**, 141115 (2006).
5. H. Ahn, Y.-P. Ku, Y.-C. Wang, C.-H. Chuang, S. Gwo, and Ci-Ling Pan, Appl. Phys. Lett. **91**, 132108 (2007).
6. R. E. Jones, K. M. Yu, S. X. Li, W. Walukiewicz, J. W. Ager, E. E. Haller, H. Lu, and W. J. Schaff, Phys. Rev. Lett. **96**, 125505 (2006).
7. H. Ahn, Y.-P. Ku, Y.-C. Wang, C.-H. Chuang, S. Gwo, and C.L. Pan, Appl. Phys, Lett. **91**, 163105 (2007).
8. H. Ahn, C.-H. Chuang, Y.-P. Ku, and C.-L. Pan, J. Appl. Phys. **105**, 023707 (2009).
9. I. Wilke, R. Ascazubi, H. Lu, and W. J. Schaff, Appl. Phys. Lett. **93**, 221113 (2008).
10. M. Fujiwara, Y. Ishitani, X. Wang, S.-B. Che, and A. Yoshikawa, Appl. Phys. Lett. **93**, 231903 (2008).
11. H. Ahn, C.-L. Pan, and S. Gwo, Proc. of SPIE, **7216**, 72160T (2009).
12. K. I. Lin,1, J. T. Tsai, T. S. Wang, J. S. Hwang, M. C. Chen, and G. C. Chi, Appl. Phys. Lett. **93**, 262102 (2008).
13. M. B. Johnston, D. M. Whittaker, A. Corchia, A. G. Davies, and E. H. Linfield, Phys. Rev. B **65**, 165301 (2002).
14. M. B. Johnston, D. M. Whittaker, A. Dowd, A. G. Davies, E. H. Linfield, X. Li, and D. A. Ritchie, Opt. Lett. **27**, 1935 (2002).
15. E. Estacio, H. Sumikura, H. Murakami, M. Tani, N. Sarukura, and M. Hangyo, Apply. Phys. Lett. **90**, 151915 (2007).
16. Y. M. Chang, Y.-L. Hong, and S. Gwo, Appl. Phys. Lett. **93**, 131106 (2008).
17. M. Reid, I. V. Cravetchi, R. Fedosejevs, Phys. Rev. B **72**, 035201 (2005).
18. R. Ascazubi, C. Shneider, I. Wilke, R. Pino, and P. S. Dutta, Phys. Rev. B **72**, 045328 (2005).
19. H. Ahn, C.-H. Shen, C.-L. Wu, and S. Gwo, Appl. Phys. Lett. **86**, 201905 (2005).

Mater. Res. Soc. Symp. Proc. Vol. 1202 © 2010 Materials Research Society   1202-I10-08

Roberto Paiella[1], Kristina Driscoll[1], Yitao Liao[1], Anirban Bhattacharyya[1], Lin Zhou[2], David J. Smith[2], and Theodore D. Moustakas[1]

[1]Department of Electrical and Computer Engineering and Photonics Center, Boston University, Boston, MA 02215, U.S.A.

[2]Department of Physics, Arizona State University, Tempe, AZ 85287, U.S.A.

ABSTRACT

Due to their large conduction-band offsets, GaN/Al(Ga)N quantum wells are currently the subject of extensive research efforts aimed at extending the spectral range of intersubband optoelectronic devices towards shorter and shorter wavelengths. Here we report our recent measurement of optically pumped intersubband light emission from GaN/AlN quantum wells at the record short wavelength of about 2 μm. Nanosecond-scale optical pulses are used to resonantly pump electrons from the ground states to the second excited subbands, followed by radiative relaxation into the first excited subbands. The intersubband origin of the measured photoluminescence is confirmed via an extensive study of its polarization properties and pump wavelength dependence.

INTRODUCTION

Intersubband (ISB) transitions in GaN/Al(Ga)N quantum wells (QWs) have been the subject of extensive research efforts for the past several years [1-8]. These heterostructures feature particularly large conduction-band offsets (up to about 1.75 eV) allowing for ISB transitions at record short wavelengths, well into the near-infrared spectral region. As a result they provide new opportunities for ISB device development, particularly in the area of all-optical switching for ultrafast fiber-optic communications [2,3] as well as short-wavelength ISB photodetectors [4,5], modulators [6], and sources [7,8]. Incidentally, it should be mentioned that nitride semiconductors are also promising for the development of THz ISB devices, owing to their particularly large LO-phonon energies [9].

Here we demonstrate optically pumped ISB light emission from a GaN/AlN multiple-QW structure at the record short wavelength of about 2 μm [8]. In these measurements the QWs are pumped with high-power near-infrared pulses from a tunable optical parametric oscillator (OPO), which allow resolving the QW ISB luminescence without the need for highly sensitive lock-in detection techniques. The output luminescence spectra are peaked at wavelengths near 2.05 μm, which represent a new record for the shortest ISB emission wavelength from any QW materials system.

EXPERIMENT

The material used in this work was grown on c-plane sapphire by rf plasma-assisted molecular beam epitaxy. Following nitridation of the substrate, a 1-μm-thick AlN buffer layer was deposited to serve as a template for growth of the subsequent layers. The multiple-QW active region consists of 200 periods of nominally 18-Å-thick GaN wells (doped n-type with Si to the level of ~ 2×10^{19} cm^{-3}) separated by 26-Å-thick AlN barriers. During deposition of the AlN layers a small flux of Ga was used as a surfactant to promote two-dimensional growth. Finally, the multiple-QW structure was terminated with a 70-nm-thick AlN cap layer. The sample microstructure was studied using a 400-keV JEM-4000EX transmission electron microscope (TEM).

In order to identify the relevant ISB transitions, the sample infrared absorption spectrum was measured at room temperature using a Fourier transform infrared (FTIR) spectrometer and a liquid-nitrogen-cooled HgCdTe detector. For this measurement the substrate side of the sample was mirror polished and two opposite facets were lapped at 45° to obtain a multi-pass waveguide geometry. Light from the FTIR internal source was then polarized TM (in accordance with the polarization selection rules of ISB transitions) and coupled in and out at normal incidence through the lapped facets.

For the light emission measurements, a pulsed OPO was used to optically pump electrons from the QWs ground states $|1\rangle$ to the second excited subbands $|3\rangle$, followed by relaxation into the first excited subbands $|2\rangle$ and corresponding luminescence at the $|3\rangle \rightarrow |2\rangle$ transition energy. The OPO output consists of a train of pulses with 5-ns width, 20-Hz repetition rate, and 0.5 mJ pulse energy. The measurements were carried out at room temperature, with the sample input and output facets polished at 45° and 90°, respectively. The pump light was delivered to the sample with a multi-mode optical fiber, and then focused at normal incidence onto the 45° input facet with a two-lens imaging system, producing an estimated spot diameter of about 300 μm. Broadband quarter- and half-wave retarders placed before the fiber and a polarizer before the sample were used to control the pump light polarization. Another two-lens imaging system was used to collect and focus the light exiting the 90° output facet onto the entrance slit of a monochromator, with a germanium plate in place to reject the residual pump light. The monochromator output light was finally measured using a room-temperature extended-range InGaAs detector with 1.2-2.6 μm spectral response and 45 MHz bandwidth. To increase the measurement sensitivity, gated detection was performed using a box-car integrator.

RESULTS AND DISCUSSION

A cross-sectional TEM image of the multiple-QW active region is shown in Fig. 1(a), revealing high structural quality with smooth interfaces and good control of the structure periodicity. Some period-to-period and in-plane thickness variations are visible, but they are limited to about one monolayer (ML), with the GaN well and AlN barrier thicknesses estimated to be in the range of 6-8 MLs and 9-10 MLs (with 1 ML ≈ 2.6 Å), respectively.

A measured absorbance spectrum normalized with respect to the system response is plotted in Fig. 1(b), where the observed feature is ascribed to ISB transitions between the ground states $|1\rangle$ and the first excited subbands $|2\rangle$. It should be noted that the $|1\rangle \rightarrow |3\rangle$ transitions are also allowed in these QWs, due to the built-in electric fields of nitride heterostructures which partially break the symmetry of the bound-state envelope functions. However, the resulting absorption is still too weak to be unambiguously resolved in our measured FTIR transmission spectra. The multi-peaked nature of the absorbance spectrum of Fig. 1(b) is attributed to the monolayer-scale variations in the QW thicknesses which give rise to fluctuations in the ISB transition energies.

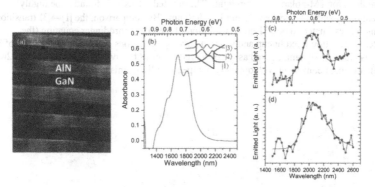

Figure 1. (a) Cross-sectional TEM image of the QW sample used in this work. (b) Room-temperature ISB absorbance spectrum of the same sample. The negative dip near 1300 nm is an artifact of the system response and is not related to the optical properties of the sample. The inset shows the calculated conduction-band diagram of one QW in the active region. (c) and (d) Emission spectra measured at two different places in the QW sample (dotted lines), and modified-Gaussian fits of the experimental data (solid lines).

In Figs. 1(c) and 1(d) we show two emission spectra measured at different positions of the QW sample using a TM-polarized pump at $\lambda_p = 910$ nm (1.362 eV). The solid lines in these figures are fits to a modified Gaussian peak, superimposed on a broad pedestal in the case of Fig. 1(c). The photon energies of peak emission and the spectral full-widths-at-half-maximum obtained from these fits are 608 meV (2.04 µm) and 99 meV in Fig. 1(c), and 602 meV (2.06 µm) and 105 meV in Fig. 1(d). The comparison between the two spectra illustrates the reproducibility of well-resolved luminescence features across the sample area. The measured emission photon energies are in good agreement (well within the luminescence linewidth) with the difference between the pump photon energy used in the measurements and the experimental value of 735 meV for the $|1\rangle\rightarrow|2\rangle$ transition energy estimated from Fig. 1(b). This observation confirms that the measured emission is indeed due to optical pumping from $|1\rangle$ to $|3\rangle$ followed by radiative relaxation from $|3\rangle$ to $|2\rangle$.

Further evidence of the ISB nature of the measured luminescence was obtained by studying its polarization properties and pump wavelength dependence, as summarized in Fig. 2. Figure 2(a) shows two spectra measured at the same spot under identical conditions except for the pump polarization, which is rotated from TM to TE. Essentially no emission is detected in the latter case, consistent with the polarization selection rules of ISB transitions. In Fig. 2(b) we plot two spectra obtained using the same TM-polarized pump wave, and with a polarizer added to the collection optics after the sample. As expected for ISB light emission, the luminescence is largely TM-polarized, with a strong reduction in the integrated power when the polarizer transmission axis is rotated from TM to TE. The small fraction of detected TE light is attributed to polarization scrambling of the TM-polarized ISB emission as it traverses the sample, due to the residual roughness in the bottom surface and sidewalls.

Finally, Figs. 2(c) and 2(d) show the evolution of the luminescence spectrum versus pump wavelength λ_p for TM-polarized pump light. The emitted intensity is found to be highly dependent on λ_p and to be peaked around $\lambda_p = 910$ nm. By comparison, the $|1\rangle \rightarrow |3\rangle$ transition wavelength is estimated to be 931 nm via solution of the QW Schrödinger equation (the calculated conduction-band lineup and squared envelope functions are shown in the inset of Fig. 1(b)). This close agreement, as well as the observed resonant behavior in the data of Figs. 2(c) and 2(d), are again consistent with optically pumped ISB light emission.

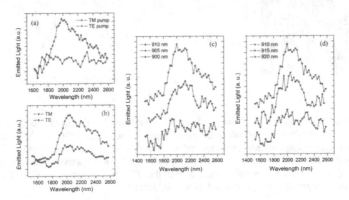

Figure 2. (a) Emission spectra measured with otherwise identical TM- (circles) and TE- (squares) polarized pump waves. (b) Emission spectra measured with the same TM-polarized pump wave and passed through a polarizer with transmission axis oriented along the TM (circles) and TE (squares) directions. (c) Dependence of the emission spectrum on the pump wavelength for $\lambda_p \leq 910$ nm. (c) Dependence of the emission spectrum on the pump wavelength for $\lambda_p \geq 910$ nm. In (c) and (d) the spectra are shifted relative to one another vertically for the sake of illustration clarity.

The peak integrated power of the detected light pulses is on the order of a few hundred nanowatts, with typical values of about 300 nW. Besides the low collection efficiency of the measurement setup, these values are mainly limited by the very inefficient $|1\rangle \rightarrow |3\rangle$ pump absorption, which is only weakly allowed in these uncoupled QWs by the parity-breaking action of the internal electric fields. As a result, only a negligible fraction of the incident pump power is actually absorbed in the sample (too small in fact to be resolved in our FTIR transmission measurements). Much higher pumping efficiencies, as well as the possibility of robust population inversion, can be obtained in more complicated systems such as asymmetric coupled QWs.

CONCLUSIONS

In summary, we have demonstrated optically pumped ISB emission of short-wave (2 μm) infrared radiation from GaN/AlN QWs. These results show that, despite the complexities of nitride heterostructures, sufficiently large electronic populations can be established in excited subbands for ISB luminescence. This is promising for the future demonstration of nitride-based short-wavelength ISB lasers. Further progress in that direction will require the development of more advanced QW structures providing sufficiently large pumping efficiency and the possibility of population inversion.

ACKNOWLEDGMENTS

This work was supported by the National Science Foundation under grant ECS-0622102. We acknowledge use of facilities in the John M. Cowley Center for High Resolution Electron Microscopy at Arizona State University.

REFERENCES

1. C. Gmachl, H. M. Ng, S.- N. G. Chu, and A. Y. Cho, Appl. Phys. Lett. **77**, 3722 (2000).
2. N. Iizuka, K. Kaneko, and N. Suzuki, IEEE J. Quantum Electron. **42**, 765 (2006).
3. Y. Li, A. Bhattacharyya, C. Thomidis, T. D. Moustakas, and R. Paiella, Opt. Express **15**, 5860 (2007).
4. E. Baumann, F. R. Giorgetta, D. Hofstetter, H. Lu, X. Chen, W. J. Schaff, L. F. Eastman, S. Golka, W. Schrenk, and G. Strasser, Appl. Phys. Lett. **87**, 191102 (2005).
5. A. Vardi, G. Bahir, F. Guillot, C. Bougerol, E. Monroy, S. E. Schacham, M. Tchernycheva, and F. H. Julien, Appl. Phys. Lett. **92**, 011112 (2008).
6. L. Nevou, N. Kheirodin, M. Tchernycheva, L. Meignien, P. Crozat, A. Lupu, E. Warde, F. H. Julien, G. Pozzovivo, S. Golka, G. Strasser, F. Guillot, E. Monroy, T. Remmele, and M. Albrecht, Appl. Phys. Lett. **90**, 223511 (2007).
7. L. Nevou, M. Tchernycheva, F. H. Julien, F. Guillot, and E. Monroy, Appl. Phys. Lett. **90**, 121106 (2007).
8. K. Driscoll, Y. Liao, A. Bhattacharyya, L. Zhou, D. J. Smith, T. D. Moustakas, and R. Paiella, Appl. Phys. Lett. **94**, 081120 (2009).
9. E. Bellotti, K. Driscoll, T. D. Moustakas, and R. Paiella, Appl. Phys. Lett. **92**, 101112 (2008).

Light Emitting Diodes

Mater. Res. Soc. Symp. Proc. Vol. 1202 © 2010 Materials Research Society   1202-I02-06

On the Light Emission in GaN Based Heterostructures at High Injection

X. Ni, X. Li, J. Lee, H. Y. Liu, N. Izyumskaya, V. Avrutin, Ü. Özgür, and H. Morkoç[*]
Dept of Electrical and Computer Engineering
Virginia Commonwealth University
601 W. Main Street, Richmond, VA 23284

T. Paskova, G. Mulholland, and K.R. Evans
Kyma Technologies, Inc., Raleigh, NC 27617

ABSTRACT

For light emitting diodes (LEDs) to be used for general lighting, high efficiencies would need to be retained at high injection levels to meet the intensity and efficiency requirements. In this regard, it is imperative to overcome the observed drop in LED efficiency at high injection levels beyond that would be expected from junction temperature. The suggested genesis of efficiency degradation includes electron overflow or spillover, also suggested to be aided by polarization induced electric field, Auger recombination, current crowding, and elevated junction temperature. Setting the junction temperature aside, the degree to which or even whether each of these mechanisms plays a role is still under debate. We have undertaken a series of experiments to isolate, whenever possible, the aforementioned processes in an effort to determine the causes of efficiency loss at high injection levels. By using 1 μs pulsed electrical injection with 0.1% duty cycle, we were able to minimize the effect of the junction temperature. By changing the design of the multiple quantum well region as well as by employing or not employing electron blocking layers, we demonstrated the important role that electron overflow plays on efficiency. Furthermore, by also exploring the same on non polar surfaces and observing any lack of dispersion in terms of the effect of the electron blocking layer we can conclude that the polarization induced field does not seem to play a major role. LEDs on non polar surface with no notable efficiency degradation, up to current densities of about 2250 Acm^{-2} used for measurements, have been obtained which seems to imply that Auger recombination up to these injection levels is not of major importance, at least in the structures investigated. The effect of current crowding on efficiency droop was investigated by comparing semitransparent Ni/Au p-contacts and transparent conducting oxide contacts (Ga-doped ZnO). Because the latter showed notably reduced efficiency degradation at high injection levels, we can conclude that current crowding plays a role as well.

INTRODUCTION

Compared to conventional lighting technology, solid-state lighting shows great potential for energy saving, not to mention the Hg free nature. As one of the key components of lighting technology, InGaN light emitting diodes (LEDs) have been widely used as blue, violet, and green as well as white light emitters when used in conjunction with a yellow fluorescent dye. Over the past decade, the performance of InGaN-based LEDs has improved significantly to warrant the aforementioned applications. Critical to insertion of LEDs, however, is to retain efficiency at high injection current levels, the loss of which has been dubbed as the "efficiency

---

[*] Electronic mail: hmorkoc@vcu.edu

droop" [1]. The external quantum efficiency (EQE) has been reported to reach its peak value at current densities lower than 50 A/cm^2 and monotonically decreases thereafter even under short pulsed current [2] It is, therefore, imperative to understand the genesis of this problem before InGaN LEDs could be widely used in the abovementioned applications. It has also been reported that the Auger process alone cannot explain the efficiency droop and very likely might not even be the dominant cause at the current levels of interest [3]. In the present work, we will show that the electron overflow (or spillover) caused by unequal carrier injection (lower number of holes) and poor hole transport represent the most likely reason for the loss of efficiency retention at current densities investigated which in our experiment far exceed those needed for lighting applications.

EXPERIMENTAL PROCEDURES

The InGaN/InGaN MQW LED samples studied (emitting at ~400-410 nm) were grown on bulk GaN in a vertical low-pressure metalorganic chemical vapor deposition (MOCVD) system. It should be noted that the efficiency retention becomes increasingly more challenging as the wavelength of emission is increased. In our case, the wavelength used emanated from the need to keep the wavelenghth near the same as the earlier experiments which arbitrarily used the near 400 nm emission. Trimethylgallium (TMGa), trimethylaluminum (TMAl), trimethylindium (TMIn), silane (SiH$_4$), Cp$_2$Mg, and ammonia (NH$_3$) were used as sources for Ga, Al, In, Si, Mg and N, respectively. The LEDs on c- and m-plane orientations investigated here have the same structure (schematic is shown in Figure 1): 6 period 2 nm In$_{0.14}$Ga$_{0.86}$N quantum wells with 12 nm In$_{0.01}$Ga$_{0.99}$N barriers and a 60 nm Si-doped (2×10^{18} cm^{-3}) In$_{0.01}$Ga$_{0.99}$N underlayer just beneath the active region for improved quality. A ~10 nm p-Al$_{0.15}$Ga$_{0.85}$N electron blocking layer (EBL) was deposited on top of the active quantum well region, which was then capped with a 100 nm thick p-GaN layer. After mesa etching (250 μm diameter), Ti/Al/Ni/Au (30/100/30/30 nm) metallization annealed at 800°C for 60 seconds was used for n-ohmic contacts, and 5nm/5nm semitransparent Ni/Au layer was used for p-contacts. To investigate the effect of current crowding a few of the investigate structures featured Ga-doped ZnO (GZO) transparent conducting oxide p-contacts (~400nm) for increased lateral conductivity and thus reduced current filamentation (or current crowding). Finally, a 30nm/30nm Ti/Au contact pad was deposited on part of the top of the mesa (albeit with opacity) for probe contacts. The internal quantum efficiency (IQE) was determined by resonant photoluminescence (PL) measurements on the LED samples at room temperature using a frequency-doubled 80 MHz repetition rate femtosecond Ti:Sapphire laser. The excitation laser wavelength was 375 nm, below the bandgap of the quantum barriers and top GaN. This form of resonant excitation ensures equal number of electron and hole generation only within the quantum wells with ensuing recombination again only within the quantum wells due to large barrier heights.

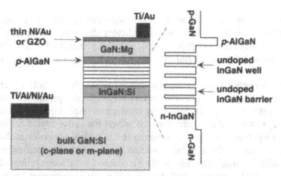

Figure 1. Schematic for the LEDs with EBL investigated in this work. As described in the text for some LEDs, the EBL is removed to investigate the carrier overflow effect, while other parts of the LEDs are kept the same as the ones with EBL.

RESULTS AND DISCUSSION

To set the stage and place our discussion in context with what has so far been reported, it is fair to state that at the moment the physical origin of the efficiency droop in InGaN LEDs is not sufficiently understood and is open to further investigations. Carrier loss due to Auger recombination has been proposed early on for the efficiency loss [4,5]. The Auger coefficient deduced through a fit to PL data in an earlier effort is $1.4\text{-}2.0 \times 10^{-30} \mathrm{cm}^6/\mathrm{s}$ for quasi-bulk InGaN layers [4]. A recent calculation reported the Auger recombination coefficient to be as large as $2 \times 10^{-30} \mathrm{cm}^6/\mathrm{s}$ from first-principle density-functional and many-body theory [6]. This notable coefficient is said to result from the upper conduction band which would be in resonance for LEDs emitting at 2.5 eV which implies that on either side of this wavelength, the Auger coefficient due to the aforementioned resonance should drop off precipitously. In the absence of this resonance effect, the direct Auger recombination coefficient decreases exponentially with the bandgap energy and the carrier losses due to Auger effect are expected to be very small in InGaN if we were to use the size of the forbidden gap alone [7]. Further credence to a small Auger coefficient is provided by fully microscopic many body models [8]. If an inherent process such as Auger recombination were to be solely responsible for the efficiency degradation, one would surmise that it would have prevented laser action in InGaN, which requires high injection levels, which is not the case. The so called phonon assisted extrinsic Auger recombination, however, must be examined, albeit very difficult, to determine whether this phenomenon could be in play here. To make matters more complex, through fitting the EL efficiency data from commercial LEDs to the rate equation, which includes Auger recombination in addition to the Shockley-Read-Hall (SRH) recombination and bimolecular radiative recombination, Ryu *et al.* [9] found unreasonably large Auger recombination coefficients in the range of $10^{-27}\text{-}10^{-24} \mathrm{cm}^6\mathrm{s}^{-1}$ when the non-radiative lifetime τ_{SRH} is assumed to be between 5-50 ns. The deduced Auger coefficients are at least three orders of magnitude higher, if not more, than the other reported values which are also extracted from the efficiency dependence on injection [4,6,10,11] implying that Auger recombination alone if any is not enough to explain the efficiency drop in InGaN LEDs. On the other hand, in below the barrier resonant photo excitation experiments (photons absorbed only in the quantum wells with ensuing generation of equal number of electron and hole generation followed by either radiative or non radiative recombination only), the efficiency

161

degradation has not been detected at carrier generation rates comparable to electrical injection (confirmed in our experiments as well) which indicates that efficiency degradation might be related to the carrier injection, transport, and leakage processes [12,13]. Aggravating the situation is the omission of the potential effects such as injection dependence of non radiative processes.

Given the arguments above, we suggest that the relatively low hole injection and/or poor hole transport (due to large hole effective mass) represent the prominent cause of electron overflow/spillover with ensuing efficiency degradation at high injection currents. The term "spillover electrons" refers to the electrons which escape the active region without contributing to the generation of light, and end up recombining in the p-GaN region or at the p-contact. The aforementioned suggestion could be further justified by our electroluminescence (EL) efficiency data from LEDs with coupled quantum wells in which the efficiency degradation has been reduced [14]. The multiple quantum well InGaN LEDs we investigated have varying InGaN barrier thicknesses (3-12 nm) and emit at ~ 400-410 nm. The EL data (under short pulsed drive current, 1KHz, 0.1% duty cycle) suggests that with 3nm InGaN barriers the extent of the efficiency drop could be reduced to ~5% for current densities up to 2000 Acm^{-2}, as compared to ~45% of drop for the LED with 12nm InGaN barriers in the same current range. This again is consistent with the above discussion about the poor hole injection and transport model.

Furthermore, in an effort to investigate whether the carrier spillover might be responsible for the efficiency drop as opposed to the Auger recombination, Li et $al.$ [3] performed fitting of the LED EL efficiency as a function of injection current with a modified recombination rate equation where the Auger recombination term can be replaced by a empirical current spillover term if desired, albeit with injection independent coefficients:

$$(J - J_{spillover})/qd = An + Bn^2 + Cn^3 + J_{spillover}/qt \qquad (1)$$

where d is the active region thickness, and t is the thickness within which the spillover current could be assumed to be linear. When an empirical dependence of the spillover current on the injection current is chosen as $J_{spillover}=kJ^b$ with k and b being fitting parameters, and $t = d$ is chosen. If one was to entirely neglect the Auger process, a good fit to the data can be obtained for the particular LEDs tested with n^3 like dependence of the leakage current [15]. While n^3 dependence is the same as Auger, others show n^2 dependence. Power dependence of the carrier overflow term on the carrier densities as high as 4.8 (ref. [16]) and 5.5 (ref. [17]) have been reported for InGaN and InGaAsP light emitters, respectively.

If we now subscribe to the concept that the electron overflow/spillover plays a major role in the efficiency droop in InGaN LEDs, the follow up question is then what causes the electron overflow. Kim et $al.$ [12] suggested that the polarization-induced electric field inside the active region and the EBL causes the electron overflow to the p-GaN layer. If this is indeed the case, then the non-polar LEDs would exhibit much smaller electron overflow. In this regard, we investigated the LED resonant PL and EL efficiency dependence on injection level (relative external quantum efficiency in the latter) as a function of current for LEDs on c-plane GaN, which is polar and possesses polarization induced field, and on $(1\overline{1}00)$ m-plane GaN, which is nonpolar and therefore free of polarization induced field. In addition, we also either inserted or not inserted EBLs in both varieties to interrogate the role of or the lack thereof polarization induced electric field on carrier spillover.

The IQE values extracted from the excitation-dependent PL method (see ref. [18] for a detailed description of the method) for the m-plane and c-plane InGaN multiple quantum well (MQW) LEDs are shown in Figure 2. First, the IQE values of c-plane LEDs with and without an

EBL layer are very similar in the entire range of photo-generated carrier concentrations up to 1.2×10^{18} cm^{-3} where a value of ~60 % is reached. This similarity in the IQE values is indicative of the fact that the resonant carrier excitation causes electron-hole pair generation only in the wells of MQWs which is also where they undergo radiative and or non radiative recombination. Second, in general the IQE values of the m-plane LEDs in our laboratory are ~30% higher than those of their c-plane counterparts at relatively high carrier concentrations, reaching slightly above 85% for the structure without the EBL and a comparable value of ~80% for the one with EBL at steady state carrier concentrations of 7×10^{17} and 1×10^{18}cm^{-3}, respectively. The higher internal quantum efficiency in the m-plane variety might probably be due to relatively larger optical matrix elements predicted for m-plane compared to c-plane of GaN [19]. We also note that for the two m-plane LEDs, the main difference in terms of IQE is that the one without EBL exhibits a smaller nonradiative SRH recombination, manifested by a steeper increase with increasing carrier density. This is attributed to the variation in the quality of the two structures, arising possibly from the variations in the quality of the m-plane bulk substrates.

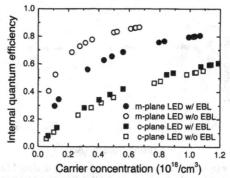

Figure 2. Extracted IQE values for the m- and c-plane LEDs on bulk GaN substrates from excitation power dependent resonant excitation PL measurements. All LEDs have the same structure: 6 period 2 nm In$_{0.14}$Ga$_{0.86}$N quantum wells with 12 nm In$_{0.01}$Ga$_{0.99}$N barriers either with or without the 10 nm p-Al$_{0.15}$Ga$_{0.85}$N EBL. During the calculation of carrier concentrations for the samples, the B coefficient was assumed to be 1×10^{-11} cm^{3}s^{-1}.

Figure 3 illustrates that the LEDs without EBL show reduced EL under the same current levels as compared to those with EBLs in both c-plane [4-5 times lower without EBL; Figure 3(a)] and m-plane cases [2-2.5 times lower without EBL; Figure 3(b)]. Recall that excitation-power dependent PL measurements discussed above show that the LEDs with and without EBL have essentially the same IQE values under optical excitation (for m-plane, the one without EBL is even slightly higher possibly due to the variation of substrates), even though under electrical injection the former group has higher EL intensity. These results suggest that the lower EL intensity for the LED without EBL is related to carrier injection (*i.e.*, more electron overflow caused by poor hole injection and hole transport inside the quantum wells for the LEDs without EBL), which do not apply under resonant optical excitation where equal number of holes and electrons are generated solely inside the quantum wells and no carrier transport is involved in the process. Moreover, substantial carrier spillover in both polarities w/o EBL also suggests that the polarization charge is not a major factor responsible for the efficiency degradation observed, particularly at high injection levels. It should also be mentioned that the m-plane variety with EBL did not show any discernable efficiency degradation up to a maximum current density of 2250 Acm^{-2} while that on c-plane showed a reduction by ~20%, indicating better ability of efficiency retention for the LED in m-plane direction. The fact that the m-plane LED with EBL shows no notable efficiency degradation seems to imply that Auger recombination up these injection levels is not of major importance, at least in the structure investigated.

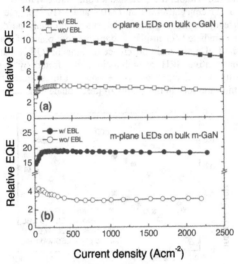

Current density (Acm^{-2})

Figure 3. Relative EQE for (a) c-plane LEDs on bulk c-GaN and (b) m-plane LEDs on bulk m-GaN with and without EBL measured under pulsed current with 0.1% duty cycle and a frequency of 1 kHz. Except the inclusion of the EBL, all the LEDs have the same structure: MQWs with 6 periods of 2nm In$_{0.14}$Ga$_{0.86}$N wells and 12nm In$_{0.01}$Ga$_{0.99}$N barriers. All of the LEDs have an emission wavelength ~400-410 nm. For both c-plane and m-plane varieties the EL intensity from the LEDs with EBLs is higher than that from respective LEDs without EBLs. The m-plane LED with EBL shows almost negligible efficiency droop (~5%) up to a current density of 2250 Acm^{-2}, compared to 20% droop for the c-plane LED with EBL.

As another potential cause of efficiency droop the detrimental effect of current crowding on the LED efficiency degradation should not be overlooked. In some LEDs where a lateral contact configuration is employed, the current density is distributed non-uniformly across the device area: concentrated near the edge of the thicker metal around the mesa and decreases exponentially with distance from the mesa edge. This locally concentrated current distribution can result in efficiency degradation due to locally increased injection levels and local heating. The particulars of current crowding in lateral contacts, although an age old problem in many devices including bipolar transistors, has been discussed in the context of nitride based LEDs in numerous papers, see, *e.g.*, ref. [20]. In addition to contact configuration design, a proper choice of p-contact materials also plays an important role to relieve the current crowding effect, thereby reduce the magnitude of efficiency drop according to our comparative studies of LED structures which employed GZO [21] and semi-transparent Ni/Au as p-contact materials. The GZO p-contact was grown by molecular beam epitaxy on a MQW-LED sample on c-plane sapphire with six-period 2nm $In_{0.15}Ga_{0.85}N$ wells and 12nm $In_{0.01}Ga_{0.99}N$ barriers. In the meantime, semitransparent Ni/Au (5nm/5nm) p-contact was used on another piece of the same LED wafer as a reference sample. The results show that in addition to an almost doubled light extraction due to improved transmittance of GZO over the thin Ni/Au contact, the magnitude of efficiency droop was also reduced from ~37% for the LED associated with thin Ni/Au contact to ~27% for the one with GZO contact in tests up to current levels of 3500 Acm^{-2}. The reduced efficiency drop in the GZO case is attributable to a partial elimination of current crowding effect, judging from the fact that a much more uniform emission intensity distribution at high current levels is achieved for the LED with GZO contact as compared to the case with thin Ni/Au contact which exhibited non uniform light emission at high injection levels.

Last but not least, junction Joule heating effect on LED efficiency deserves close attention and it has also been proposed to contribute to the InGaN LED efficiency drop, indicating a obvious suggestion that optimal heat removal should be employed [22]. The junction heating issue is very important due to several reasons. First, the junction heating would affect the internal quantum efficiency due to enhanced non-radiative processes including the SRH recombination with the increase of temperature [15]. Second, excessive junction heating could also degrade the active layer as well as contacts which would manifest itself as increased series resistance, thereby degrading efficiency as well as the power conversion efficiency. At high currents, the dominant heat source is from the Joule heating of p-type GaN and p-contact, having a dependence of I^2R. Additional heat may also come from an increased non-radiative recombination due to the efficiency drop at high operating current and low photon extraction efficiency. Kim et al. [12] proposed that the junction heating only reduce the overall efficiency of the LEDs, but not the cause of the efficiency drop. According to our measured EL data, at high injection levels thermally induced drop is observed even at 1% duty cycle as judged from the fact that the drop can be partially eliminated by employing 0.1% duty cycle (as shown in Figure 4).

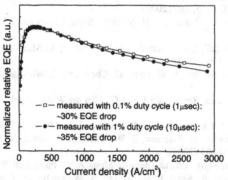

Figure 4. Relative EQE data of a MQW-LED (six period of 2nm $In_{0.15}Ga_{0.85}N$ with 12nm $In_{0.01}Ga_{0.99}N$ barriers) on sapphire measured with two different duty cycles (1% and 0.1% duty cycles) to test the thermal effect on the LED efficiency behavior. During measurement, the sample was placed on a heat sink. The frequency of the pulse current is 1KHz. 1% and 0.1% duty cycles correspond to 10 μsec and 1

μsec, respectively. By switching the duty cycle from 1% to 0.1%, the efficiency drop was reduced from ~35% to ~30%.

CONCLUSIONS

We elaborated on possible physical origins (electron overflow/spillover, Auger recombination, current crowding, and junction heating) for the efficiency loss observed in InGaN LEDs at high electrical injection levels. Our experimental results indicate that the electron overflow seriously reduces the LED EL performance and might be the major reason for the efficiency drop in LEDs at high current levels, at least at 400 nm like wavelengths investigated. Moreover, this electron overflow is present in both c-plane and non-polar m-plane LEDs, with the latter being free of polarization-induced electric field. We also interrogated the importance of the current crowding effect on the LED efficiency by a comparative study in which semitransparent Ni/Au and nearly fully transparent conducting GZO contacts were used.

ACKNOWLEDGMENTS

The work at VCU is supported by grants from the Air Force Office of Scientific Research and National Science Foundation. Partial support by ARO under Phase II W911NF-07-C-0099 contract for non-polar bulk development at Kyma Technologies, Inc., is acknowledged. The study of GZO contact is partially supported by a grant from the Department of Energy, Basic Energy Sciences, through a subgrant from the University of Wisconsin. HM would like to acknowledge useful discussions with Dr. C. Tran of SemiLEDs.

REFERENCES

1. H. Morkoç, "Handbook of Nitride Semiconductors and Devices", Chapter 1, Volume 3, Wiley-VCH (2008).
2. M. R. Krames, O. B. Shchekin, R. Mueller-Mach, G. O. Mueller, L. Zhou, G. Harbers, and M. G. Craford, J. Display Tech. 3, 160 (2007).
3. X. Li, H. Liu, X. Ni, Ü. Özgür, and H. Morkoç, Superlattices and Microstructures, in press, 2009.
4. Y. C. Shen, G. O. Mueller, S. Watanabe, N. F. Gardner, A. Munkholm, and M. R. Krames, Appl. Phys. Lett. 91, 141101 (2007).
5. N. F. Gardner, G. O. Müller, Y. C. Shen, G. Chen, and S. Watanabe, Appl. Phys. Lett. 91, 243506 (2007).
6. Kris T. Delaney, Patrick Rinke, and Chris G. Van de Walle, Appl. Phys. Lett. 94, 191109 (2009).
7. A. R. Beattie and P. T. Landsberg, Proc. R. Soc. Lond. A. 249, 16 (1958).
8. J. Hader, J. V. Moloney, B. Pasenow, S. W. Koch, M. Sabathil, N. Linder, and S. Lutgen, Appl. Phys. Lett. 92, 261103 (2008).
9. Han-Youl Ryu, Hyun-Sung Kim, and Jong-In Shim, Appl. Phys. Lett. 95, 081114 (2009).
10. G. Chen, M. Craven, A. Kim, A. Munkholm, S. Watanabe, M. Camras, W. Götz, and F. Steranka, Phys. Status Solidi A 205, 1086 (2008).
11. K. A. Bulashevich and S. Y. Karpov, Phys. Status Solidi C 5, 2066 (2008).
12. M. H. Kim, M. F. Schubert, Q. Dai, J. K. Kim, E. F. Schubert, J. Piprek, and Y. Park, Appl. Phys. Lett. 91, 183507 (2007).

13. J. Xie, X. Ni, Q. Fan, R. Shimada, Ü. Özgür, and H. Morkoç, Appl. Phys. Lett. 93, 121107 (2008).
14. X. Ni, Q. Fan, R. Shimada, Ü. Özgür, and H. Morkoç, Appl. Phys. Lett., 93, 171113 (2008).
15. Ü. Özgür, H. Morkoç, H. Liu, X. Li, and X. Ni, Special issue of Proc. of IEEE, 2010, in press.
16. Petr G. Eliseev, Appl. Phys. Lett., 75, 3838, (1999).
17. C. B. Su, J. Schafer, J. Manning, and R. Olshansky, Electron. Lett. 18, 1108 (1982).
18. Q. Dai, M. F. Schubert, M. H. Kim, J. K. Kim, E. F. Schubert, D. D. Koleske, M. H. Crawford, S. R. Lee, A. J. Fischer, G. Thaler, and M. A. Banas, Appl. Phys. Lett., 94, 111109 (2009).
19. A. Niwa, T. Ohtoshi, and T. Kuroda, Appl. Phys. Lett. 70, 2159 (1997).
20. X. Guo and E. F. Schubert, Appl. Phys. Lett., 78, 3337 (2001).
21. H.Y. Liu, V. Avrutin, N. Izyumskaya, M. A. Reshchikov, Ü. Özgür, and H. Morkoç, Applied Physics Letters, Submitted.
22. A. A. Efremov, N. I. Bochkareva, R. I. Gorbunov, D. A. Larinovich, Yu. T. Rebane, D. V. Tarkhin, and Yu. G. Shreter, Semiconductors 40,605 (2006).

Mater. Res. Soc. Symp. Proc. Vol. 1202 © 2010 Materials Research Society   1202-I10-01

Development of Milliwatt Power AlGaN-Based Deep UV-LEDs by Plasma-Assisted Molecular Beam Epitaxy

Yitao Liao, Christos Thomidis, Anirban Bhattacharyya, Chen-kai Kao, Adam Moldawer, Wei Zhang and Theodore D. Moustakas

Department of Electrical and Computer Engineering and Photonics Center, Boston University, Boston, MA 02215, U.S.A.

ABSTRACT

In this paper, we report the development of AlGaN-based deep ultraviolet LEDs by rf plasma-assisted molecular beam epitaxy (MBE) emitting between 277 and 300 nm. Some of these devices were evaluated after fabrication at bare-die and some at wafer-level configurations. Devices with total optical output of 1.3 mW at injection current of 200 mA were produced, with maximum external quantum efficiency (EQE) of 0.16%. These performance values are equivalent to those reported for deep UV-LEDs grown by the Metalorganic chemical vapor deposition (MOCVD) method and measured at bare-die configuration. In parallel, we have evaluated the internal quantum efficiency (IQE) of AlGaN quantum wells, and found that such wells emitting at 250 nm have an IQE of 50%. From the analysis of these data, we concluded that the efficiency of deep UV LEDs is not limited by the IQE but by the light extraction efficiency, injection efficiency or a combination of both.

INTRODUCTION

Deep ultraviolet light emitting diodes (UV-LEDs) are expected to find a number of applications such as water purification, epoxy curing, non-line-of-sight communications, fluorescence identification of biological/chemical agents and surface sterilization. However, despite the intense effort, deep UV-LEDs based on AlGaN alloys have been reported to be highly inefficient. Specifically, at wavelength below 300 nm, the reported maximum external quantum efficiency (EQE) is typically equal or less than 1% [1-9]. This result can be accounted for by low internal quantum efficiency (IQE), current injection efficiency, light extraction efficiency, or a combination of all three factors. In the majority of the previous reports the deep UV-LEDs were produced by the MOCVD method. Only limited reports exist on producing such LED structures by the MBE method [10-14]. Contrary to the earlier reports that AlGaN alloys have a relatively poor IQE [15], our group has demonstrated recently that AlGaN multiple quantum wells (MQWs) produced by plasma-assisted MBE have IQE as high as 50% at 250 nm [16]. In this paper, we report the development of AlGaN-based deep UV-LEDs by rf plasma-assisted molecular beam epitaxy emitting in the wavelength range of 277 and 300 nm. Devices with total optical output of 1.3 mW at injection current of 200 mA were produced, with maximum external quantum efficiency (EQE) of 0.16%.

EXPERIMENTAL DETAILS

The deep UV-LED structures were grown by plasma-assisted molecular beam epitaxy on to (0001) sapphire substrates in a Varian Gen II MBE system. Nitrogen is activated by an Applied Epi Uni-bulb rf plasma source. The Ga and Al metals as well as Si and Mg dopants were supplied from standard effusion cells. The epilayer structure for these devices is schematically shown in Figure 1. Prior to the growth of the AlN template, the sapphire substrates were nitridated at high temperature by exposure to active nitrogen in order to convert the surface layers from Al_2O_3 to AlN [17, 18]. The AlN template was 2 μm thick and grown under stoichiometric conditions (the ratio of Al/N fluxes is equal to one). A 10 period AlGaN/AlN strain management superlattice (SL) was grown on to the AlN template before the growth of the Si-doped AlGaN cladding layer. Such strain management SL has been proved to be effective in accommodating the elastic strain induced by Si incorporation into the AlGaN lattice and to prevent cracking [19, 20]. The AlGaN layer in the SL has the same Al mole fraction to that of the n-AlGaN cladding layer. The average electron concentration across the entire n-AlGaN cladding layer was on the order of mid-10^{18} cm^{-3}. The active region consists of an asymmetric AlGaN/AlGaN quantum well and a 10 nm AlGaN electron blocking layer (EBL) heavily doped with Mg. The thickness of the barrier close to the n-AlGaN cladding is 10 nm, while the other barrier next to EBL is 3 nm. Such an asymmetric design was chosen in order to balance the electron and hole injection into the quantum well due to differences in diffusion length of holes and electrons [21]. The composition and thickness of the EBL were optimized in order to best block electron overflow into the p-type region [22]. The AlGaN quantum well is 3 nm thick, and was grown under Ga-rich conditions to allow for the formation of compositional inhomogeneities due to atomic ordering, which has been shown to improve the IQE of the quantum wells [16, 23, 24, 25]. The final structure is capped by a p-AlGaN layer and p-GaN contact layer. Table 1 shows the Al composition in the active region and EBL of the LED structures investigated here. Some of the devices were grown on (0001) sapphire substrates coated first with approximately 1 μm AlN by the hydride vapor phase epitaxy (HVPE).

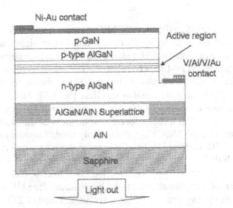

Figure 1: Schematic of the epilayer structure of the investigated deep UV LEDs

Table 1: Al composition in the active region and EBL of the investigated LED structures

Emission wavelength(nm)	AlN mole fraction in the quantum well	AlN mole fraction in the barriers	AlN mole fraction in the EBL
300	50%	65%	75%
294	60%	65%	75%
277	65%	75%	80%

The deep UV-LED devices were fabricated using standard photolithography techniques. The samples were etched in a Cl-plasma to form 300 μm × 300 μm square mesa structures. As n-contact we had employed a multilayer consisting of V (2 nm)/Al (80 nm)/V (20 nm)/Au (100 nm) stack annealed to 1000 °C. Such a contact was reported previously to form a good Ohmic contact to n-AlGaN alloys with AlN mole fraction up to 70% [26]. A thin Ni/Au layer was deposited on the top of the mesa to form a contact to p-GaN and a thicker Au pad was deposited in the corner of the mesa.

The devices were not packaged but instead were evaluated by probing at the bare-die configuration. In this method the sapphire was rested on the top of the aperture of a calibrated integrated sphere, and the emitted light was collected from the rough surface of the sapphire.

RESULTS AND DISCUSSIONS

Figure 2 (a) shows the electroluminescence (EL) spectra of a single diode emitting at 300 nm under ac injection current varying from 10 mA to 200 mA. The device was evaluated under pulsed mode of 5% and 2% duty cycle (5 μs and 2 μs pulse width, 10 KHz frequency). The EL spectra consist of a single peak with a full width at half maximum of 10 nm. Up to the investigated injection current of 200 mA the peak emission remains the same. The total optical power of the device described in Figure 2 (a) as a function of injection current is shown in Figure 2 (b). Thus, at 200 mA injection current, the integrated optical power of the device was 1.3 mW, which as discussed earlier is only a lower limit of the performance of this device. Using the data of Figure 2 (b) we evaluated the EQE of this device and the results are shown in Figure 2 (c). Thus, the EQE of the device reaches a value of 0.16 % at injection current of 90 mA. This value is of the same order as in the majority of the MOCVD grown deep UV LEDs and measured the same way [1].

Figure 3 (a) shows the electroluminescence (EL) spectra of a single diode emitting at 294 nm under ac injection current varying from 10 mA to 200 mA. The device was evaluated under pulsed mode of 5% duty cycle (5 μs pulse width, 10 KHz frequency). Up to the investigated injection current of 200 mA the peak emission remains the same. The total optical power of the device described in Figure 3 (a) as a function of injection current is shown in Figure 3 (b). Thus, at 200 mA injection current, the integrated optical power of the device was 0.7 mW. Using the data of Figure 3 (b) we evaluated the EQE of this device and the results are shown in Figure 3 (c). Thus, the EQE of the device reaches a value of 0.09% at injection current of 100 mA. We attribute the reduction in the performance to this device compared to that described in Figure 2 to the shallower quantum well employed as indicated in Table 1.

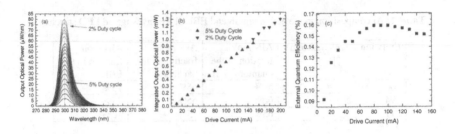

Figure 2. (a) Electroluminescence spectra of the deep UV-LED emitting at 300 nm; (b) Integrated optical power versus drive current; (c) calculated external quantum efficiency versus drive current.

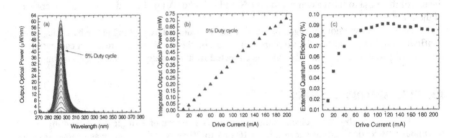

Figure 3. (a) Electroluminescence spectra of the deep UV-LED emitting at 294 nm; (b) Integrated optical power versus drive current; (c) calculated external quantum efficiency versus drive current.

Figure 4 shows the EL spectra and integrated output power of deep UV-LED emitting at 277 nm. This particular LED structure was not fabricated into a device and it was only evaluated by probing at the wafer level. Even though the spectra and optical output power are not indicative of the actual device performance, nevertheless they show these devices upon fabrication are going to demonstrate equivalent performance as those described in Figures 2 and 3.

The performance values of the devices described in Figures 2 and 3 are of the same order as in the majority of the MOCVD grown deep UV LEDs measured at bare-die configuration. Therefore, the issue remains why deep UV LEDs based on AlGaN alloys produced by both MOCVD and MBE are relatively inefficient comparatively to violet-blue LEDs based on InGaN alloys? As discussed previously, the EQE of an LED is the product of the IQE, extraction efficiency and injection efficiency.

Our group has recently [16] reported the photoluminescence spectra of AlGaN multiple quantum wells emitting in the wavelength region between 220 to 250 nm, as shown in Figure 5. By measuring the photoluminescence as a function of temperature we also calculated the room-temperature IQE of these structures and the results are shown in Figure 6.

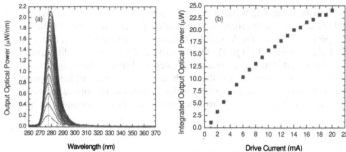

Wavelength (nm)   Drive Current (mA)

Figure 4. (a) Electroluminescence spectra of the deep UV LED structure emitting at 277 nm; (b) Integrated optical power versus drive current. This LED structure was evaluated by probing at the wafer level.

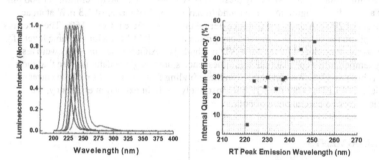

Wavelength (nm)   RT Peak Emission Wavelength (nm)

Figure 5. (a) Room-temperature luminescence from AlGaN/AlN MQWs with identical well and barrier width; (b) IQE for all $Al_{0.7}Ga_{0.3}N/AlN$ MQWs investigated. The IQE of these nominally identical MQWs' varies from 5% at 220 nm to 50% at 250 nm. [Ref.16]

Based on the results of Figure 5, we conclude that IQE of AlGaN as discussed in detail in Ref [16] is not a limiting factor in the performance of deep UV LEDs. Thus, the efficiency performance must be limited by the extraction efficiency, injection efficiency or a combination of both. As discussed by Jiang, et al., [27] the light extraction efficiency of AlGaN alloys and multiple quantum wells is affected by the change of the band structure of AlGaN alloys as the mole fraction of AlN increases. More specifically due to the difference in the crystal-field splitting in AlN (Δ_{CF} = -219meV) and GaN (+38 meV) there is a gradual reversal of the order of the valence bands as the AlN mole fraction increases; at 25% AlN the three valence bands are degenerate and at higher AlN mole fraction there is a complete reversal of the order of the valence bands. This change in the valence band structure imposes certain selection rules in the fundamental optical transitions, which influences the emission properties of the AlGaN system as the AlN mole-fraction increases. Specifically, the recombination between the conduction band electrons and the holes in the top valence band is polarized along the direction $E \parallel c$ in AlN and E

173

$\perp c$ in GaN. The implication is that LEDs based on polar (0001) AlGaN alloys with high Al concentration should be designed in the form of edge emitters rather than surface emitters.

In our devices specifically, since the AlN mole fraction in the active region is over 50%, a significant amount of the emitted light is parallel to the surface of the die. In addition, only a small fraction of the surface emitted light arrives at the integrated sphere due to light coupling into the sapphire substrate and scattering from the rough surface in the back of the sapphire substrate. Thus, the evaluation of our device greatly underestimates its real optical power. Furthermore, the lack of heat sinking prevents us from evaluating the devices under dc current injection.

CONCLUSIONS

In conclusion we report the growth and fabrication of deep UV LEDs emitting in the wavelength region from 277 to 300 nm by plasma-assisted MBE. The device emitting at 300 nm evaluated at a bare-die configuration were found to have an optical power of 1.3 mW at an ac injection current of 200 mA and a maximum external quantum efficiency of 0.16%. These results show that the plasma-assisted MBE method can produce deep UV LEDs with the same level of performance as those produced by the MOCVD method. The AlGaN quantum wells were also evaluated by temperature-dependant photoluminescence studies and the data indicate that their IQE can be as high as 50%. We conclude from these finding that the limitation of the current generation deep UV LEDs is not the IQE but most likely the light extraction efficiency, the injection efficiency or a combination of both.

ACKNOWLEDGMENTS

This work was supported by the BU-ARL cooperative agreement. We would like to thank Enrico Bellotti, Roberto Paiella, Alexey Nikiforov and Leah Ziph-Schatzberg for illuminating discussions.

REFERENCES

1. A. Khan, K. Balakrishnan, and T. Katona, Nat. Photonics **2**, 77 (2008).
2. H. Hirayama, et al, Phys. Status Solidi A **206**, 203 (2009).
3. K. H. Kim, Z. Y. Fan, M. Khizar, M. L. Nakarmi, J. Y. Lin, and H. X. Jiang, Appl. Phys. Lett. **85**, 4777 (2004).
4. K. Mayes, A. Yasan, R. McClintock, D. Shiell, S. R. Darvish, P. Kung, and M. Razeghi, Appl. Phys. Lett. **84**, 1046 (2004).
5. A. A. Allerman, M. H. Crawford, A. J. Fischer, K. H. A. Bogart, S. R. Lee, D. M. Follstaedt, P. P. Provencio, and D. D. Koleske, J. Cryst. Growth **272**, 227 (2004).
6. Z. Ren et al., Phys. Status Solidi C **4**, 2482 (2007).
7. L. Zhou, J. E. Epler, M. R. Krames, W. Goetz, M. Gherasimova, Z. Ren, J. Han, M. Kneissl, and N. M. Johnson, Appl. Phys. Lett. **89**, 241113 (2006).
8. V. Adivarahan, A. Heidari, B. Zhang, Q. Fareed, M. Islam, S. Hwang, K. Balakrishnan and A. Khan, Appl. Phys. Express **2**, 092102 (2009).

9. K. Kawasaki, C. Koike, Y. Aoyagi and M. Takeuchi, Appl. Phys. Lett. **89**, 261114 (2006).
10. S. Nikishin, B. Borisov, V. Kuryatkov, M. Holtz, G. A. Garrett, W. L. Sarney, A. V. Sampath, H. Shen, M. Wraback, A. Usikov and V. Dmitriev, J Mater Sci: Mater Electron **19**, 764 (2008).
11. G. Kipshidze, V. Kuryatkov, K. Zhu, B. Borisov, M. Holtz, S. Nikishin, and H. Temkin, J. Appl. Phys. **93**, 1363 (2003).
12. G. Kipshidze, V. Kuryatkov, B. Borisov, S. Nikishin, M. Holtz, S. N. G. Chu and H. Temkin, Phys. Status Solidi A **192**, 286 (2002).
13. A. V. Sampath, G. A. Garrett, C. J. Collins, W. L. Sarney, E. D. Readinger, P. G. Newman, H. Shen, and M. Wraback, J. Electron. Mater. **35**, 641 (2006).
14. V. Jmerik et al., J. Cryst. Growth **311**, 2080 (2009).
15. S. F. Chichibu et al., Nature Mater. **5**, 810 (2006).
16. A. Bhattacharyya, T. D. Moustakas, L. Zhou, D. J. Smith, and W. Hug, Appl. Phys. Lett. **94**, 181907 (2009).
17. T. D. Moustakas, T. Lei, and R. J. Molnar, Physica B **185**, 36 (1993).
18. Y. Wang, A. Özcan, G. Özaydin, K. Ludwig Jr, , A. Bhattacharyya, T. D. Moustakas, H. Zhou, R. Headrick, and D. P. Siddons, Phys. Rev. B. **74**, 235304 (2006).
19. J. P. Zhang, H. M. Wang, M. E. Gaevski, C. Q. Chen, Q. Fareed, J. W. Yang, G. Simin, and M. Asif Khan, Appl. Phys. Lett. **80**, 3542 (2002).
20. L. T. Romano, C. G. Van de Walle, J. W. Ager III, W. Gotz, and R. S. Kern, J. Appl. Phys. **87**, 7745 (2000).
21. K. Mayes, A. Yasan, R. McClintock, D. Shiell, S. R. Darvish, P. Kung, and M. Razeghi, Appl. Phys. Lett. **84**, 1046 (2004).
22. K. B. Lee, P. J. Parbrook, T. Wang, J. Bai, F. Ranalli, R. J. Airey and G. Hill, J. Cryst. Growth **311**, 10 (2009).
23. G. A. Garrett, A. V. Sampath , C. J. Collins, E. D. Readinger, W. L. Sarney , H. Shen , M. Wraback , V. Soukhoveev , A. Usikov , V. Dmitriev, Phys. Status Solidi C **3**, 2125 (2006).
24. Min Gao, S. T. Bradley, Yu Cao, D. Jena, Y. Lin, S. A. Ringel, J. Hwang, W. J. Schaff, and L. J. Brillson, J. Appl. Phys. **100**, 103512 (2006).
25. Yiyi Wang, Ahmet S. Ozcan, Karl F. Ludwig, Jr., Anirban Bhattacharyya, T. D. Moustakas, Lin Zhou, and David J. Smith, Appl. Phys. Lett. **88**, 181915 (2006)
26. Ryan France, Tao Xu, Papo Chen, R. Chandrasekaran, and T. D. Moustakas, Appl. Phys. Lett. **90**, 062115 (2007).
27. K. B. Nam, J. Li, M. L. Nakarmi, J. Y. Lin, and H. X. Jiang, Appl. Phys. Lett. **84**, 5264 (2004).

Mater. Res. Soc. Symp. Proc. Vol. 1202 © 2010 Materials Research Society   1202-I10-02

Reliability and Performance of Pseudomorphic Ultraviolet Light Emitting Diodes on Bulk Aluminum Nitride Substrates

Shawn R. Gibb[1] James R. Grandusky[1], Yongjie Cui[1], Mark C. Mendrick[1], and Leo J. Schowalter[1]
[1]Crystal IS Inc., 70 Cohoes Avenue, Green Island, NY 12183 U.S.A.

ABSTRACT

Low dislocation density epitaxial layers of $Al_xGa_{1-x}N$ can be grown pseudomorphically on c-face AlN substrates prepared from high quality, bulk crystals. Here, we will report on initial characterization results from deep ultraviolet (UV) light emitting diodes (LEDs) which have been fabricated and packaged from these structures. As reported previously, pseudomorphic growth and atomically smooth surfaces can be achieved for a full LED device structure with an emission wavelength between 250 nm and 280 nm.

A benefit of pseudomorphic growth is the ability to run the devices at high input powers and current densities. The high aluminum content $Al_xGa_{1-x}N$ (x~0.7) epitaxial layer can be doped n-type to obtain sheet resistances < 200 Ohms/sq/μm due to the low dislocation density. Bulk crystal growth allows for the ability to fabricate substrates of both polar and non-polar orientations. Non-polar substrates are of particular interest for nitride growth because they eliminate electric field due to spontaneous polarization and piezoelectric effects which limit device performance. Initial studies of epitaxial growth on non-polar substrates will also be presented.

INTRODUCTION

Ultraviolet disinfection is becoming very important as an efficient means of providing disinfection to water, air and surfaces without the use of potentially harmful chemicals. This requires a light source that is emitting in the ultraviolet-C (UVC) range (<300 nm) for efficient disinfection due to a peak in DNA absorption at ~ 265 nm[1]. Traditionally, mercury lamps are used, however these suffer the disadvantages of short lifetimes, slow start up, and the use of toxic mercury which can be a hazard if the bulb is broken and leads to problems with disposal. Light emitting diodes (LEDs) have the capability of high efficiencies, fast start up which can be synchronized with water flow, long lifetimes, variable wavelengths, and no toxic materials. However, currently devices are fabricated from $Al_xGa_{1-x}N$ layers on sapphire substrates which lead to a high dislocation density and thus low efficiencies and short lifetimes[2]. Bulk AlN substrates offer several advantages over growth on sapphire, including low lattice and thermal mismatch between the substrate and the device layers. In addition, pseudomorphic growth of $Al_xGa_{1-x}N$ with x>0.6 can be obtained resulting in device layers with low dislocation densities, low resistivities, and atomically smooth surfaces[3].

EXPERIMENT

Epitaxial growth was carried out via OMVPE in a Veeco D180 vertical rotating disc reactor on low dislocation density bulk AlN substrates of both polar (c-plane) and non-polar (m-plane) orientations. The growths utilized trimethylgallium, trimethylaluminum, 100 ppm silane and ammonia as precursors in hydrogen ambient. Prior to epitaxial growth the AlN substrates are etched to remove surface oxides and an *in-situ* high temperature annealing step is performed. For polar growth 500 nm of homoepitaxial AlN was grown post anneal and the remainder of the device structure consisted of an n-type $Al_{0.7}Ga_{0.3}N$ layer, a 5 period multiple quantum well (MQW), an electron blocking layer, and a p-type GaN contact layer. X-ray diffraction was used to characterize the strain state of the layers using both the symmetric (0002) and asymmetric (10-12) reflections. Aluminum content and strain were calculated as described previously[3]. Growth on non-polar substrates was performed in the same hardware utilizing the precursors described above.

Devices on c-plane AlN substrates were processed using standard LED processing with a mesa diameter of 350 μm. The devices are flip chip mounted to an AlN submount and packaged in TO-8 and surface-mounted design (SMD) packages. For testing of the diodes, the packaged devices were mounted on a pin fin anodized Al heat sink (40 mm x 40 mm) with forced convection cooling which had an experimentally validated theoretical thermal resistance of 2.5 °C/W. With an applied forward bias, the output power was measured in a calibrated integrating sphere. For lifetime measurements a photodiode was used to measure the output power of each device.

DISCUSSION

Growth on c-plane AlN:

Previous work has demonstrated[3] that 70% AlGaN layers can be grown pseudomorphically up to a thickness of 1 μm. The main advantage to the pseudomorphic growth is that since there are no misfit dislocations generated, no new threading dislocations will need to be generated and it is possible to grow thick layers with a threading dislocation density (TDD) comparable to the starting substrate. This phenomenon is demonstrated in the cross sectional scanning transmission electron (STEM) micrograph in Figure 1a which highlights the lack of defects generated through the active region of the 265 nm LED device structure. Threading dislocations are observed in the p-GaN layer due to the large lattice mismatch and an array of misfit dislocations are observed along the p-AlGaN/p-GaN interface. Additional benefits of the pseudomorphic growth technique are atomically smooth surfaces of the n-type $Al_xGa_{1-x}N$ which provides very sharp interfaces in the p-n junction and thin active region. Figure 1b shows the typical step flow growth pattern with atomically smooth surfaces.

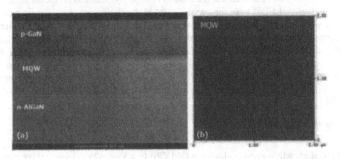

Figure 1: (a) Bright field scanning transmission electron micrograph highlighting the lack of defects generated in the active region of a 265 nm LED during pseudomorphic epitaxial growth. (b) 2X2 μm AFM micrograph highlighting atomically smooth surfaces achieved via pseudomorphic growth after 5 period MQW.

One challenge of incorporating a 70% $Al_xGa_{1-x}N$ layer into an LED structure is the conductivity. As the Al concentration is increased, the conductivity is typically decreased owing mainly to the reduced mobilities[4] and deeper level of the donor in the conduction band[5]. We have optimized the AlGaN growth conditions and routinely achieve a mean sheet resistance value of 320 Ohm/sq for 0.5 μm layers. Hall measurements were performed in the Van der Pauw geometry and gave conductivity values suitable for an LED structure, a resistivity of 0.01892 Ohm-cm with a mobility of 70 cm^2/V-s at a carrier concentration of 4.5E18 cm^{-3}. These results are a direct result of the high quality, pseudomorphic growth on native AlN substrates and are an integral part of the formation of reliable, low contact resistance Ohmic contacts.

Device performance and reliability:

Initial device measurements were carried out at input currents ranging from 20 mA to 150 mA. At each step, the system (device and heat sink) was allowed to achieve thermal equilibrium and the output power was stable. In each device, the increase in power at currents up to 150 mA is linear with no roll off or thermal effects. The forward voltage of the devices is quite high and typically 20 V at 150 mA forward current resulting in an input power of 3 W. The forward voltage is expected to become lower as the n and p contact metallization is improved.

In order to test the potential of the devices at elevated currents, devices were tested at currents greater than 150 mA without letting the device reach thermal equilibrium (quasi-cw). Figure 2 shows the results with an output power of 1.3 mW at 400 mA of current with a peak wavelength of 258 nm. Direct comparison of these results to those grown on non-native substrates is somewhat difficult since the optical power behaves in a non linear fashion with decreasing wavelength, however we feel that these results compare favorably to device results on non-native substrates reported elsewhere[6,9,10].

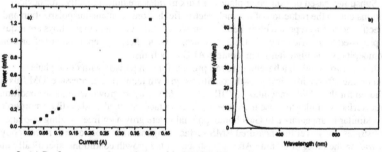

Figure 2: (a) Performance of pseudomorphic UVC LEDs on bulk AlN substrates when driven to a forward current of 400 mA and (b) the spectra of the device at 400 mA with a peak wavelength of 258 nm.

In addition to improved performance of the devices, an important feature of the pseudomorphic LED is the reliability. It is expected that the low dislocation density in the substrates will lead to increased reliability and increased lifetimes over devices fabricated on sapphire substrates. Lifetime measurements were carried out at input currents of 20 mA, 100 mA, and 150 mA and shown in Figure 3a. As expected, the device reliability is much improved over devices grown on foreign substrates. Devices have currently been operating for 3500 hours at 20 mA and 2000 hours at 100 mA without reaching L50 and are currently still operating. Further analysis of these measurements and associated decay mechanisms used to predict the L50 lifetimes have been reported on elsewhere[8]. At this juncture, it appears that fitting the data to a log fit after only a few days worth of data gives the best method of estimating the L50 with limited data as shown in Figure 3b; however we are continuing to collect data and analyze the fitting functions to determine the best way to predict the L50.

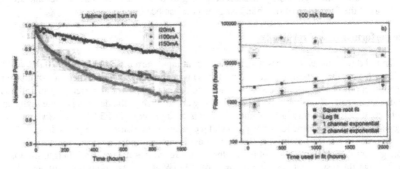

Figure 3: (a) Power decay during the first 1000 hours of testing for different drive currents, and (b) fitted L50 as a function of time used in fit for 4 different models at 100 mA.

Homoepitaxial growth on m-plane AlN:

Substrates based on non-polar planes of bulk nitrides are quite attractive for device applications due to the reduction of internal electric fields from spontaneous polarization and piezoelectric effects as reported elsewhere[6,7]. Growth of AlN, GaN and their alloys are required for the proposed high efficiency UV LED structure, however here we report on our efforts on AlN homoepitaxy and high aluminum content $Al_xGa_{1-x}N$ films.

In order to establish a baseline growth process for comparison with our c-plane pseudomorphic efforts, identical growth conditions (growth temperature, pressure, OM flux, etc) were chosen for the AlN homoepitaxy (V/III ratio ~ 460). Layers grown with these conditions exhibited surfaces similar to those in Figure 4a. The surfaces showed a "slate-like" morphology which is similar in appearance to GaN homoepitaxial layers grown on free standing m-plane GaN substrates[7], however the measured RMS surface roughness via atomic force microscopy (AFM) was on the order of 30 nm. After adjustment of the growth conditions, specifically the V/III to 20 we measured more than a four-fold decrease in the RMS surface roughness to 7 nm as is shown in Figure 4b. We believe that these smoother AlN surfaces are advantageous as a foundation for growing smooth, high quality $Al_xGa_{1-x}N$ layers for use in high performance device structures.

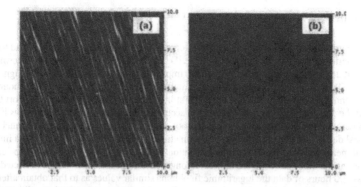

Figure 4: 10x10 μm AFM micrographs highlighting AlN homoepitaxy surface morphology for V/III ratios of (a) 460, and (b) 20.

$Al_xGa_{1-x}N$ growth on m-plane AlN:

Next, utilizing the AlN homoepitaxial growth conditions above, we evaluated the growth of $Al_xGa_{1-x}N$ layers on these templates. Again, to establish a baseline for comparison with our c-plane efforts nearly identical growth conditions were chosen (growth temperature, pressure, etc). Aluminum compositions of 60% and 70% were targeted and XRD analysis confirmed films of 55% and 62% were demonstrated. Figure 5 highlights the surfaces of these films and the RMS surface roughness was determined to be approximately 15 nm for the 55% sample while the 62% sample had an RMS roughness of 8 nm. It is worth noting that the this result is consistent with our previous work on c-plane pseudomorphic growth where it was found that layers begin relaxing by a surface roughening mechanism, where initially the surface appears to buckle to reduce the compressive strain in the layers[3].

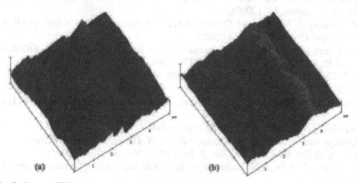

Figure 5: 5x5 μm AFM micrographs of 0.5 μm thick AlGaN layers on m-plane AlN with (a) 55% and (b) 62% Al content.

181

CONCLUSIONS

The use of pseudomorphic layers grown on c-plane bulk AlN substrates has lead to the improved performance and reliability of UVC LEDs. In addition to the improved performance at high current densities, device reliability is much improved over devices grown on foreign substrates. Devices have currently been operating for 3500 hours at 20 mA and 2000 hours at 100 mA without reaching L50 and are currently still operating. This has lead to an effort to be able to fit the decay of the power and be able to accurately predict the L50 value with as little data as possible. The data was fitted to square root, logarithmic, and one and two channel exponential decay functions to obtain the best function to predict the L50 values with a limited amount of measured data. The logarithmic function was chosen as the best based on the current predictions as the other fits continue to predict longer L50 values as more data is collected. With as little as 100 hours of data the logarithmic fit obtains similar values as to that obtain after 2000 hours at 100 mA. More data is currently being collected and analyzed to further validate this after the devices reach their L50 values.

Additionally we have demonstrated the ability to grow relatively smooth, crack free $Al_xGa_{1-x}N$ layers with compositions up to 62% at 0.5 μm thickness on m-plane bulk AlN substrates. While further development is required to produce functioning UVC LED devices on these substrates we have shown that high Al content $Al_xGa_{1-x}N$ films are achievable.

ACKNOWLEDGMENTS

This material is based upon work partially supported by an Advanced Technology Project (ATP) grant which is administered by the National Institute of Science and Technology (NIST) and by the Department of Energy (DOE) under Award No. DE-FC26-08NT01578.

REFERENCES

1. EPA Document # 815-D-03-007
2. A. Khan, S. Hwang, J. Lowder, V. Advirahan, and Q. Fareed, *Reliability Physics Symposium, 2009 IEEE International*, 89, (2009).
3. J. R. Grandusky, J. A. Smart, M. C. Mendrick, L. J. Schowalter, K. X. Chen, and E. F. Schubert, *J. Cryst. Growth* 311, 2864 (2009).
4. A.A. Allerman, M.H. Crawford, A.J. Fischer, K.H.A. Bogart, S.R. Lee, D.M. Follstaedt, P.P. Provencio, D.D. Koleske, *J. Crystal Growth* 272 (2004) 227.
5. T. Xu, C. Thomidis, I. Friel, T.D. Moustakas, *Phys. Status Solidi (c)* 2 (2005) 2220.
6. A. Khan,K. Balakrishnan, *Materials Science Forum Vol. 590 (2008) pp 141-174.*
7. S.C. Cruz, S. Keller, T.E. Mates, U.K. Mishra, S.P. DenBaars, *J. Cryst. Growth* 311, 3817 (2009).
8. J.R. Grandusky et al, 2009 MRS Fall Meeting, B3-4.
9. Y.Bilenko, A.Lunev, X.Hu, T.M. Katona, J.Zhang, R. Gaska, M.S.Shur, W.Sun, V.Adivarahan, M.Shatalov, A.Khan, *Jpn. J. Appl. Phys., Vol. 44, No. 3 (2005).*
10. H.Hirayama, T.Yatabe, N.Noguchi, T.Ohashi, N.Kamata, *Applied Physics Letters 91, 071901 (2007).*

Mater. Res. Soc. Symp. Proc. Vol. 1202 © 2010 Materials Research Society

1202-I10-03

Short Period p-Type AlN/AlGaN Superlattices for Deep UV Light Emitters

S. Nikishin[1], B. Borisov[1], V. Mansurov[1], M. Pandikunta[1], I. Chary[1], G. Rajanna[1], A. Bernussi[1], Yu. Kudryavtsev[2], R. Asomoza[2], K. A. Bulashevich[3], S. Yu. Karpov[3], S. Sohal[1], and M. Holtz[1]
[1]Nano Tech Center, Texas Tech University, Lubbock, TX 79401, USA
[2] SIMS Laboratory of SEES, CINVESTAV, Mexico D.F.07300, Mexico
[3]STR Group – Soft-Impact, Ltd., St. Petersburg, 194156, Russia

ABSTRACT

The Mg doped AlN/Al$_x$Ga$_{1-x}$N ($0.03 \leq x \leq 0.05$) short period superlattices (SPSLs) were grown by gas source molecular beam epitaxy on (0001) sapphire substrates. The average AlN mole fraction is ~ 0.7 and the hole concentration is ~ 7×10^{17} cm^{-3}. Contacts formed to the SPSLs using Ni/Au bilayer are found to have specific contact resistance ~ 5×10^{-5} Ωcm^2 near room temperature and to show weak temperature dependence attributed to activation of Mg acceptors in the AlN barriers of SPSLs. These p-SPSLs are attractive for fabrication of transparent low resistive ohmic contacts for deep UV LEDs.

INTRODUCTION

Al$_x$Ga$_{1-x}$N alloys with energy gaps from 3.4 in GaN to 6.2 eV in AlN can be grown epitaxially in the wurtzite structure across the full composition range using several methods, including gas source molecular beam epitaxy (GSMBE) with ammonia. This provides a wide range of possible emission wavelengths for photonic devices. However, device preparation is

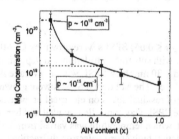

Fig. 1. Mg incorporation in AlGaN grown at ~ 830 °C. III-N ratio and Mg flux are constant.

fundamentally limited by the difficulties of preparing p-type layers of wide-bandgap Al$_x$Ga$_{1-x}$N. Figure 1 summarizes our prior investigations on efficiency of Mg incorporation in Al$_x$Ga$_{1-x}$N ($0 \leq x \leq 1$) when all layers are grown at the same temperature, III/N ratio, and Mg flux [1]. The resulting hole concentration obtained using Hall measurements is also shown for two samples. Reduction in the hole concentration, when x increases, is due to increasing Mg activation energy and reducing Mg incorporation in Al-rich alloys. These factors have driven us to develop AlN/AlGaN short period superlattices (SPSLs) allowing high hole concentrations in wide bandgap alloys with average AlN concentration up to 70% [2 – 4]. These structures, consisting of 3-6 monolayer (ML) thick barriers of AlN and 2-3 ML thick wells of Al$_x$Ga$_{1-x}$N, have been shown to have optical bandgaps in the deep UV suitable for light emitting diodes (LEDs) operating down to ~ 250 nm [5]. However, until now there has been no systematic investigation of the influence of Mg incorporation on growth rate and polarity of the well and barrier materials. Furthermore, the mechanism of current injection in (Ni-Au)/(p-AlGaN) ohmic contacts has been discussed [6, 7], although the temperature behavior of such contacts for p-SPSL is unknown.

In this paper, we report on the structural, compositional, and electrical properties of p-type Mg-doped SPSLs with a range of Mg concentrations, N_{Mg}, from 10^{18} to 10^{20} cm^{-3}. We have found the Mg excess at the initial stage of p-SPSL growth to be capable of inducing formation of N-polar material and reducing Mg incorporation into solid. Using the results of temperature varied Hall measurements (TVHM) we propose a model describing the process of Mg activation in SPSLs. The new approach takes into account a complex valence band structure and elastic strain in both the well and barrier materials in SPSL. The temperature dependence of specific contact resistance of Au/Ni contacts to p-SPSL is investigated. Current-voltage-temperature measurements allow us to propose the carrier injection mechanism in ohmic contacts to p-SPSL.

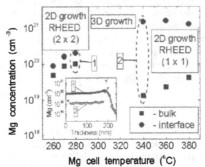

Fig.2. Growth mode, Mg incorporation, and typical SIMS profiles versus Mg flux.

EXPERIMENT

AlN/Al$_x$Ga$_{1-x}$N (0.03 ≤ x ≤ 0.05) SPSLs were grown by GSMBE on 2" (0001) sapphire substrates. The growth started with substrate nitridation and deposition of thin AlN nucleation layer, followed by an undoped ~ 30 nm thick SPSL with an average AlN content ~ 70% or undoped ~ 0.4 µm thick AlN layer. The Mg doped SPSLs were grown on different buffers in order to investigate influence of residual stress on electrical properties of these structures. The average composition of p-type SPSL was controlled by well and barrier thickness. Mg was evaporated from effusion cell, which temperature was varied from 260 to 380°C. Growth mode was *in situ* controlled by reflection high energy electron diffraction (RHEED). The average SPSL compositions and barrier/well thicknesses were obtained using X-ray diffraction (XRD) [8, 9]. Transmission electron microscopy (TEM) was used for validation of XRD data. Atomic force microscopy was used for surface morphology studies. The secondary ion mass spectroscopy (SIMS) was used to obtain the Mg concentration. The details of growth procedure and measurements are described elsewhere [2 – 5, 8 – 10].

Three sets of experiments were carried out to understand: i) the influence of Mg flux on growth mode, Mg incorporation, and surface polarity of SPSLs; ii) the temperature dependence

184

of hole concentration in p-SPSLs with average AlN content near 70%; and iii) the temperature dependence of specific contact resistance of p-SPSLs.

Table I. Parameters of the samples used for Hall and contact resistance measurements

Sample number	Average AlN conc. (%)	RT Hole density (cm^{-3})	X-ray Thickness (nm)		Mg concentration, N_{Mg} (cm^{-3})		Screw dislocation density (cm^{-2})
			Well	Barrier	Average SIMS	in AlN (fitting)	
1	65%	6.9×10^{17}	0.52	0.90	1.0×10^{19}	4.3×10^{18}	3.3×10^9
2	67%	6.6×10^{17}	0.60	1.08	1.5×10^{19}	3.1×10^{18}	2.9×10^9
3	72%	5.6×10^{17}	0.56	1.30	8.0×10^{18}	2.8×10^{18}	2.9×10^9

Figure 2 summarizes the results on growth mode and Mg distribution as a function of Mg flux (temperature of Mg effusion cell). Note all samples in Fig. 2 were grown under the same conditions except for Mg flux. Three regions are distinguished in this figure. At low Mg flux we observed two-dimensional (2D) growth mode and surface structure (2×2) at low temperatures. The Mg concentration at the interface between SPSL/buffer and the average Mg concentration in SPSL are in very close proximity and both increase following the temperature of Mg effusion cell. This corresponds to formation of (group III)-polar surface when samples are grown by GSMBE with ammonia [11]. Curve 1 in the inset shows a typical SIMS profile for one of the samples grown at low Mg flux. The average Mg concentration in this SPSL is $(1.5\pm0.5)\times10^{19}$ cm^{-3}. The 2D growth of SPSL can also be reached when Mg cell temperature is in the range from 340 to 380°C. In these samples, the Mg concentration at the SPSL/buffer interface and its value in the SPSL are very different. Although Mg concentration increases with its evaporation temperature, one can see that average Mg concentration in this SPSL is significantly lower compared to the structure grown at lower Mg fluxes. The curve 2 in inset shows a rapid change of Mg concentration from the interface to the bulk for a typical sample grown under these conditions. Note the surface structure of these SPSLs remains (1 x 1) at the growth and low temperatures. Moreover, wet etching of these samples in diluted KOH yielded formation of very rough surface. These effects and ~30% decrease of average growth rate of Mg doped layers grown with high effusion cell temperature are attributed to changes in the crystal polarity during exposure of the buffer layer surface to large Mg flux. This conclusion is in very good agreement with reported N-polar GaN growth at high Mg fluxes [11]. Note 3D growth mode was observed for the samples grown at intermediate Mg cell temperatures. In this region both N-polar and (group-III)-polar domains are formed simultaneously yielding formation of extremely rough surfaces and unreliable doping properties of the SPSLs. As a result, no electrical studies were carried out on these samples.

We used GSMBE conditions yielding (group III)-polar SPSL surfaces for the next set of experiments. The parameters of these SPSLs are summarized in Table I. The SPSLs consist of 300 AlN/Al$_{0.03}$Ga$_{0.97}$N pairs. The well thickness was hold at ~ 2 MLs and barrier thickness was varied in order to change an average AlN content from 65 to 72 %. All SPSLs were coated by 5

185

nm thick p-Al$_{0.03}$Ga$_{0.97}$N. Average SIMS Mg concentration in all samples was ~ 10^{19} cm^{-3}. These samples were characterized by the temperature-dependent Hall and transmission line method (c-TLM) measurements. The details of Au/Ni contact fabrication are described elsewhere [7, 12].

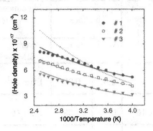

Fig. 3. Average hole density in various SPSLs as a function of inverse temperature calculated, assuming either AlN layers in SPSL to be relaxed (lines for samples #2 and #3 and dotted line for sample #1) or AlGaN layer to be relaxed (line for sample #1). Symbols are the Hall data.

DISCUSSION

Figure 3 (symbols) shows the results of temperature-dependent Hall characterization of the above SPSLs. The hole density here corresponds to the hole concentration averaged over the full SPSL thickness. Compared to bulk GaN:Mg and AlGaN:Mg materials [7, 12], the SPSL layers exhibit much weaker dependence on temperature.

In order to understand the temperature dependence of the hole density, we consider the valence band edge profile in a SPSL (sample #1) computed by the BESST 2.0 simulator [13]. The horizontal line in Fig. 4a at zero energy corresponds to the Fermi level position in the SPSL, while the dashed line indicates the positions of the acceptor levels in Al$_{0.03}$Ga$_{0.97}$N-well (activation energy of ~170 meV) and AlN-barrier (activation energy of ~750 meV) layers. The acceptor levels were estimated by neglecting the influence of the individual acceptor position in a quantum well/barrier and hole confinement in the wells on acceptor activation energy.

One can see that the acceptor levels in the Al$_{0.03}$Ga$_{0.97}$N quantum wells (QWs) lie well above the Fermi level, so that these acceptors are not activated. In contrast, the Fermi level crosses the acceptor energy levels in the AlN barriers. Therefore, acceptors in the barriers are partially activated. The activation degree depends on temperature, producing hole transfer from barriers to wells, which results in the dependence shown in Fig. 3. The hole concentration is found to depend essentially on the acceptor concentration N_{Mg} in the AlN barriers. In order to get the theoretical dependence of the hole density on temperature we fitted N_{Mg} to reproduce experimentally observed hole density at room temperature. Then the variation of the hole density with temperature was calculated with this value of N_{Mg}. Fitting parameters are shown in Table I and compared with average Mg concentration from SIMS. The fitting result is consistent with the results for bulk AlN shown in Fig. 1. The simulation results are found to depend on the strain in both well and barrier layers via the polarization charges at the SPSL interfaces. For sample #1, Fig. 3 compares the hole densities computed by assuming either QW or barrier to be completely relaxed. Though the band-edge profiles are not very different in these two cases, the small shift

186

in the hole miniband energy with respect to the acceptor levels produces quite different temperature dependences of the hole density.

Figure 4b shows the distributions of the heavy, light, and split-off holes in the SPSL obtained by self-consistent solution of the Poisson and Schrödinger equations [13] accounting for the complex valence band structure of III-nitride materials [14]. One can see that heavy holes are primary contributors to the total hole density. The contribution of the split-off holes can be neglected due to the fact that this subband in the superlattice is far below the heavy and light subbands. The holes in the SPSL are accumulated in the wells, partially penetrating into the AlN barriers.

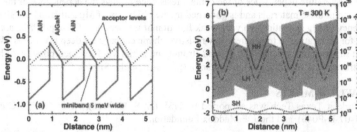

Fig. 4. Valence band layout in the sample #1 computed for room temperature (a) and band diagram and heavy (HH), light (LH), and split-off (SH) hole concentrations in the sample (b). In (a), dotted lines indicate the acceptor levels estimated by neglecting the effect of the hole confinement in the quantum wells.

Figure 5 shows the temperature dependence of the specific contact resistances, ρ_c, for three SPSLs with average AlN content of ~70% and for the "bulk" (~300 nm thick) p-$Al_{0.03}Ga_{0.97}N$. The Mg concentration in all samples is ~ 10^{19} cm^{-3} and room-temperature ρ_c is ~ 4×10^{-5} $\Omega\cdot$cm^2.

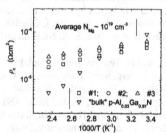

Fig. 5. Temperature dependences of specific contact resistance for SPSLs and p-$Al_{0.03}Ga_{0.97}N$.

We have already reported that thermionic emission is the dominant mechanism of carrier injection in the (Ni/Au) ohmic contacts to p-$Al_xGa_{1-x}N$ (0.05 < x < 0. 0.08) [7, 12]. The temperature dependence of the contact resistance was found to be primarily controlled by the activation energy of Mg acceptors in $Al_xGa_{1-x}N$ [7]. The ρ_c of metal/SPSL ohmic contact, as seen in Fig. 5, also follows the temperature dependence of hole concentration in these structures. As we mentioned above, the density of acceptors activated in AlN barriers is the primary factor influencing the hole concentration in the well material, producing a weaker temperature dependence compared to bulk materials. Thus, a weak dependence of ρ_c on temperature is expected even when thermionic mechanism of current injection dominates in the metal/SPSL ohmic contact. This expectation agrees with the results shown in Fig. 5.

CONCLUSIONS

We report fabrication of low-resistance ohmic contacts to p-SPSLs with a high (up to 72%) average Al content. We compare our results with our prior work on AlGaN alloys. In both cases, the temperature dependence of the contact resistance follows the dependence of the hole concentration in these materials and is governed by the efficiency of Mg activation. In SPSLs, the efficiency of Mg activation is a very weak function of temperature. These p-SPSLs are attractive for fabrication of transparent low resistive ohmic contacts for deep UV LEDs. We found that surface polarity of SPSLs depend strongly on initial Mg flux. Incorporation of Mg in N-polar structures is significantly lower than in (group-III)-polar SPSLs.

ACKNOWLEDGMENTS
Work at Texas Tech was supported by NSF (ECS-0609416), Army CERDEC contract (W15P7T-07-D-P040), and the J. F Maddox Foundation.

REFERENCES

1. S. Nikishin, V. Kuryatkov, B. Borisov, G. Kipshidze, A. Chandolu, K. Zhu, M. Holtz, S. N. G. Chu, Yu. Kudryavtsev, R. Asomoza, and H. Temkin, Presentation at International Workshop on Nitride Semiconductors, Aachen, July 22 – 25, 2002 (unpublished).
2. G. Kipshidze, V. Kuryatkov, B. Borisov, S. Nikishin, M. Holtz, S. N. G. Chu and H. Temkin, Phys. Status Solidi A **192** (2002) 286.
3. S. A. Nikishin, V. V. Kuryatkov, A. Chandolu, B. A. Borisov, G. D. Kipshidze, I. Ahmad, M. Holtz, and H. Temkin, Jpn. J. Appl. Phys., **42**, L1362 (2003).
4. V. Kuryatkov, K. Zhu, B. Borisov, A. Chandolu, I. Gheriasou, G. Kipshidze, S. N. G. Chu, M. Holtz, Yu. Kudryavtsev, R. Asomoza, S. A. Nikishin and H. Temkin, Appl. Phys. Lett., **83** 1319 (2003).
5. S. A. Nikishin, M. Holtz, and H. Temkin, Jpn. J. Appl. Phys., **44**, 7221 (2005).
6. T. V. Blank and Yu. A. Gol'dberg, Semiconductors, **41**, 1263 (2007) *and references therein*.
7. S. Nikishin, I. Chary, B. Borisov, V. Kuryatkov, Yu. Kudryavtsev, R. Asomoza, S. Yu. Karpov, and M. Holtz, Appl. Phys. Lett., **95**, 163502 (2009).
8. A. Chandolu, S. Nikishin, M. Holtz, and H. Temkin, J. Appl. Phys., **102**, 114909 (2007).
9. M. Pandikunta, M.S. Thesis, Texas Tech University, 2009.
10. M. Holtz, G. Kipshidze, A. Chandolu, J. Yun, B. Borisov, V. Kuryatkov, K. Zhu, S. N. G. Chu, S. A. Nikishin, and H. Temkin, Mat. Res. Soc. Symp. Proc., **744**, M10.1.1 (2003).
11. N. Grandjean, A. Dussaigne, S. Pezzagna, and P. Vennegues, J. Cryst. Growth, **251**, 460 (2003), and references therein.
12. I. Chary, B. Borisov, V. Kuryatkov, Yu. Kudryavtsev, R. Asomoza, S. Nikishin, and M. Holtz, Mat. Res. Soc. Symp. Proc., **1108**, 1108-A09-30 (2009).
13. http://www.str-soft.com/products/BESST/
14. S. L. Chuang and C. S. Chang, Phys. Rev. B **54**, 2491 (1996).

Growth and Processing

Mater. Res. Soc. Symp. Proc. Vol. 1202 © 2010 Materials Research Society   1202-I05-04

AlN periodic multilayer structures grown by MOVPE for high quality buffer layer

V. V. Kuryatkov, W. Feng, M. Pandikunta, D. Rosenbladt, B. Borisov, S. A. Nikishin and M. Holtz
Nano Tech Center, Texas Tech University, Lubbock, Texas 79409, USA

ABSTRACT

High crystal quality crack-free AlN on sapphire was grown by low pressure metal organic vapor phase epitaxy (MOVPE). Growth experiments combine two recent approaches: the ammonia pulse-flow method and ammonia continuous-flow growth mode by varying the V/III ratio. The detailed aspects of MOVPE, employing the periodic multilayer approach at low, intermediate, and high temperatures are described. This method yields significant reduction of screw dislocation density and provides very smooth surface for thin AlN layers.

INTRODUCTION

Research into III-nitride semiconductors is important due to applications in photonics and electronics. The applications rely on the wide range of energy gaps possible with this material class. AlN is often used as a buffer layer which initiates epitaxy on various substrates and is transparent to ~ 210 nm, i.e., throughout the deep UV spectral region. Therefore, improvements in AlN growth are critical in obtaining high quality growth for devices based on wide-bandgap AlGaN [1]. In general, the difficulties of AlN growth are related to the low migration mobility of Al atoms and to significant pre-reaction of Al and ammonia.

Khan et al. have proposed low pressure metalorganic pulsed atomic layer epitaxy (PALE) for the growth of AlN [2] on (0001) sapphire substrate. Imura et al. applied two- and three dimensional (2D and 3D) multi-growth mode, using various III-V ratios and high temperature (HT) to obtain AlN with low threading dislocation density [3]. Okada et al. have compared crystal quality of AlN grown at high temperature in multi-growth mode on low temperature (LT) and HT buffer layers [4] and found that combined LT buffer grown in continuous-flow mode and HT buffer grown in multi-growth mode yield better crystal perfection of crack-free thick AlN layers. Hirayama et al. have proposed the ammonia pulse-flow multilayer approach at onset of the low pressure metal organic vapor phase epitaxy (MOVPE) of HT AlN buffer layer [5]. In order to reduce the surface roughness of AlN grown in pulse-flow mode, they additionally used the high-growth-rate continuous-flow mode for the following epitaxial layer grown at high temperature. Repetition of pulse- and continuous-flow mode yielded crack-free layers of thick AlN with smooth surfaces.

Although the ammonia pulse-flow/continuous-flow (APF/ACF) approach at HT results in high crystal quality, the application of this method at LT for the growth of high quality thin AlN layers has not been thoroughly investigated. In this paper we summarize our systematic study of APF/ACF applied for the growth of LT and HT thin AlN buffer layers. The structural quality of the epilayers was characterized by X-ray rocking curve (XRC) analysis. The surface morphology of the epilayers was evaluated by atomic force microscopy (AFM) along with scanning electron microscopy (SEM). Influence of the thickness of LT AlN grown in APF mode, thickness of HT AlN grown in CPF mode, ammonia interruption time (τ_{in}), and period of APF/ACF HT AlN on threading dislocation density and surface roughness is presented and discussed.

EXPERIMENT

AlN growth was carried out on sapphire (0001) substrates by low-pressure MOVPE. Prior to growth, substrates were annealed at 1100 °C in a 20 Torr H_2 environment and nitridated at the same temperature for 2 minutes in NH_3 flow of 1250 sccm. Trimethylaluminum (TMAl) precursor was used with H_2 as carrier gas and NH_3 as nitrogen source for the growth of AlN. The TMAl and NH_3 flow rates were varied from 2 to 10 μmol/min and from 0.1 to 1.0 slm, respectively, for both LT and HT growth. The growth pressure was varied in the range from 20 to 200 Torr for experiments at both low and high temperatures.

Table I. MOVPE growth parameters for different stages of AlN growth.

Process step	Substrate temp. (°C)	Time (min)	Reactor gases	Flow rates (sccm)	NH_3 mode τ_p/τ_{in} (s/s)
(I) Cleaning	1100	10	H_2	5000	NA
(II) Nitridation	1100	2	H_2/NH_3	3800/1200	NA
(III) Ramp	890	3	H_2/NH_3	4500/500	NA
(IV) AlN (LT)	890	15	$H_2/NH_3/TMAl$	4500/500/4.1	APF (3/3)
(V) Ramp	1110	3	H_2/NH_3	4500/500	NA
(VI) AlN (IT)	1110	15	$H_2/NH_3/TMAl$	4500/500/4.1	APF (3/3)
(VII) Ramp	1160	1	H_2/NH_3	4500/500	NA
(VIII) AlN (HT)	1160	10	H_2/NH_3	4500/500/4.1	ACF
(IX) AlN (HT)	1160	60	$H_2/NH_3/TMAl$	4500/500/4.1	APF (1/5)

First, we investigated the growth of LT AlN using separately ACF and APF approaches. For the III-V flux ratios, TMAl and NH_3 flow rates, and growth pressures investigated, we found that ACF leads to formation of 3D islands in the temperature range from 550 to 1000 °C. The APF approach yields significant improvement of surface smoothness and the best surface morphology of LT AlN buffer was obtained when layer was grown at 890 °C and duration of the ammonia pulse was $\tau_p = 3$ s with interruption time $\tau_{in} = 3$ s. Application of combined APF/ACF mode did not improve surface morphology of LT layers. The detailed parameters of the onset of MOVPE are summarized in the rows 1, 2, 3, and 4 of Table I and in the left part of Fig. 1.

In order to reduce dislocation density, a LT AlN buffer layer is usually annealed at intermediate

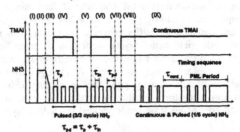

Figure 1. The timing sequence of TMAl and NH_3 varying over the growth process of AlN at LT (left) and HT (right), where τ_p and τ_{in} are duration and interruption times in APF mode, $\tau_{pd} = \tau_p + \tau_{in}$, and τ_{cont} is growth time in ACF mode. PLM period corresponds to APF/ACF growth mode at high temperature.

192

temperature before HT growth. However, high ramping rate from LT to intermediate temperature (IT) and annealing could significantly rough a surface of LT buffer [6]. To avoid this undesirable

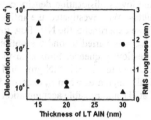

Figure 2. Screw dislocation density (triangles) and RMS surface roughness (circles) vs. thickness of LT AlN.

effect, we applied APF approach for IT growth also. Rows 5 and 6 of Table I summarize optimal parameters of APF approach at IT yielding the best crystal and morphological quality of AlN grown on LT buffer.

After LT and IT deposition, the TMAl and H_2 fluxes were turned off and substrate was heated to the high temperature (HT), as illustrated in Fig. 1. HT AlN layers were grown at 1160 °C. Following this, ACF growth mode was applied to begin HT deposition. Then, APF/ACF approach was used, as illustrated in Fig. 1. Finally, the desired HT AlN, ~ 150 nm thick, was grown. The best growth conditions of HT AlN are summarized in the Table I.

DISCUSSION

Figure 2 shows the dependences on LT AlN thickness of the screw dislocation density and (AFM) RMS surface roughness. Dislocation density (N_D) was calculated from the full-width at half-maximum (FWHM = β) of X-ray ω-scan of (0002) AlN peak using standard formula $N_D = \{\beta/(2\pi ln2xb^2)\}$ [7], where b is the length of the Burger vector. For screw dislocations b equals to c lattice parameter of AlN. The dislocation density decreases when thickness of LT AlN increases, but roughness is seen to rise above thickness 20 nm. This effect could be attributed to

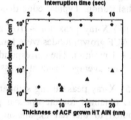

Figure 3. Dislocation density as a function of ACF grown thickness (triangles) and interruption time, τ_{in}, (circles) during APM growth of HT AlN.

low surface mobility of Al atoms at low deposition temperature and the resulting formation of 3D islands even in APF growth mode. Note all LT layers grown in ACF mode had 3D surface structure, i.e., similar to island formation. The ~ 20 nm LT AlN grown in APF mode was used for the next set of experiments. The dependence of dislocation density on thickness and interruption time during ACF/APF approach is shown in Fig. 3. Our experiments show that it is important to begin HT growth using ACF mode. The duration of ACF mode affects the extent of lateral growth. We found that the lowest dislocation density corresponds to the growth of ~ 10 nm ACF layer within ~10 minutes. We investigated the influence of pulse NH_3 flow on the quality of grown AlN. During these experiments the NH_3 flow time, τ_p, was kept constant for 3 s and the NH_3 interruption time, τ_{in}, was varied to find the optimum condition. TMAl flow was continuous during the NH_3 pulse-flow sequence. Similar to the LT growth, the optimal NH_3 interruption time of 3-5 sec was found for the HT AlN growth. Filled circles in Fig. 3 show that screw dislocation density drastically increases when τ_{in} exceeds 5 s. We tentatively attribute this to high evaporation rate of AlN at HT and possible accumulation of Al at interstitials of epitaxial film.

The last set of experiments was carried out in order to optimize thicknesses of ACF/APF grown layers of periodic multilayer (PML) structure. The ratio ~ 3/2 of pulsed/continuous growth time was found optimal and it was kept the same for all runs. The growth period of PML structure

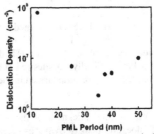

Figure 4. The dislocation density vs. period of HT PML.

was varied by changing both APF and ACF duration times holding their ratio close to 1.5. The density of screw dislocation as a function of PML structure period is shown in Fig. 4. Note that although ACF/APF approach yields significant reduction of screw dislocation density, it almost does not affect the density of edge dislocation. Typical edge dislocation density of 5×10^{11} cm^{-2} was estimated from FWHM of $(10\bar{1}5)$ X-ray peak for ~ 100 nm thick AlN layers [8]. The optimum conditions for the APF and ACF growth modes were found to be 15 ± 2 min and 10 ± 2 min, respectively, for the growth rate ~ 1 nm/min. The corresponding number of periods used in the APF growth was typically 150. The average residual stress in AlN was calculated from angular position of (0002) and $(10\bar{1}5)$ X-ray peaks using standard procedure [9]. The value of compressive stress, − 0.64 GPa, was obtained for the optimal period ~ 30 ± 5 nm of PML structure. This stress can be attributed to thermal expansion mismatch between AlN and sapphire.

The typical $2\Theta - \omega$ scan of (0002) reflection for 130 nm thick AlN layer grown using ACF/APF approach is shown in Fig. 5. Sharp Pendellosung fringes correspond to a high degree of top layer flatness with AFM RMS roughness of 0.23 nm. One interesting feature of a (0002) peak is its asymmetric shape. We believe the asymmetric shape of (0002) peak is due to varying lateral lattice parameter *"a"* in the PML layers. To verify this, we have simulated PML AlN layers at different relaxations using Philips Epitaxy software [8]. The percentage of relaxation is expressed in terms of stressed and relaxed lattice constants in AlN layers by following formula

$$\%relaxation = \frac{a'_{N+1} - a_N}{a_{N+1} - a_N} \times 100\%$$

where N+1 is the successive layer above layer N, a'_{N+1} is the stressed lattice constant of N+1 layer, while a_{N+1} and a_N are respective relaxed lattice constants of N+1 and N layers. The best fit,

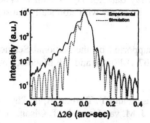

Figure 5. Experimental 2Θ – ω scan (solid curve) and its simulation (dashed curve) for 130 nm thick AlN grown using ACF/APF approach.

shown as dashed curve in Fig. 5, corresponds to LT AlN buffer layer on substrate at 98% relaxation, followed by gradual increase of relaxation in PML layers from 0% relaxation to 100% relaxation. The top HT AlN ~ 40 nm is at 100% relaxation in this simulation.

The presence of a successively relaxed layers was checked using similar ACF/APF growth for the growth of thin crack-free AlN layers on Si(111) substrates. This approach was used to reduce charging during SEM for AlN on sapphire. SEM cross-section of typical AlN/Si sample is shown in Fig. 6. We observe slight contrast, corresponding to strain relaxation steps. The width of the contrast layers corresponds with the thickness of AlN layers grown in ACF and APF mode. This

Figure 6. Cross-section SEM image of AlN on Si(111). 1 – Silicon wafer; 2 – LT & IT AlN layers; 3 – PML AlN layers; 4 – 40 nm AlN grown in APF mode.

corroborates the identification of strain relaxation layers identified by the asymmetry in our X-ray diffraction.

CONCLUSIONS

Detailed aspects of MOCVD AlN layer growth, employing the periodic multilayer approach, are described. Dependence on sapphire preparation and growth conditions is presented, including the influence of the pre-growth annealing, the sapphire nitridation, and both LT and HT growth conditions are discussed. MOVPE growth of AlN on c-plane sapphire substrates was investigated as a function of the PML AlN period. This method leads to a reduction in the screw-type dislocation density of AlN. The screw-type dislocation densities of AlN layer was ~ 6×10^5 cm^{-2}, as determined from the X-ray data for layers with thickness ~130 nm. The AFM RMS surface roughness was 0.23 nm. AlN layers with atomically flat surfaces are crack free.

ACKNOWLEDGMENTS

This work at TTU was partly supported by the National Science Foundation (ECS-0609416), U.S. Army CERDEC (W15P7T-07-D-P040), and J. F Maddox Foundation.

REFERENCES

1. H. Amano, N. Sawaki, I. Akasaki and Y. Toyoda, Appl. Phys. Lett. **48**, 353 (1986).
2. M. A. Khan, R. A. Skogman, J. M. van Hove, D. T. Olson, and J. N. Kuznia, Appl. Phys. Lett. **60**, 1366 (1992).
3. M. Imura, K. Nakano, N. Fujimoto, N. Okada, K. Balakrishnan, M. Iwaya, S. Kamiyama, H. Amano, I. Akasaki, T. Noro, T. Takagi, and A. Bandoh, Jpn. J. Appl. Phys. **45**, 8639 (2006).
4. N. Okada, N. Kato, S. Sato, T. Sumii, T. Nagai, N. Fujimoto, M. Imura, K. Balakrishnan, M. Iwaya, S. Kamiyama, H. Amano, I. Akasaki, H. Maruyama, T. Takagi, T. Noro, and A. Bandoh, J. Cryst. Growth, **298**, 349 (2007).
5. H. Hirayama, T. Yatabe, N. Noguchi, T. Ohashi, and N. Kamata, App. Phys. Lett., **91**, 071901 (2007).
6. D. D. Koleske, M. E. Coltrin, K. C. Cross, C. C. Mitchell, and A. A. Allerman, J. Cryst. Growth, **273**, 86 (2004).
7. J. E. Ayers, J. Cryst. Growth, **135**, 71 (1994).
8. M. Pandikunta, M.S. Thesis, Texas Tech University, 2009.
9. F. Wright, J. Appl. Phys. **82**, 2883 (1997).

Mater. Res. Soc. Symp. Proc. Vol. 1202 © 2010 Materials Research Society   1202-I05-02

Growth of high quality c-plane AlN on a-plane sapphire

Reina Miyagawa, Jiejun Wu, Hideto Miyake and Kazumasa Hiramatsu
Department of Electrical and Electronics Engineering, Mie University, 1577 Kurimamachiya, Tsu 514-8507, Japan
re_miyagawa@opt.elec.mie-u.ac.jp

Abstract

c-plane (0001) AlN layers were grown on (11-20) a-plane and (0001) c-plane substrates by hydride vapor phase epitaxy (HVPE) and metal-organic vapor phase epitaxy (MOVPE). The growth temperature was adjusted from 1430 to 1500 ^{0}C and the reactor pressure was kept constant at 30 Torr. Mirror and flat c-plane AlN were obtained on both a-plane and c-plane sapphire. The crystalline quality and surface roughness, evaluated by high-resolution X-ray diffraction (HRXRD) and atomic force microscopy (AFM), respectively, improved with increasing growth temperature. The full width at half maximum (FWHM) values of (10-12) diffraction were 519 and 1219 arcsec for c-plane AlN grown on a-plane sapphire and c-plane sapphire, respectively. This indicates that the a-plane sapphire substrate has the advantage of providing decreased dislocation density.

Introduction

The strong demand for high-efficiency deep-ultra violet light-emitting diodes (UV-LEDs) and sensors has led to renewed interest in the growth of high-quality AlN. However, the growth of high-quality, thick, and crack-free AlN is very difficult owing to the low surface migration of Al atoms, the large lattice and thermal mismatches in the absence of native substrates, the strong parasitic reaction at elevated temperatures, and the narrow growth window. Some methods, such as maskless lateral epitaxy overgrowth (LEO) [1], and similar techniques such as the use of a patterned AlN template [2] or a patterned substrate [3], pulsed atomic layer epitaxy (PALE) and similar methods [4, 5], and a V/III ratio modulation multi-interlayer method [6], have proven to be effective at decreasing the threading dislocation (TD) density, particularly that of edge-type dislocations, and improving the crystalline quality. Chen *et al.* used a combination of PALE and a modulation growth (3D-2D) method to reduce the FWHM value of (10-12) diffraction on a SiC substrate to 363 arcsec [4]. However, LEO requires *ex situ* pattern fabrication and PALE places a critical demand on the shutters of equipment. These methods are also complicated from a technological viewpoint, and a simple method for growing high-quality AlN is desirable.

Although c-plane sapphire is normally used as a substrate, a-plane sapphire is actually preferable for some applications such as edge-emitting lasers [7] owing to the easy cleavage along the r-plane. Despite the fact that improved GaN has been grown on a-plane sapphire [8], the growth of AlN on a-plane sapphire by MOVPE or HVPE has seldom been reported.

In this study, we report on a simple method of AlN growth on a-plane and c-plane sapphire substrates by HVPE and MOVPE without the use of LEO, PALE, or a multi-interlayer.

Experimental procedure

A c-plane AlN layer was grown on c-plane and a-plane sapphire substrates by low-pressure (LP) HVPE and MOVPE. For the growth of AlN by HVPE, ammonia (NH$_3$), Al, and HCl were used as source materials. c-plane AlN layers were grown at a V/III ratio of 600 at 1450 and 1500 °C at a pressure of 30 Torr. Prior to the growth of AlN, an AlN buffer layer was grown at a V/III ratio of 1800 at 1150 °C for 1 min after 10 min thermal cleaning in H$_2$ ambient at 1150 °C. Growth parameters, such as the growth temperature, and NH$_3$ and HCl flow rates, were optimized for the growth of the AlN film. For the growth of AlN by MOVPE, NH$_3$ and trimethylaluminum (TMA) were used as the precursors for N and Al, respectively. c-plane AlN layers with a thickness of 1 µm were grown at a V/III ratio of 584 at 1430 °C at a pressure of 30 Torr. Prior to the growth of AlN at 1430 °C, a medium-temperature (MT) AlN buffer layer was grown at a V/III ratio of 2763 at 1000 °C after 10 min thermal cleaning in H$_2$ ambient at 1000 °C. During the growth, the growth mechanism was observed by *in situ* optical reflectance monitoring. The dependence of the crystalline quality and surface morphology on the thickness of MT-AlN buffer layer was investigated. Growth parameters such as growth temperature, reactor pressure, and V/III ratio were also optimized for the growth of the AlN film. The crystalline quality was characterized by high-resolution X-ray diffraction (HR-XRD), and the surface morphology was characterized by optical microscopy and atomic force microscopy (AFM).

Results and discussion

Figure 1 shows optical microscopy images of c-plane AlN grown by HVPE. Cracks were only observed in the <11-20> direction. For the sapphire substrate, the thermal expansion coefficient (TEC) parallel to its <0001> c-axis is known to be larger than that perpendicular to the c-axis [8]. Thus, the thermal strain in AlN should be higher along one of the three equivalent <10-10> m-axes [marked x$_c$ in Fig. 1, which is parallel to the c-axis of sapphire] than along the other two m-axes [marked x$_a$ in Fig. 1, which is at an angle to the c-axis of sapphire]. This can explain why cracks were only observed perpendicular to the x$_c$ direction. The crack density was lower when a-plane sapphire was used as a substrate. Then, the in-plane epitaxial relationships between the film and substrate were determined by X-ray phi-scans of AlN (10-12) and sapphire (11-23). Results indicated that AlN [-1120] is parallel to sapphire [0001] when a-plane sapphire was used. Such a relationship is inconsistent with the epitaxial relationship found in HVPE-GaN layers on a-plane sapphire but is consistent with that for GaN grown by MOVPE [9]. It is possible that the high temperature in our HVPE experiments resulted in an AlN epitaxial orientation similar to that obtained by MOVPE. The FWHM values of the X-ray rocking curve (XRC) obtained from (0002) AlN diffraction were 324 and 404 arcsec, and those from (10-12) AlN diffraction were 519 and 1219 arcsec for c-plane AlN grown at 1500 °C on a-plane sapphire and c-plane sapphire, respectively. The results show that the asymmetric (10-12) diffraction peaks of AlN are markedly narrowed upon using the a-plane sapphire substrate. This suggests that the in-plane twisting of AlN islands was restricted when a-plane sapphire was used as the substrate, preventing the formation of edge dislocations. Thus, the growth window for AlN on the a-plane sapphire substrate was found to be wider than that on the c-plane sapphire substrate.

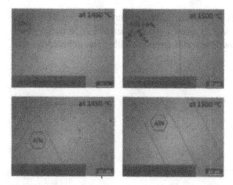

Fig. 1 Optical microscopy images of c-plane AlN grown by HVPE.

Next, we show the XRC FWHM of AlN grown by MOVPE in Figure 2, which indicates the dependence of the crystalline quality on the thickness of the MT-AlN buffer layer grown at 1100 °C. The in-plane epitaxial relationships between the film and substrate were similar to those for AlN grown by HVPE. The results indicate that the XRC FWHM for (10-12) diffraction decreased with decreasing thickness of the MT-AlN buffer layer, perticularly on the c-plane sapphire substrate, and when an AlN layer was grown without an MT-AlN buffer layer, the crystalline quality improved, as indicated by the value of FWHM for symmetric (0002) and asymmetric (10-12) diffraction. The FWHM values of XRC obtained from (0002) AlN diffraction were 455 and 190 arcsec, and those from (10-12) AlN diffraction were 688 and 461 arcsec for c-plane AlN grown without an MT-AlN buffer layer on a-plane and c-plane sapphire substrates, respectively.

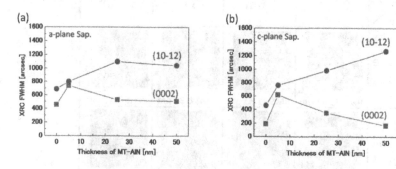

Fig. 2 The XRC FWHM of AlN grown on (a) a-plane and (b) c-plane sapphire substrates by MOVPE.

199

Figure 3 shows the radius of curvature of AlN layers. The radius of curvature is small, i.e., the AlN layers are highly curved when c-plane AlN is grown without an MT-AlN buffer layer. Therefore, it is suggested that AlN was grown on each sapphire substrate coherently, thus improving the crystalline quality.

c-plane Sap.
a-plane Sap.
Radius of curvature [m]
Thickness of MT-AlN [nm]
0
10
20
30
40
50

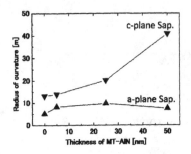

Fig. 3 Radius of curvature of AlN layers grown by MOVPE.

AFM images of AlN layers grown on a-plane and c-plane sapphire substrates by MOVPE are shown in Figure 4. The surface morphology improved with increasing thickness of the MT-AlN buffer layer. When AlN was grown on an MT-AlN buffer layer with a thickness of 25 or 50 nm, atomic steps were clearly observed and the root-mean-square (RMS) roughness values were small. The results suggest that the surface morphology deteriorated because island growth was dominant and that the AlN grains became larger when AlN was grown directly on the sapphire substrates or on a thin MT-AlN buffer layer.

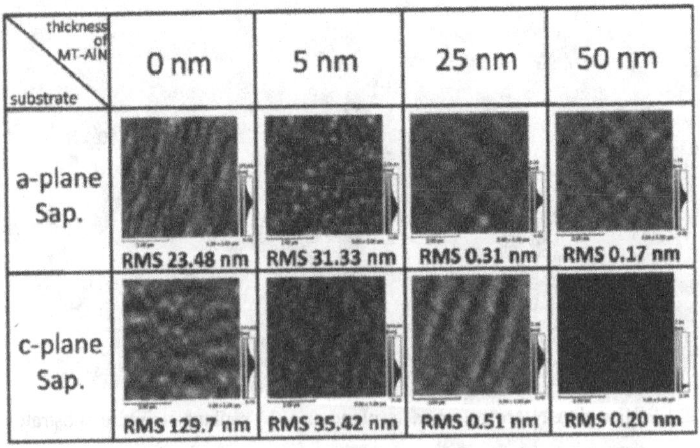

thickness of MT-AlN
substrate
0 nm
5 nm
25 nm
50 nm
a-plane Sap.
RMS 23.48 nm
RMS 31.33 nm
RMS 0.31 nm
RMS 0.17 nm
c-plane Sap.
RMS 129.7 nm
RMS 35.42 nm
RMS 0.51 nm
RMS 0.20 nm

Fig. 4 AFM images of AlN grown by MOVPE.

Conclusions

A c-plane AlN layer was grown on a-plane and c-plane sapphire substrates by HVPE and MOVPE. As a result of AlN growth by HVPE, the XRC FWHM of asymmetric (10-12) diffraction and the crack density decreased when a-plane sapphire was used as a substrate. As a result of AlN growth by MOVPE, the surface morphology improved with increasing thickness of the MT-AlN buffer layer. On the other hand, the crystalline quality improved when AlN layers were grown without an MT-AlN buffer layer on both a-plane and c-plane sapphire substrates.

Acknowledgments

This work was partially supported by Akasaki Research Center of Nagoya University, Grants-in-Aid for Scientific Research (No. 21360007 and No. 21560014) and Scientific Research on Priority Areas (No. 18069006) of MEXT, and Research for Promoting Technological Seed of JST.

References

[1] S. A. Newman, D. S. Kamber, T. J. Baker, Y. Wu, F. Wu, Z. Chen, S. Nakamura, J. S. Speck, and S. P. DenBaars: Appl. Phys. Lett. 94 (2009) 121906.

[2] M. Imura, K. Nakano, T. Kitano, N. Fujimoto, G. Narita, N. Okada, K. Balakrishnan, M. Iwaya, S. Kamiyama, H. Amano, and I. Akasaki: Appl. Phys. Lett. 89 (2006) 221901.

[3] J. Mei, F. A. Ponce, R. S. Q. Fareed, J. W. Yang, and M. A. Khan: Appl. Phys. Lett. 90 (2007) 221909.

[4] Z. Chen, S. Newman, D. Brown, R. Chung, S. Keller, U. K. Mishra, S. P. DenBaars, and S. Nakamura: Appl. Phys. Lett. 93 (2008) 191906.

[5] M. Takeuchi, S. Ooishi, T. Ohtsuka, T. Maegawa, T. Koyama, S. F. Chichibu, and Y. Aoyagi: Appl. Phys. Express 1 (2008) 021102.

[6] M. Imura, N. Fujimoto, N. Okada, K. Balakrishnan, M. Iwaya, S. Kamiyama, H. Amano, I. Akasaki, T. Noro, T. Takagi, and A. Bandoh: J. Cryst. Growth 300 (2007) 136.

[7] S. Nakamura, M. Senoh, S. Nagahama, N. Iwasa, T. Yamada, T. Matsushita, H. Kiyoku, and Y. Sugimoto: Jpn. J. Appl. Phys. 35 (1996) L217.

[8] H. K. Chauveau, P. D. Mierry, H. Cabane, and D. Gindhart: J. Appl. Phys. 104 (2008) 113516.

[9] T. Paskova, V. Darakchieva, E. Valcheva, P. P. Paskov, B. Monemar, and M. Heuken: J. Cryst. Growth 257 (2003) 1.

Mater. Res. Soc. Symp. Proc. Vol. 1202 © 2010 Materials Research Society 1202-I05-01

TEM Analysis of Microstructures of AlN/Sapphire Grown by MOCVD

Bo Cai, and Mim L. Nakarmi[a]

Department of Physics, Brooklyn College of the City University of New York, Brooklyn, NY 11210, USA, and
the Graduate Center of the City University of New York, New York, NY 10031, USA

a) mlnakarmi@brooklyn.cuny.edu

ABSTRACT

We report on microstructure analysis of aluminum nitride (AlN) epilayers by transmission electron microscopy (TEM). AlN epilayer samples were grown on sapphire substrates by metal organic chemical vapor deposition. Cross section and plan view images were taken by TEM to investigate the threading dislocations. Edge type threading dislocations dominate the total dislocation density. The threading dislocations are greatly reduced by inserting an intermediate layer prior to the growth of high temperature AlN epilayer. The dislocations further reduce with increasing thickness. Our results correlate with the dislocation density estimated from x-ray diffraction analysis.

INTRODUCTION

Aluminum nitride (AlN) that has a direct band gap ~ 6.1 eV, has emerged as a promising deep ultra violet (DUV) material with the demonstration of AlN-based light emitting diode (LED) and photodetector [1,2]. AlN possesses outstanding properties such as high temperature stability, hardness and high thermal conductivity that makes it a good candidate for high temperature/power/radiation devices [3]. Because of the piezoelectric property, it has also application in surface acoustic wave (SAW) devices [4]. Due to the growth of AlN epilayers on foreign substrates such as sapphire, it suffers from a high density of threading dislocation (TD) density on the order of $\sim 10^{10}$ cm^{-2}. AlN with low defects and dislocation density is sought to improve the device performance as well as to design advanced devices such as avalanche photodiodes. High quality, low dislocation density AlN/sapphire can also be used as templates to grow nitride based device structures especially AlGaN based DUV optoelectronic devices such as LEDs and photodetectors. DUV LEDs have many applications in the area of water/air purification, equipment decontamination, florescence detection of bio-chemical agents, general-purpose solid-state lighting, and medical and health care [5,6]. DUV photodetectors with solar-blind characteristics (cut-off wavelength below 280 nm) have important applications from UV astronomy, combustion engineering, flame detection, early missile plume detection, secure space-to-space communications and pollution monitoring [7]. Thus, the characterization of threading dislocations and techniques to reduce the threading dislocations are crucial. In this paper, we present a transmission electron microscopy (TEM) investigation of AlN epilayers grown on sapphire substrates. Both cross-section and plan view TEM images were taken to investigate the threading dislocations. The samples were grown with an intermediate layer that acts as a dislocation filter layer reducing the threading dislocations.

EXPERIMENTAL DETAIL

AlN epilayer samples investigated were grown on c-plane (0001) sapphire substrate by metal organic chemical vapor deposition (MOCVD) at Kansas State University. Trimethyl aluminum (TMA) and ammonia (NH_3) were used as aluminum and nitrogen sources respectively. Hydrogen was used a carrier gas. As usual process of nitride growth, sapphire substrate was baked with hydrogen at high temperature prior to the growth. The growth was initiated with a thin buffer layer. An intermediate layer of thickness ~ 120 nm was then grown followed by a thick high temperature (HT) AlN layer. Total thickness of the AlN epilayers were ~1 – 4 µm. The samples were characterized by atomic force microscopy (AFM), photoluminescence, x-ray diffraction (XRD) etc. For TEM imaging, the samples were prepared by using Fischione ultrasonic disk cutter, mechanical grinding and polishing followed by dimple grinding and Ar^+ ion milling. Samples for both cross section and plan view imaging were prepared. The TEM characterization was carried out on a JEOL 2010 TEM system operating at 200 kV.

RESULTS AND DISCUSSION

Figure 1 shows a bright field cross section TEM image of an AlN epilayer grown on sapphire substrate. The image was taken near zone axis$[2\bar{1}\bar{1}0]$. Threading dislocations are generated at the interface of AlN and sapphire. From the TEM images, we observed that the most of the threading dislocations are annihilated within the thickness of ~ 300 nm.

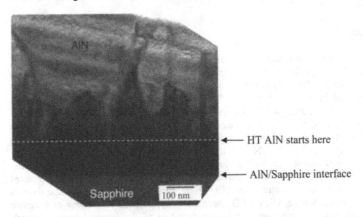

Figure 1. Bright field cross section TEM image of an AlN epilayer grown on sapphire substrate

The samples were grown with an intermediate layer with thickness ~ 120 nm following the buffer. We believe that three dimensional island growth dominates during the growth of the intermediate layer. The arrow in the Fig.1 indicates where the high temperature epilayer starts. The high temperature epilayer growth was carried out at ~ 1300°C. Two dimensional growth dominates during the high temperature AlN layer growth as evident from the smooth surface observed by AFM. In the beginning of the high temperature layer, the threading dislocation starts bending and annihilating. Within the thickness of ~ 300 nm, most of the threading dislocations are annihilated. Thus, we believe that insertion of the intermediate layer plays an important role in diminishing the threading dislocations during the high temperature growth. Figure 2 shows a cross section TEM image taken at far away from the interface. It is also clearly visible that the threading dislocations are further annihilated as the thickness increased forming large loops as depicted in Fig. 2. However, some dislocations propagate vertically.

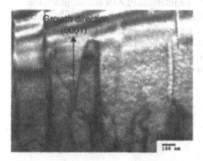

Figure 2. Bright field cross section image taken in HT AlN layer

Figure 3 shows a bright field plan view TEM image of an AlN epilayer taken with g = (11$\bar{2}$0) and the specimen was tilted ~18° from (0001) zone axis. All three types of dislocations- edge, screw, and mixed can be detected by tilting the sample in the plan view image using g = (11$\bar{2}$0) [8]. Based on the plan view images taken in different tilting angles, we observed the dominant threading dislocations are edge type.

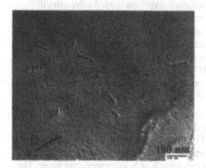

Figure 3. Bright field plan view TEM images of AlN epilayer taken with g = (11$\bar{2}$0) and tilted 18° from (0001) zone axis.

205

The average total dislocation density is ~ 2.7 x 10^9 cm^{-2}. Our results agree well with the dislocation densities estimated from the analysis of XRD rocking curves [9]. The full width at half maximum (FWHM) of XRD rocking curves of the (0002) and (10$\bar{1}$2) reflection planes decreases with increasing thickness, and were 63 and 437 arcsec respectively for AlN epilayer with thickness of 4 µm. The FWHM of rocking curves measured at (0002) and (10$\bar{1}$2) planes are related to the density of screw and edge type threading dislocations respectively [10].

Figure 4 shows a high resolution TEM image near the interface of AlN and sapphire substrate. The image was taken in the [2$\bar{1}$$\bar{1}$0] zone. At the interface of AlN and sapphire, the contrast modulations due to the strain near misfit dislocations (MD) can be observed. Edge type dislocations are normally generated from the misfit dislocations. Apparently, the stacking fault is not readily evident. AlN structure has the highest stacking fault energy among all nitride-based wurtize materials [11]. Screw dislocations are normally generated from the stacking faults. This is consistent with our observation of dominant edge type dislocations density.

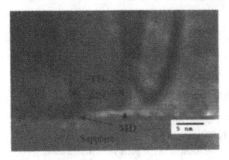

Figure 4. High resolution TEM image taken at the interface of sapphire an AlN.

Initial stage of the growth is very influencing to reduce the dislocations. As the growth temperature of AlN is higher than GaN, the growth mechanism of AlN is different from GaN. A high temperature buffer at (850 – 950°C) was used to grow AlN on sapphire substrate. High temperature buffer reduces the stacking fault dislocation at the interface leading to a very small FWHM of rocking curve measured at symmetric plane (0002) indicating reduced tilt (out of plane rotation). Other groups also reported small FWHM of rocking curve measured at symmetric plane [12,13]. However, the FWHM of rocking curves measured at asymmetric plane (10$\bar{1}$2) in AlN grown on sapphire is on the order of 10^3 arcsec. That means AlN grown on sapphire substrate suffers from a large twist (in-plane rotation) resulted from the large density of edge dislocations. Thus, reducing edge type threading dislocation is an issue to improve the material quality of AlN grown on sapphire substrate. For the reduction of the edge dislocations, an intermediate layer is inserted that was grown relatively at lower temperature. This layer allows bending the propagation of dislocations during the growth of high temperature layer. The dislocations with large deviation angles annihilate faster. Thus, the intermediate layer acts as a dislocation filter layer. As the thickness increases the dislocations further annihilate forming a large loops. Annihilation of the edge type dislocation could be by forming a loop with similar neighboring dislocations having opposite Burger vectors [14].

CONCLUSION

In conclusion, we have investigated AlN epilayers grown on sapphire substrates by TEM. From the cross section view images, the threading dislocations are significantly diminished within 300 nm thickness. Screw dislocations are reduced by using high temperature buffer. Edge type dislocations are reduced by inserting an intermediate layer between the buffer and high temperature AlN layers. From plan view TEM images, the average threading density is ~ 2.7 x 10^9 cm^{-2}, which agrees well with the dislocation density estimated by XRD rocking curve analysis. The threading dislocations further decreases with increasing the thickness. Thus, a thick AlN/sapphire template could be used to grow DUV optoelectronic device structures for improved performance.

ACKNOWLEDGEMENTS

We would like to acknowledge PSC-CUNY for supporting this project on TEM analysis. We would like to thank Profs. H. X. Jiang and J.Y. Lin, currently in the department of Electrical and Computer Engineering at Texas Tech University, Lubbock, Texas for the samples.

REFERENCES

[1] Y. Taniyasu, M. Kasu, and T. Makimoto, Nature (London) **441**, 325 (2006).
[2] J. Li, Z. Y. Fan, R. Dahal, M. L. Nakarmi, J. Y. Lin, and H. X. Jiang, Appl. Phys. Lett. **89**, 213510 (2006).
[3] O. Madelung, *Semiconductor- Basic Data*, Springer, New York, (1996) p.69.
[4] M. E. Levinshtein, S.L. Ramyantev, and M. S. Shur, *Properties of Advanced Semiconductor Materials*, Wiley, New York, (2001), p31.
[5] A. Bergh, G. Craford, A. Duggal, and R. Haitz, *Physics Today*, December 2001, pp42.
[6] J. R. Lakowicz, *Pronciples of Fluorescence Spectroscopy*, 2nd edition, (Kluwer Academic Publishers, New York, 1999).
[7] M. Razeghi, and A. Rogalski, J. Appl. Phys. **79**, 7433 (1996).
[8] D. M. Follstaedt, N. A. Missert, D. D. Koleske, C. C. Mitchell, and K. C. Cross. Appl. Phys. Lett. **83**, 4797 (2003).
[9] B. N. Pantha, R. Dahal. M. L. Nakarmi, N. Nepal, J. Li, J. Y. Lin, and H. X. Jiang, Appl. Phys. Lett. **90**, 241101 (2007).
[10] S. R. Lee, A. M. West, A. A. Allerman, K. E. Waldrip, D. M. Follstaedt, P. P. Provencio, D. D. Koleske, and C. R. Abernathy, Appl. Phys. Lett. **86**, 241904 (2005).
[11] K. Dovidenko, S. Oktyabrsy, J. Narayan, J. Appl. Phys. **82**, 4296 (1997).
[12] M. Asif Khan, M. Shatalov, H. P. Maruska, H. M. Wang, and E. Kuokstis, Jpn. J. Appl. Phys. **44**, 719 (2005).
[13] Y. A. Xi, K. X. Chen, F. Mont, J. K. Kim, C. Wetzel, E. F. Schubert, W. Liu, X. Li, And J. A. Smart, Appl. Phys. Lett. **89**, 103106 (2006).
[14] H. Klapper, Mater. Chem. Phys. 66, 101 (2000).

Mater. Res. Soc. Symp. Proc. Vol. 1202 © 2010 Materials Research Society   1202-I05-05

Properties of Digital Aluminum Gallium Nitride Alloys Grown via Metal Organic Vapor Phase Epitaxy

L. E. Rodak[1] and D. Korakakis[1,2]

[1]Lane Department of Computer Science and Electrical Engineering, West Virginia University, PO Box 6109, Morgantown, WV 26506, USA

[2]National Energy Technology Laboratory, National Energy Technology Laboratory, 3610 Collins Ferry Road, Morgantown, WV 26507, USA

ABSTRACT

Deep Ultra Violet (UV) emitters are of particular interest for applications including, but not limited to, biological detection and sterilization. Within the III-Nitride material system, Aluminum Gallium Nitride ($Al_xGa_{1-x}N$) alloys are the most promising for UV device fabrication due to the wide, direct band gap. The growth of high quality $Al_xGa_{1-x}N$ alloys via Metal Organic Vapor Phase Epitaxy (MOVPE) is challenging due to large sticking coefficient of the Al species compared to that of Ga and also the high reactivity of Al precursors. As a result, films are often characterized by large dislocation densities, cracks, and poor conductivity. Digital alloy growth, or Short Period Superlattices (SPS), consisting of layers of binary or ternary alloys with a period thickness of a few monolayers has been shown to be a viable means of growing high quality ternary alloys via Metal Organic Vapor Phase Epitaxy (MOVPE). In certain materials, such as AlGaInP, the electronic properties of digitally grown alloys differ considerably from the equivalent random alloy. Specifically, the bandgap has been shown to differ significantly from the equivalent random alloy. As a result, digital alloy growth presents the potential to further engineer material properties. However, the influence of digital growth on the electronic properties of III-Nitride alloys has not been extensively characterized. This study focuses on Aluminum Gallium Nitride ($Al_xGa_{1-x}N$) alloys grown using a digital technique via MOVPE. The influence of the growth technique over a wide range of compositions is reported along with the electronic properties.

INTRODUCTION

Aluminum Gallium Nitride ($Al_xGa_{1-x}N$) alloys are of particular interest for Ultra Violet (UV) applications due to the wide direct bandgap characteristic of III-Nitride alloys. The bandgap can range from 3.4 eV to 6.2 eV as the Al concentration is varied from $x = 0$ to $x = 1$ and as a result the emission spans through the near UV and into the deep UV regime. When grown via Metal Organic Vapor Phase Epitaxy (MOVPE), Al containing alloys are inherently difficult to grow due to the short diffusion length of the Al species and also the high reactivity of the Al containing precursors[1]. This requires high growth temperatures and low reactor pressures when compared GaN growth conditions. As such there has been considerable interest in the development of alternative growth techniques which yield high quality $Al_xGa_{1-x}N$ films. Modulated Precursor Expitaxial Growth (MPEG) for example has been one technique reported to yield high quality films in which the metal-organic and ammonia precursor are introduced into the growth chamber at separate times[1]. This work investigates the uses of a digital alloy technique as a viable growth method for $Al_xGa_{1-x}N$ alloys and also its influence on the electrical properties of the films. Digital alloy growth, often called short period superlattices, consists of layers of binary or ternary alloys with a period thickness of a few monolayers[2]. It has been demonstrated as a suitable means of growing ternary and quaternary alloys via Molecular Beam Expitaxy (MBE) in nitride[3], phosphide[4], antimonide[5], and arsenide[6] growth. In the antimonide and phosphide systems, material properties have been shown to depend heavily on the modulation period. At very small periods the bandgap of digital alloys is similar to that of random alloys. InAsGaSb using $(InAs)_n(GaSb)_{3n}$ digital alloys resulted in a bandgap within 15 meV of the analog equivalent[7] but showed decreasing bandgap with increasing period. The bandgap of $(Al_{0.5}Ga_{0.5})_{0.51}In_{0.49}P$ was shown also shown to decrease with increasing period[2]. III-Nitride growth via MOVPE using this technique has been far less studied. Early reports on AlN/GaN digital alloy growth via MOVPE showed it was an effective means of growing $Al_xGa_{1-x}N$ alloys and modulating the bandgap using the period thickness[8]. More recent studies have employed the MPEG technique and observed non uniform composition for alloys with large Ga concentrations[1]. In this work, we are targeting AlN/GaN digital superlattices grown on sapphire substrates. The relationship between the period thickness and the bandgap is reported and further characterization using X-Ray Diffraction measurements reveals this technique is suitable for the growth of coherent $Al_xGa_{1-x}N$ on GaN.

EXPERIMENT

In this work, the digital alloys consisted of a 300 period AlN/GaN superlattice grown on c-plane sapphire substrates as shown in figure 1. All materials were grown in an AIXTRON 200/4 RF-S MOVPE system using trimethylgallium, trimethylaluminum, and ammonia as the precursors. The structures consisted of ~40 nm AlN nucleation layer followed by a ~50 nm GaN layer and finally the 300 period superlattice. In these structures, the GaN layer thickness was fixed at approximately 4 ML and the thickness of the AlN layer was varied from approximately 1 ML to 7 ML. The superlattice growth was conducted at 1000 °C and a reactor pressure of 60 mbar. The ammonia flow rate was 1.2 slpm.

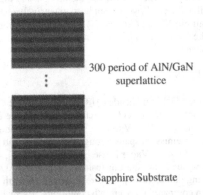

300 period of AlN/GaN superlattice

Sapphire Substrate

Figure 1. Schematic of the 300 period AlN/GaN digital alloy

X-Ray Diffraction (XRD) and transmission measurements were used to characterize all the films under investigation. XRD measurements were preformed with a Bruker D8-Discover four circle diffractometer using Cu-Kα radiation and a 4 bounce Ge (022) monochromator. The periodicity of the superlattice was extracted from the XRD measurements while the bandgap was extracted from the transmission measurements.

DISCUSSION

Figure 2a shows a typical 2-theta/omega scan of the (0002) reflection of a representative 300 period sample. The first order satellite peaks resulting from the periodic structure are visible and indicate abrupt interfaces within the superlattice. From this data, the period of the superlattice was measured and agreed well with the targeted values. As additionally shown in figure 2b, the period of the structure was linearly dependent on the AlN growth time as expected. The y-intercept of the points in figure 2b confirms the growth of ~4 ML of GaN.

Figure 3 shows the bandgap obtained from transmission measurements as a function of the period thickness. There is an observable saturation of the bandgap as the AlN thickness in the superlattice structure increases. Using the AlN thickness as obtained from the XRD measurements, the *expected Al mole fraction*, that is the AlN thickness/period thickness, was calculated. The bandgap as a function of the expected Al mole fraction for the digital alloy is also shown in figure 3b. The dotted line corresponds to the bandgap of a random $Al_xGa_{1-x}N$ alloys as a function of the Al mole fraction calculated from $Eg_{AlxGa1-xN} = Eg_{AlN} + (1-x)Eg_{GaN} + bx(1-x)$ with a bowing parameter of 1 [9]. For low Al concentrations, the measured bandgap is higher than the expected value however it saturates for large Al concentrations.

Additional characterization was performed using off-axis reciprocal space scans. The intensity contour plot around the $(10\bar{1}4)$ diffraction peak is shown in figure 4. The primary superlattice peak and the underlying 50 nm GaN layer are observable. Clearly, the $Al_xGa_{1-x}N$ layer is coherent with the underlying GaN. All samples grown in this work showed the same coherent nature however it was observed in samples with the largest period thickness (26 Å), the $Al_xGa_{1-x}N$ tends to shift slightly toward a smaller c-lattice parameter.

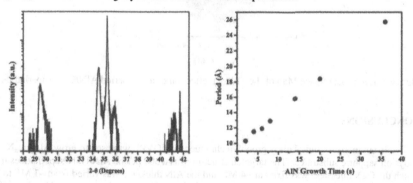

Figure 2. (a) 2-theta/omega scan of the (0002) reflection and (b) the period of the superlattice as a function of AlN growth time

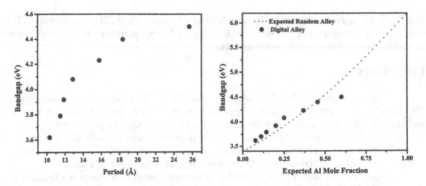

Figure 3. (a) Bandgap of the digital alloy as a function of the superlattice period and (b) bandgap of the digital alloy as a function of the *expected Al mole fraction* (AlN thickness/period thickness)

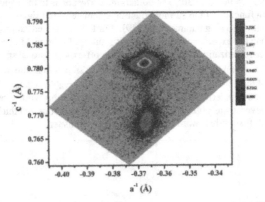

Figure 4. Reciprocal Space Map of the $(10\bar{1}4)$ reflection of a 300 period AlN/GaN superlattice

CONCLUSIONS

In summary, a digital alloy growth technique via MOVPE was used to grow $Al_xGa_{1-x}N$ alloys on sapphire substrates. The targeted structure was a 300 period AlN/GaN superlattices in which the GaN thickness was fixed at ~4 ML and the AlN thickness was varied from ~1 ML to 7 ML. XRD and transmission measurements were used to characterize the period and bandgap of the films. For structures with thin AlN layers, the bandgap was observed to be larger than expected but as the AlN thickness approached 7 ML, the bangap began to saturate. The saturation can be speculated to be due to slight relaxation of the $Al_xGa_{1-x}N$ for large periods as observed in the RSM. Calculations using the QWProject software package were performed in order to determine the transitions occurring in a multiple quantum well with these dimensions. Since the bandgap measured did not correspond to the calculated transitions, especially for small periods, it was concluded that these superlattice structures are acting as alloys not quantum wells.

ACKNOWLEDGMENTS

This work was supported in part by AIXTRON and a grant from the West Virginia Graduate Student Fellowships in Science, Technology, Engineering and Math (STEM) program to LER.

REFERENCES

[1] M. E. Hawkridge, Z. Liliental-Weber, H. Jin Kim, S. Choi, D. Yoo, J. Ryou, and R. Dupuis. Appl. Phys. Lett. **94**, 071905 (2009).

[2] J. Son, D. Heo, W. Choi, I. Han, J. Lee, J. Kim, K. Chang, and Y. Lee. IEEE International Semiconductor Device Research Symposium, Dec. 7-9, 430 (2005).

[3] P. K. Kandaswamy, C. Bougerol, D. Jalabert, P. Ruterana, and E. Monroy. J. Appl. Phys. **106**, 013526 (2009).

[4] O. Kwon, Y. Lin, J. Doeckl, and S. Ringel. J. Electronic Materials, **34**, 1301 (2005).

[5] C. Mourd, D. Gianardi, K. Malloy, and R. Kaspi. J. Appl. Phys. **88**, 5543 (2000).

[6] P. Newman, J. Pamulapati, H. Shen, M. Taysing-Lara, J. Liu, W. Chang, G. Simonis, B. Koley, M. Dagenais, S. Feld, and J. Loehr. J. Vac. Sci. Technol. B., **18**, 1619 (2000)

[7] R. Kasapi, and G. Donati. "J. Cryst. Growth **251**, 515 (2003).

[8] M. Asif Khan, J. N. Kuznia, D. T. Olson, T. George, and W. T. Pike, Appl. Phys. Lett. **63**, 3470 (1993).

[9] O. Ambacher, J. Phys. D: Appl. Phys. **31** 2653 (1998).

Mater. Res. Soc. Symp. Proc. Vol. 1202 © 2010 Materials Research Society   1202-I05-06

Studies of Ni and Co Doped Amorphous AlN for Magneto-Optical Applications

W. M. Jadwisienczak[1], H. Tanaka[1], M. E. Kordesch[2], A. Khan[3,4], S. Kaya[1] and R. Vuppuluri[1]

[1] School of EECS, Ohio University, Athens, Ohio, U.S.A.

[2] Department of Physics and Astronomy, Ohio University, Athens, OH, U.S.A.

[3] Department of Chemistry and Biochemistry, Ohio University, Athens, OH, U.S.A.

[4] Department of Physics, University of Peshawar, Peshawar, Pakistan.

ABSTRACT

Magneto-optical properties of Ni- and Co-doped amorphous AlN thin films were investigated as a function of post grown annealing temperature using magneto-optical Kerr effect (MOKE) spectroscopy. The x-ray diffraction spectra confirmed that the as-grown material is amorphous and retained its morphology after thermal treatment; however the sample morphology strongly depends on the concentration of incorporated transition metals. We observed with help of transmission electron microscopy and atomic force microscopy that the films surface containing TMs with concentrations larger than ~10 at.% undergo morphological changes suggesting possible Ni and Co atom clustering. Significant enhancement of the polar Kerr rotation signal was observed for Ni- and Co-doped a-AlN materials annealed above 300 °C in nitrogen. The studied materials have shown strong magnetic isotropy in polar geometry whereas the MOKE measurements in longitudinal geometry did not show an explicit signal for the transition metals doped a-AlN studied.

INTRODUCTION

Transition metals (TM) doped group-III nitride semiconductors (III-Ns) are well known for their potential applications in spintronics [1]. The room-temperature magneto-optic Kerr effect in the magnetic semiconductor materials combines the magnetic and optical properties of the materials together, making them attractive for the development of magnetic recording media, as well as for multifunctional photo-magneto-electronics applications including sensors where the strong light-matter coupling is essential [2]. There are a number of reports about ferromagnetism in TM-doped III-Ns [3-6], respectively, although more work focused on the structural and magnetic properties of these materials is necessary before practical applications will become feasible. In spite of the difficulty and controversies surrounding the understanding of the origin of the magnetization in magnetic semiconductors, the formation of magnetic III-Ns has sparked widespread interest due to the possibility of making highly efficient spin injectors for spintronics. It is known that diluted magnetic semiconductors for practical applications require two essential features. One, Curie temperature must remain higher than room temperature; second, the basis should be on a typical semiconductor in which the carrier control technique is well established [3,4]. Among the most promising candidates for this task are some diluted ferromagnetic nitrides and oxides [3]. Nickel and Cobalt were already studied in polycrystalline and crystalline AlN materials and their magnetic properties were assessed [6-11]. In this paper we report on preliminary MOKE spectroscopy results of Ni-and Co-doped amorphous AlN (a-AlN) not studied before.

EXPERIMENTAL DETAILS

Typical a-AlN films grown for this project were produced by *rf*-sputtering on Si (0001) substrate at low temperature using 120 W *rf* power in a pure nitrogen atmosphere of 8 mTorr.

The films were grown using a 50 mm diameter aluminum target with a few 6 mm diameter pieces of Ni or Co mounted on the target surface for *in situ* doping with varied concentrations. The nominal thickness of the deposited films was measured with a quarts crystal thickness monitor to be 50 nm (for TEM analysis) and 120 nm. Selected samples were thermally annealed in a resistive tube furnace for changing time durations in temperature up to 1000 °C in nitrogen ambient. The samples were characterized with the x-ray diffraction (XRD) (*Rigaku Geigerflex*, 2000 Watts) with CuKα (0.154 nm), energy dispersive x-ray (EDX) attached to the *JEOL JSM-6390* scanning electron microscope, transmission electron microscopy (TEM) (*JEOL 1010*) and atomic force microscope (*Agilent 5500LS*), respectively. The nominal concentrations of Ni and Co in a-AlN sputtered using targets with different Al/TMs ratios determined by EDX ranged from a small (less than 10 at.%), to medium (~20 at.%) and large (more than 30 at.%), respectively. The specimens for TEM study were grown on a glass slide covered with a thin film of soap. The film was dissolved from the substrate using a drop of distilled water then floated onto a copper TEM grid. Atomic force microscope (AFM) was operated in a contact mode with the frequency ranging between 320 kHz to 330 kHz and amplitude set-point between 2.5V to 3.5V, respectively. Magnetic characterization was carrier out using home build MOKE spectrograph working in polar and longitudinal configurations at room temperature [12]. Specimens were mounted on a sample holder made of PVC material and placed between an electromagnet poles with a variable gap. The HeNe laser (632.8 nm) having linearly polarized beam (both *p*- and *s*-polarization were exercised) and stabilized intensity (15 mW) was used as a probe. The maximum incident angle was less than 4° (polar) and ~65° (longitudinal), respectively. The laser probe polarization was achieved by passing laser probe through a set of crossed out Glan-Thompson polarizers (*Oriel 25706*). The light intensity detected by a high-speed Silicon detector (*Thorlabs Inc., DET200*) was proportional to the laser probe polarization change induced by a sample subjected to magnetic field. The signal was analyzed using lock-in technique operating between 2 kHz to 3 kHz. The data collection process was computerized and multiple scans were performed to improve Kerr rotation signal to noise ratio. Collected MOKE hysteresis loops were corrected for the response of the MOKE spectrograph [12].

RESULTS AND DISCUSSION

Figure 1 shows representative XRD spectra of Ni-doped a-AlN films. All spectra show no peaks related to AlN or Ni suggesting that the as-grown film is amorphous. This morphology is retained in all studied samples annealed for short time duration (a few minutes) disregard of TMs concentrations with annealing temperature up to ~600 °C; however films annealed for an extended time duration, even at lower annealing temperatures, shown evidences of the material morphology change (polycrystalline phases). We should emphasize here that XRD may not resolve very small TMs agglomeration due to the very thin thickness (120 nm) of the studied films. We observed the similar XRD results for Co-doped a-AlN (not shown here). In general all as-grown TMs-doped a-AlN samples reveal very weak material magnetization as determined by the MOKE characterization in this study.

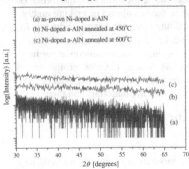

Figure 1. (left) X-ray diffraction patterns (θ-2θ scan mode) of a large Ni concentration doped (a) as-grown a-AlN, (b) annealed and 450 °C for 15 min and (c) annealed at 600 °C for 15 min, respectively.

216

To enhance the magnetic properties of as-grown material selected samples were subjected to thermal annealing treatment resulting in significant enhancement of observed Kerr rotation signal which is proportional to the overall magnetization of material subjected to magnetic field. We investigated how the magnetic moment induced by the Ni and Co in a-AlN host increases with the TM concentration and depends on the post growth processing. Figure 1 shows results of magneto-optical characterization of Ni- and Co-doped a-AlN films using MOKE spectrograph operating in polar and longitudinal configurations at room temperature.

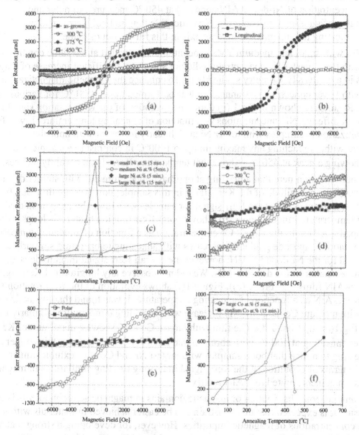

Figure 2. (a) Evolution of polar Kerr rotation for a large Ni at.% in a-AlN as a function of annealing temperature. (b) Comparison between polar and longitudinal Kerr rotation signals for a large Ni at.% doped a-AlN annealed at 450 °C. (c) Change of Kerr rotation signal as a function of annealing temperature observed for Ni-doped a-AlN films. (d) Evolution of polar Kerr rotation signal for a large Co at.% in a-AlN as a function of annealing temperature. (e) Comparison between polar and longitudinal Kerr rotation signals for a large Co at.% doped a-AlN annealed at 400 °C. (f) Change of Kerr rotation as a function of annealing temperature observed for Co-doped a-AlN films.

Figure 1(a) shows polar Kerr rotation hysteresis loops recorded for Ni-doped a-AlN with a large Ni concentration as a function of annealing temperature. The change in the shape of hysteresis loops is evident, and a gradual variation in the saturation Kerr signal is observed as

217

a function of annealing temperature. The Kerr signal amplitude is three orders of magnitude stronger for sample annealed at 450 °C than for as-grown material. Also, all samples almost approach saturation at 8 kOe magnetic field; however different saturation Kerr rotation is expected for each individual film. The width of hysteresis loops changes with annealing indicating increase of the magnetic coersivity.

The polar Kerr rotation is in principle proportional to the material magnetization; moreover it strongly depends on the morphology. Figure 1(b) shows Kerr rotation signals for a largest concentration of Ni-doped AlN annealed at 450 °C measured in polar and longitudinal configurations, respectively. No noticeable longitudinal Kerr signal was observed indicating that magnetization ease axis in this film is out of plane.

Samples with small and medium TMs concentrations annealed at the same conditions as samples with the largest TMs concentration did not show any significant increase of magnetic properties in any MOKE geometry studied here. To enhance the magnetic magnetization in samples with lower concentrations annealing was performed up to 1000 °C for various time durations. Results are shown in Fig.1 (c) depicting evolution of the Kerr rotation signal in samples with different Ni concentrations as a function of annealing temperature. It was found that magnetization in samples with lower Ni at.% linearly increases with annealing temperature without reaching any maximum up to 1000°C in contrast to a sample with largest Ni at.% showing a peak at 450 °C. Above that temperature Kerr signal abruptly reduces.

We should mention here that annealing at optimal temperature but with shorter duration time decreases the magnitude of the Kerr signal in this sample; however the time duration factor does not strongly affect the Kerr signal for samples with lower Ni at.% in studied annealing temperature range. Figure 1(d) shows polar Kerr rotation hysteresis loops recorded for Co-doped a-AlN with the largest Co concentration as a function of annealing temperature. The Kerr rotation amplitude gradually increases with annealing temperature similarly as it was observed for the Ni-doped a-AlN; however the overall Kerr signal magnitude is an order of magnitude smaller in Co-doped a-AlN. We did not observe any longitudinal Kerr signal for Co-doped a-AlN films (see Fig. 1 (e)). Figure 1(f) shows evolution of the Kerr rotation signal in Co-doped a-AlN as a function of annealing temperature. It was found that magnetization in samples with medium Co at.% linearly increase with annealing temperature (similarly to Ni case, see Fig.1(c)) up to 600 °C. The film with largest Co at.% shows a maximum of Kerr signal after annealing at 400 °C with abrupt signal intensity decrease above this temperature. It is interesting to notice that both, samples with largest Ni and Co at.% exhibit similar behavior in terms of evolution of the Kerr signal indicating existing similarities in magnetic properties when incorporated to a-AlN matrix.

The homogeneous distribution of magnetic domains in magnetic semiconductors is a critical issue. It is expected that this constrain can be easier achieved in materials with low or moderate concentrations of magnetic impurities. However, for developing a strong material magnetization it is necessary to incorporate magnetic impurities at as high as possible concentration. We have studied the surface of selected samples with TEM and AFM to elucidate if the surface of studied films changed upon annealing. Furthermore, any significant changes on the surface of a very thin film may indicate that similar transition took place in the volume of material. Figure 3 shows TEM images of a-AlN sample with largest Ni at.%. It is seen that as-grown film is smooth with some unidentified at the present topographical features. The film was annealed in Ar at 400 °C for 60 min. and a portion of the film was reexamined. It is seen that the annealed film surface underwent some morphological change as indicated by observed dark spots. TEM diffraction pattern of this film showed sharp rings, indicating polycrystalline inclusions; however, the diffraction pattern has not been analyzed yet (work in progress). One can speculate here that extensive annealing of Ni-doped AlN resulted in Ni atoms agglomeration which caused the decrease of material magnetization as indicated by measured Kerr rotation signal (see Fig.2). We observed that samples annealed at comparable temperature but for much shorter time did not show such obvious changes in material's (here surface) morphology. Figure 4 shows representative AFM images of a-AlN

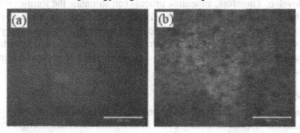

Figure 3 TEM black and white images of a-AlN film doped with largest Ni concentration: (left) as-grown film and (rigth) film annealed for 1 hr at 400 °C in Ar. Please notice that the films appear to be dark, probably due to a magnetic effect. The scale bar is 200 nm.

film doped with medium Co at.%. The roughness parameter of as-grown and annealed at 600 °C film are $Rms = 1.12$ nm and $Rms = 1.65$ nm, respectively. The change of the Rms parameter indicates that the size of grains containing Co atoms increases with annealing temperature, perhaps promoting easier Co clustering. It is expected that annealing above 600 °C will further increase the Rms parameter reaching a critical value above which material magnetization drastically decrease. It should be mentioned here that annealing time for this sample was only a few minutes. Because of that the annealing process stimulating excess Co atoms clustering agglomeration or film crystallization did not accumulate until the higher temperature and/or longer time were reached. The same conclusions apply to Ni-doped a-AlN

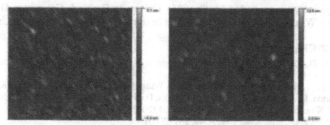

Figure 4 Selected set of two dimensional 1μm×1μm AFM images corresponding to medium Co at.% doped a-AlN film (left) as deposited, and (right) annealed at 600°C for a few minutes, respectively. The scale bar is 200 nm.

We can speculate that annealing at elevated temperature, responsible for formation of TMs clusters, modifies the ferromagnetic state of the system. This can be due to the change of the TM atom distribution in the matrix resulting in magnetic disconnection of the TM clusters or change of the "magnetic" size of the TM cluster [11]. Furthermore, it is documented in the literature that thermally annealed amorphous semiconductors experience some structural reordering (not necessarily crystallization) [13, 14]. When this takes place the dangling or broken bonds and/or the atomic rearrangements of the amorphous material may have great influence on the electronic states of amorphous host and at the same time effect the magnetic moment of embedded impurities [14]. It was shown recently that the presence of Al may quench the magnetism of Ni atom through the existence of a nonmagnetic zone at the Ni cluster surface [8]. Also, the formation of a nonmagnetic cobalt-nitride phase was suggested to reduce the magnetic response of Co-doped AlN [11]. In both cases the hybridization of the s-p electrons of Al and N with the d electrons of Ni or Co may play a critical role. Furthermore, in films annealed at high temperatures the AlNi and AlCo inclusions may develop. Therefore, annealing process can stimulate the formation of nonmagnetic inclusions or dead magnetic areas (layers or interfaces, $e.g.$ surface) resulting from the interaction of the AlN matrix and magnetic TM atoms. In the future study the interaction between a-AlN host atoms and TM atoms and their clusters will be critically addressed since an interaction between matrix and magnetic impurities depends strongly on the nature of a magnetic atom local environment.

CONCLUSIONS

We have investigated magneto-optics of Ni-and Co-doped a-AlN grown by rf sputtering. It was observed that the thermal annealing of the investigated materials strongly influence the material magnetization as observed in recorded MOKE hysteresis loops. The studied films have shown strong magnetic isotropy in polar geometry whereas the MOKE measurement in longitudinal geometry did not show a measurable signal. This indicates that the magnetization in the active Ni- and Co-doped a-AlN layers is out of plane. Furthermore, the Kerr rotation signal depends on the Ni and Co concentrations and post growth annealing conditions. It was found that polar Kerr rotation monotonically increases with annealing temperature for smaller Ni and Co at.%. For films with large Ni and Co at.% the optimal annealing temperature was in the 400 °C to 450 °C temperature range. We speculate that strong reduction of the Kerr rotation signal in samples with high TMs concentrations is probably due to the TMs cluster formation in a-AlN matrix as evidenced by observed surface morphology change for annealed films.

ACKNOWLEDGMENT

The project was supported by the Ohio University Research Committee OURC 08-27 grant. Also, WJ and HT acknowledge support from the CMSS Program at Ohio University.

REFERENCES

1. S. Wolf, D. Awschalom, R. Buhrman, J. Daughton, S. Von Molnar, M. Roukes, A. Chatchelkanova, and D. Treger, *Science*, **294**, 1488 (2002).
2. H. Song, L. Mei, Y. Ahang, S. Yan, X. Ma, Y. Wang, Z. Zhang, *et al.*, *Physica B*, .**388**, 130 (2007).
3. S. Pearton, C. Abernathy, M. Overberg, G. Thaler, D. Norton, N. Theororopouou, A. Hebard, Y. Park, F. Ren, J. Kim, and L. Boatner, *J. Appl. Phys.*, **93**, 1 (2003).
4. S. Pearton, *et al.*, *Mat. Scien. Eng. R*, **40**, 137 (2003).
5. R. Frazier, J. Stapleton, G. Thaler, C. Abernathy, S. Pearton, R. Rairigh, J. Kelly, A. Hebard, M. Nakarmi, K. Nam, J. Lin, H. Jiang, J. Zavada, and R. Wilson, *J. Appl Phys.*, **94**, 1592 (2003).
6. V. Ivanov, T. Aminov, V. Novotortsev, and V. Kalinnikov, *Russ. Chem. Bull., Int.* **53**, 2357 (2004).
7. D. Zanghi, A. Delobbe, A. Traverse, and G. Krill, *J. Phys.: Condens. Matter.*, **10**, 9721 (1998).
8. D. Zanghi, C. M. Teodorescu, F. Petroff, H. Fischer, C. Bellouard, C. Clerc, C. Pelissier, and A. Traverse, *J. Appl. Phys.*, **90**, 6367 (2001).
9. H. Li, H. Q. Bao, B. Song, W. J. Wang, and X. L. Chen, *Solid State Commun.*, **148**, 406 (2008).
10. T. Sato, Y. Endo, Yu. Shiratsuchi, and M. Yamamoto, *J. Magn. Magn. Mater.*, **310**, e735 (2007).
11. Y. Huttel, H. Gomez, C. Clavero, A. Cebollada, G. Armelles, E. Navarro, M. Ciria, L. Benito, J. I. Arnaudas, and A. J. Kellock, *J. Appl. Phys.*, **96**, 1666 (2004).
12. H. Tanaka, M.S thesis, Dept. Elect. Eng, Ohio University, Athens, OH, 2008.
13. A. Zanatta, A. Khan and M. Kordesch, *J. Phys-Condens. Mat.***19**, 436230 (2007).
14. *Charge transport in disordered solids with applications in electronics*, ed. S. Baranovski Chichester, England; Hoboken, NJ, Wiley (2006).

Mater. Res. Soc. Symp. Proc. Vol. 1202 © 2010 Materials Research Society   1202-I05-08

Thermodynamic Analysis and Purification for Source Materials in Sublimation Crystal Growth of Aluminum Nitride

Li. Du[1] and J.H. Edgar[1]
[1]Department of Chemical Engineering, Kansas State University
Manhattan, KS 66506-5102, U.S.A.

ABSTRACT

Source material purification according to a thermodynamic analysis is reported for the sublimation crystal growth of aluminum nitride in an inert reactor. OAlOH is strongly favored over all other possible oxygen containing compounds in both the Al-O-H-N and Al-O-H-C-N systems, while Al_2O the most favorable oxygen containing gas species for Al-O-N system, become secondary favorable gas species. A low temperature (<1200 °C) treatment is effective in eliminating oxygen and hydrogen from the source powder. Carbon monoxide is another important oxygen containing gas species in the Al-O-H-C-N system, and is favored over Al_2O at certain temperature and pressure. Carbothermal reduction with intentionally added carbon (graphite) can further reduce the oxygen concentration. Experiments show that high-temperature sintering minimizes the oxygen concentration and reduces the specific surface area of the source. With 5.5% of mass loss, source purification reduced impurities concentrations to $0.019\pm0.001wt\%O$, $7\pm1ppmH$, and $0.010\pm0.004wt\%C$.

INTRODUCTION

Single crystal aluminum nitride is an important substrate for UV-LEDs and laser diodes (LDs): its high thermal conductivity helps to dissipate heat generated by devices and its small lattice mismatch makes it well-suited for $Al_xGa_{1-x}N$ epitaxy. However, these properties are affected by the impurities in the AlN, specifically oxygen incorporation causes Al vacancies that degrade its thermal conductivity and can reduce its lattice constants [1].

In moist or dry air aluminum compounds form hydroxides and/or oxides [2]. Li et al. reported that amorphous AlOOH is produced in moist air at room temperature and can be further hydrolyzed to form mixtures of $Al(OH)_3$[3]. The surface oxide or hydroxide may be a mixture of $Al_2O_3/Al(OH)_3/AlOOH$[4]. Our previous study summarized the analysis by several groups on AlN hydroxides [5], and experimentally demonstrated that a heat treatment under 900°C in either vacuum or under nitrogen can decompose the aluminum hydroxides, removing oxygen via the formation of water. Carbon is another common impurity in the AlN. It can be beneficial: Fukuyama's group [6-7] showed that carbon can reduce oxide to nitride.

This study applies thermodynamic analysis to AlN crystal sublimation and recondensation growth process to predict the most stable species present, and experimentally tests source purification by annealing and sintering experiments. High temperature (>1900°C) source sintering with or without low temperature (<1200°C) annealing process are discussed and compared. The benefit of adding carbon to reduce the oxygen concentration was investigated theoretically and experimentally.

THEORY CALCULATION

The thermodynamics calculations were based on a non-reacting reactor (tungsten furnace). In the Al-O-H-N system, oxygen and hydrogen were assumed to be the main impurities, originating from the AlN source powder (0.9wt% oxygen, 0.024wt% hydrogen). In the Al-O-H-C-N system, carbon could be considered a residual impurity in the as-received powder (0.06wt%) or an impurity intentionally added to facilitate oxygen removal. Pure N_2 gas was the ambient gas. The total pressure was calculated from the ideal gas law knowing the initial temperature and pressure. Two solid-vapor equilibriums were assumed, one at the source sublimation zone, the other at the crystal growth zone. The crystal growth zone was assumed to be 5 to 50 °C lower in temperature than the source sublimation zone.

Table I: all the possible species for each system

Al-O-C-N system						**Al-O-H-N system**			
				Al-O-N system						
N-C-O	Al-C	N-C	N-O	Al-N	Al-O	Al-H-O	H-N	H-N-O		
Liquid				Al		H2O				Liquid
Solid	C,			Al,	Al_2O_3	AlOOH*				Solid
	Al_4C_3			AlN		$Al(OH)_3$*				
Gas	CO_2,	C,	CNN,	NO,	Al,	O,	H_2O,	H_3N,	HNO	Gas
	CO,	C_2,	NCN,	NO_2,	Al_2,	O_2,	H_2O_2,	H_4N_2,	$HONO_{cis}$	
	C_3O_2,	C_3,	CN,	NO_3,	N,	O_3,	HO,	H_2N_2,	$HONO_{trans}$	
	C_2O,	C_4,	C_2N,	N_2O,	N_2,	Al_2O,	AlH,	H_2N,	HONO2	
	NCO	C_5,	C_2N_2,	N_2O_3,	N_3,	AlO,	OAlH,	HN,		
		AlC	C_4N_2	N_2O_4,	AlN	Al_2O_2,	AlOH,	H_2		
				N_2O_5		AlO_2	OAlOH			
		CH	CH_2	CH_3	CH_4	C_2H	C_2H_2	C_2H_4		
			CHN	CHNO	CHO	C_2H_4O	CHNO			
Al-O-C-H-N system										

* aluminum-oxygen-hydrogen compounds, but no thermal data was found in JANAF table.

All the possible species according to the JANAF table [8] are presented in Table I. Reactions used in thermodynamic calculation can be represented as:

$$aA(s) + bB(g) = cC(g) + Dd(g) \qquad (1)$$

Then the equilibrium constant of the reaction can be expressed by:

$$K = \frac{\alpha_C^c \cdot \alpha_D^d}{\alpha_A^a \cdot \alpha_B^b} = \exp(-\frac{\Delta rG^0}{RT}) = \exp(-\frac{\sum \upsilon_i \Delta_f G_i^0}{RT}) \qquad (2)$$

Where α_i is the activity of the reactants or products i with reference to their standard states (pure substance, activities of unity and 0.1 M Pa) and ν is the stoichiometric coefficients. Free energy of formation $\Delta_f G^0{}_i$ comes from JANAF data [8].

Al-O-N system

Our previous study [9] showed that Al_2O was the most favorable oxygen containing gas species in the Al-O-N system. The reaction $Al_2O_3(s) + 4Al(g) = 3Al_2O$ has a positive LogKs; all the reactions forming other oxygen containing species have negative LogKs. Both Al and O begin to enter the gas phase at 1900°C. Much of the oxygen originally present in the source enters the gas phase: 56.3% at 2200°C and 99.5% at 2300°C. In comparison, less of the Al present in the source enters the gas phase: 6.0% at 2200°C and 17.1% at 2300°C. The stability of Al_2O at high temperature suggests that high temperature (> 1900°C) sintering of AlN source can remove the oxygen from the solid phase.

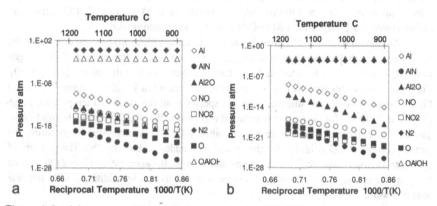

Figure 1: Partial pressure of the selected gas species in Al-O-H-N system; a, total pressure is 500 torr (0.658 atm); b, total pressure is 1 torr (0.001atm).

Al-O-H-N system

Aluminum hydroxides are easily decomposed in the AlN source annealing process. For a total pressure of 500 torr (0.658 atm, Fig1a) and 1 torr (0.001 atm, high vacuum, Fig.1b), thermodynamics calculation predict that between 900 °C to 1200 °C, the major gas-phase species present are N_2 and OAlOH. Although the Al vapor higher pressure is higher under vacuum (about 10^{-9}), it is a minor gas species. Gas species Al_2O, Al_2, AlO, AlN, Al_2O_2, AlO_2, NO, NO_2, N_2O, N_3, O, O_2 with pressure lower than 10^{-10} atm are negligible, while other gas species NO_3, N_2O_3, N_2O_4, N_2O_5, O_3, N, H_2O, H_2O_2, HO, H_2, AlH, OAlH, AlOH, H_3N, H_4H_2, H_2H_2, H_2N, HN, HNO, HONOc, HONOt, $HONO_2$, with pressure lower than 10^{-28} atm are unlikely to be present in the vapor. While Al_2O gas is only stable at high temperature (higher than 1900°C), OAlOH gas is stable at lower temperature regardless of the ambient. In addition, the lower Al/O mole ratio in OAlOH (0.5) than Al_2O (2) indicates less source mass is lost during the removal of oxygen from the solid phase.

Al-O-C-H-N system

For the system with carbon, thermodynamic analysis shows that the major gas-phase species present are still N_2 and OAlOH (with pressure of about 10^{-3}). Consider carbon as a residual impurity in the as-received powder; figure 2a represents the partial pressure for the gas species that has partial pressure higher than 10^{-13} at the calculated temperature range. Between 900 °C to 1200 °C, CO is a minor gas species (with a partial pressure between 10^{-6} -10^{-8}). The addition of carbon does not change the concentrations of those species which are negligibly small in the Al-O-H-N system. Carbon containing species C_2N_2, C_2N, C_4N_2, CNN, NCN, C, C_2, C_3, C_4, C_5, C_3O_2, C_2O, AlC, NCO, CH, CH_2, CH_3, CH_4, C_2H, C_2H_2, C_2H_4, CHN, CHNO, CHO, C_2H_4O and CHNO with pressure lower than 10^{-12} atm are also negligible. Between 1400 °C to 1600 °C (Fig. 2a), CO gas with pressure between 10^{-4} -10^{-6} become the third most major gas species. Between 1900 °C to 2100 °C (Fig.2a), both the Al and Al_2O partial pressures increased faster than that of CO gas as the temperature increased. The Al_2O partial pressure finally exceeds CO above 2050 °C. Since the partial pressure of cyanogen (CN) is lower than 10^{-10} at 1400 to 1600 °C and 10^{-8} at 1900 to 2100 °C, CO is the main carbon containing gas species in Al-O-C-H-N system in an inert reactor.

Aluminum oxides and hydroxide can be written as $(Al_2O_3)_x(H_2O)_y$ to emphasize the water they contain. By adding carbon, oxygen can be removed by transferring all the oxygen from Al_2O_3 to CO, thus the total moles of carbon are the same as that of oxygen in Al_2O_3. Then partial pressures for this new system were represented in Fig.2b. Compare Fig.2a and Fig.2b, N_2 and OAlOH are still the major gas species at all the calculated temperature range (900 °C to 2100 °C) for the system with added carbon (fig.2b) . CO is become the third major gas and has significant pressure increasing since 1400 °C. At the same temperature, the partial pressure of CO is higher for the system with added carbon (fig.2b). The pressure of cyanogen is lower than 10^{-10} in this temperature range. Between 1900 °C to 2100 °C, the sharp increasing in the Al_2O partial pressure was constrained by additional CO, and was always lower than that of CO.

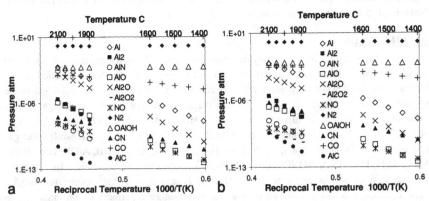

Figure 2: Partial pressure of the selected gas species in Al-O-C-H-N system for as received AlN source (a) and AlN source with additional carbon (b)

EXPERIMENT

A resistively-heated tungsten furnace with a vertical temperature gradient of about 5 °C /cm was used in the experiments. The AlN source powders were placed in a tungsten crucible (25 mm diameter) within a tungsten retort. Three sets of experiments were run to remove oxygen from the AlN source. In the first set of experiments, the AlN powder was sintered at about 1950 °C in 615 torr of nitrogen for 2, 5 and 10 hours. In the second set of experiments, the AlN was first annealed at 960–1000 °C for 2 to 10 hours in a vacuum, and then sintered at about 1920 °C, 615 torr of N_2 for 2 to 10 h; the high temperature sintering time was the same as the low temperature annealing time. In the third set of experiments, carbon was added, in the same amount as the oxygen present in $(Al_2O_3)_x$ minus the carbon moles initially present in the source. A carbothermal reduction at about 1500°C and 1050 torr followed the low temperature annealing (at about 1000 °C in vacuum). Then the source was sintered at 1920 °C, and 615 torr N_2), the time range for each step was 2 hours. The mass of the source was measured before and after the experiments to determine the percentage of source mass lost. The oxygen, carbon and hydrogen concentrations after the heat treatment were measured by inert gas fusion technique (LECO analysis).

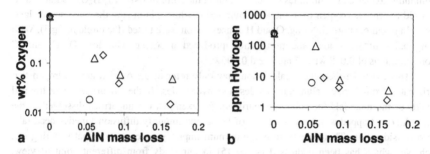

Figure 3: Oxygen (a) and hydrogen (b) concentration in the AlN source as function of the source mass loss. Symbols: triangle - source only sintering at 1950°C; diamond - source annealing at 960–1000 °C followed by sintering at 1920°C; circle - source with added carbon for carbothermal reduction during the annealing process.

DISCUSSION

Due to the temperature gradient in the furnace, the source powder lost mass due to sublimation. We previously proved that reducing the surface area of the AlN source through particle agglomeration and an increase in their average size is effective at reducing the oxygen concentration [5]. The particle size significantly increased after high temperature sintering (>1900 °C), however long time sintering at this temperature results in high source mass loss.

According to thermodynamic analysis, OAlOH, CO, and Al_2O are the major impurity containing species when the AlN source is directly heated to a temperature higher than 1900 °C. Their significantly higher partial pressure than Al and stability in the vapor helps remove impurities from the solid phase. In contrast, OAlOH is the only impurity containing species if first annealing AlN source at about 1000 °C. The oxygen and hydrogen can be removed first through OAlOH volatilization and then by Al_2O forming when the temperature is increased to 1920°C. The low Al/O ratio of OAlOH over Al$_2$O indicates the source purification with low AlN source mass loss.

Figure 3 shows the O (a) and H (b) concentration in the AlN source vesus source mass loss. Comparing the two sets of experiments (triangle stands for the source only sintering at 1950°C; diamond stands for the source annealing at 960–1000 °C followed by sintering at 1920°C), for the same AlN source mass loss, the experiments with the low temperature annealing step have lower O and H concentration in general, indicating more effective source purification.

Thermodynamic analysis predicts that CO is the major oxygen containing species and only carbon containing vapor species between 1400 °C to 1600 °C, thus a carbothermal reduction at 1500°C after the 1000°C annealing can further reduce oxygen with nearly zero source mass loss. Considering that as-received AlN powder has an impurity concentration of 0.9 wt% O, 0.024wt% H, and 0.06wt% C, the atomic ratio of C:H:O is 4:19:45. Since aluminum oxides and hydroxides can be written in the form of (Al$_2$O$_3$)$_x$(H$_2$O)$_y$, additional C is need to remove oxygen from the aluminum oxide while minimizing the mass source loss. After high temperature sintering, O and H concentration were tested (the circle in fig.3). With only 5.5% of mass loss, the purification produced a source with low O, H, and C concentrations of 0.018 wt%, 6 ppm, and 0.006wt%.

However, the theoretical calculation may only refer to the oxygen and hydrogen that originates from surface aluminum oxides and hydroxides. If the impurities were trapped inside of the materials (or volume impurities), for example, the impurities dissolved in the bulk AlN or trapped at internal surfaces of voids, the impurity diffusion kinetics need take into consideration. Therefore there is a minimum impurity concentration that sintering can achieve, which has been addressed before [5] in our study from different point of view. Meanwhile, the stability of the major impurity containing species OAlOH, CO, and Al$_2$O in the gas phase suggest that impurities trapped inside of the materials (or volume impurities) may be removed by recrystallization of aluminum nitride by, for example, completely subliming and recondensing the source material.

CONCLUSIONS

A two step process (a low temperature (<1000°C) annealing plus a high temperature (>1900°C) sintering) was shown theoretically and experimentally to be effective in reducing the oxygen in a AlN source for AlN sublimation crystal growth. High-temperature sintering minimizes the oxygen concentration by reducing the specific surface area of the source and transferring oxygen to the vapor phase as Al$_2$O. Low temperature annealing removes aluminum hydroxide with a low source mass loss compared to high temperature sintering. Intentionally adding carbon can further reduce the oxygen concentration and reduce the source total treatment time. Carbothermal reduction at a low temperature (around 1500°C) can reach the minimum AlN mass loss possible. With only 5.5% of mass loss, the purification produced a source with low O, H, and C concentrations of 0.019±0.001wt% O, 7±1ppm H, and 0.010±0.004wt% C.

ACKNOWLEDGMENTS

Support for this project from the II-VI Foundation is greatly appreciated.

REFERENCES

1. M. Miyanaga, N. Mizuhara, S. Fujiwara, M. Shimazu, H. Nakahata, T. Kawase, J. Crystal Growth **300** (2007) 45.
2. K. Wafers, C. Bell, *Oxides and hydroxides of aluminum*, Alcoa Technical Paper no. 19, Alcoa Research Laboratories.
3. J. Li, M. Nakamyra, T. Shirai, K. Matsumary, C. Ishizaki, and K. Ishizaki, J.Am.Ceram. Soc., **89** [3] 937-943 (2006)
4. M.L. Panchula, J.Y. Ying, J. Am. Ceram. Soc. **86** (2003) 1114.
5. J. Edgar, L. Du, L. L. Nyakiti , J. Chaudhuri, J. Crystal Growth **310** (2008) 4002-4006
6. H.Fukuyama, S.Kusunoki, A.Hakomori, and K.Hiraga, J. Appl. Phys. **100**, 024905 (2006)
7. Wataru Nakaoa, Hiroyuki Fukuyama, J. Crystal Growth **259** (2003) 4001-4006 302-308
8. *NIST-JANAF Thermochemical Tables*, 4th ed., M. W. Chase, Jr. Am. Chem. Soc., DC, 1998, Am. Inst. Phys., NY, 1998, the NIST, Gaithersburg, MD 1998.
9. L. Du and J.H. Edgar, Mater. Res. Soc. Symp. Proc. 955 I11-05 (2007).

Mater. Res. Soc. Symp. Proc. Vol. 1202 © 2010 Materials Research Society   1202-I05-12

Facet control in selective area growth (SAG) of a-plane GaN by MOVPE

Bei Ma, Reina Miyagawa, Hideto Miyake and Kazumasa Hiramatsu
Department of Electrical and Electronic Engineering,
Graduate School of Engineering, Mie University,
1577 Kurimamachiya, Tsu 514-8507, Japan

ABSTRACT

The selective area growth (SAG) of a-plane GaN grown on r-plane sapphire with a stripe pattern oriented along the <1-100> was investigated. The key technology of facet control is optimizing the growth temperature and reactor pressure. Our experiments reveal that the growth temperature determined the facet form: in samples grown at 1000 °C, the structure consists of {11-22} and (000-1) planes; upon increasing the growth temperature to 1050 °C, the area of the {11-22} facet gradually decreases, and facets in two new planes, (0001) and {11-20}, facets form. In samples grown at 1000 °C, the {11-22} facet eventually completely disappears, whereas (0001) and {11-20} facets continue to increase in area to form a rectangular cross section. The reactor pressure determines the ratio of the lateral growth rate to the vertical growth rate, upon decreasing the reactor pressure from 500 Torr to 100 Torr, the rectangular structure gradually decreases in height and increases in width, and the volume of the structure remains nearly constant.

INTRODUCTION

It is well known that [0001] c-axis-oriented GaN exhibits a strong electric field produced by spontaneous and strain-induced piezoelectric field [1], which causes reduced radiative efficiency and a redshift of optical transitions [2]. These problems can be overcome by growing nonpolar GaN, such as {11-20} a-plane GaN, on a r-plane sapphire [3] and on a-plane SiC [4], which has been demonstrated by metalorganic vapor phase epitaxy (MOVPE). Unfortunately, owing to the lack of a suitable substrate, nonpolar nitrides still suffer from high densities of thread dislocations (TDs) and basal stacking faults (BSFs): the density of TDs is usually in the range of $\sim 10^{10}$ cm^{-2} and that of BSFs is $\sim 10^5$ cm^{-1} [3].

For the growth of c-plane nitrides growth, selective area growth (SAG) [5,6], epitaxial lateral overgrowth (ELO) [7] and facet control ELO (FACELO)[8,9] using MOVPE have been explored and are now recognized as a mature technology for improving crystal quality and for fabricating novel devices, such as facet light-emitting diodes, facet lasers, field emitters, waveguides and nanostructures. In the case of a-plane GaN on r-plane sapphire, SAG and ELO has also been developed to reduce TD density [10-12]. The facet structures on a-plane GaN grown by SAG with MOVPE have been investigated [3, 13, 14]. However, in in-depth knowledge is currently lacking on the facet structure; such knowledge is required to achieve FACELO or for the realization of novel devices using facet structures on nonpolar-based III-nitrides. In the present study, we have investigated the facet structure of a-plane GaN during SAG and studied the growth mechanism for each facet.

EXPERIMENT

Prior to SAG, by a nitridation process, high-quality a-plane GaN templates were grown on r-plane sapphire substrates by MOVPE [15]. Then, a ~ 100 nm-thick SiO$_2$ mask was deposited on the template. Using a conventional photolithographic process, stripe patterns consisting of a 5 μm window and 5 μm wing were then fabricated along the <1-100> axis of a-plane GaN. The patterned a-plane GaN template was then reloaded into the MOVPE apparatus for SAG. Trimethylgallium (TMGa) and NH$_3$ were used as the group III and V sources, respectively, and H$_2$ was used as the carrier gas. The first series of samples were grown at a pressure of 500 Torr with various growth temperatures from 1000 °C to 1100 °C and the second series of samples were grown under various reactor pressures from 100 Torr to 500 Torr with a growth temperature of 1100 °C. Scanning electron microscopy (SEM) was used to characterize the morphology.

RESULTS AND DISCUSSION

Cross-sectional SEM images of samples grown at the different growth temperatures are shown in Fig.1. The sample grown at 1000°C has a roughly triangular shape, and the main facets include an inclined {11-22} and a vertical (000-1) facet. However, upon increasing the growth temperature to 1050 °C, the area of the {11-22} facets (inclined planes) gradually decreases, and facets in two new planes (0001) (vertical plane) and {11-20} (horizontal plane) form. In samples grown at 1000 °C, the {11-22} facet eventually completely disappear, and (0001) and {11-20} facets continue to increase in area, resulting in a rectangular cross section at a higher temperature of 1100 °C. Thus, our experiments reveal that the growth temperature determined the facet form.

Fig.1.SEM images of the cross section of SAG experiment at a pressure of 500 Torr. (a) 1000 °C (b) 1050 °C and (c) 1100 °C.

In addition, the SEM images also show that Ga-polar (0001) plane and N-polar (000-1) plane have different growth rates in each sample as shown in Fig.2. With increasing the growth temperature, the (000-1) plane growth rate does not change distinctly; in contrast, the (0001) direction growth rate decreases linearly from 4.3 µm/h to 1.7 µm/h.

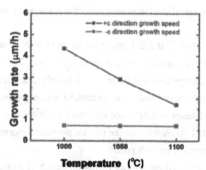

Fig.2. growth rates in [0001] and [000-1] directions at 500 Torr at different growth temperatures.

231

The growth mechanism is attributed to the stability of different facets at different temperatures; at a low temperature of 1000 °C, the inclined {11-22} facet is more stable owing to the relatively rough surface covered with N atoms [16], resulting in the triangular cross section; on the other hand, at a high temperature of 1100 °C, the vertical (0001) and horizontal {11-20} facets are more stable. Thus, with increasing temperature, the less stable inclined {11-22} facet gradually dissolves; in contrast, the stable (0001) and {11-20} facets gradually increase.

Similar phenomena have been reported in c-plane GaN [8,9] as shown in Fig.3 : at a low temperature, an inclined {11-22} facet, can be observed; however, at a high temperature, a vertical {11-20} plane, gradually replaces the {11-22} facet.

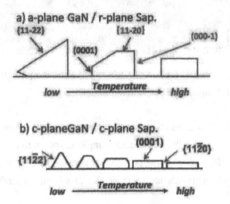

Fig.3. schematic diagram of the Facet structure of a-plane and c-plane GaN [9]

Figure. 4 shows cross-sectional SEM images of samples grown under different reactor pressures. All samples exhibit a similar rectangular cross section. This means changing the reactor pressure does not change the facet form, which is only determined by the growth temperature. In addition, we noticed that with decreasing reactor pressure, the rectangular cross section gradually decreases in height and increases in width, and the volume of the structure remains constant. This means that the reactor pressure determines the ratio of the lateral growth rate to the vertical growth rate.

Many models have revealed the surface diffusion length of atoms is inversely proportional to the reactor pressure. Thus, decreasing the pressure increases the lateral growth rate. Under a similar quasi-equilibrium condition, the chemical reaction rate is determined by the sources supply, particularly that of TMGa. Thus, the mass and volume of the product grown GaN remain nearly constant.

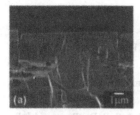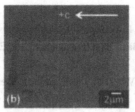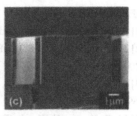

Fig.4. SEM images of the cross section of the SAG experiment at 1100 °C and reactor pressures of (a) 100 Torr (b) 300 Torr and (c) 500 Torr

CONCLUSIONS

We studied the SAG of a-plane GaN grown on r-plane sapphire by MOVPE. The growth temperature determined the facet form: at a low temperature of 1000 °C, the structure consists of {11-22}and (000-1) facet; upon increasing the growth temperature to 1050 °C, the area of the {11-22} facet gradually decreases, and facets in two new planes, (0001) and {11-20} form. In samples grown at 1000 °C, the {11-22} facet eventually completely disappears, whereas (0001) and {11-20} facets continue to increase in area to form the rectangular cross section. The reactor pressure determines the ratio of the lateral growth rate to the vertical growth rate: upon decreasing the reactor pressure 500 Torr to 100 Torr, the rectangular structure gradually decreases in height and increases in width, and the volume of the structure nearly remains constant.

Thus, we found that facet control in a-plane GaN can be achieved by optimizing the growth temperature and reactor pressure. This technology can be used to develop the FACELO technique or to fabricate novel devices, such as field emitters, waveguides, facet lasers and nanostructures.

ACKNOWLEDGMENTS

This work was partially supported by the Akasaki Research Center of Nagoya University, Grants-in-Aid for Scientific Research (No. 21360007 and No. 21560014) and Scientific

Research on Priority Areas (No.18069006) of the MEXT, and A Research for Promoting Technological Seeds of JST.

REFERENCES

[1] T. Takeuchi, H. Amano, and I. Akasaki, Jpn. J. Appl. Phys. 39, 413 (2000).

[2] J. Piprek, R. Farrell, S. DenBaars, and S. Nakamura, IEEE Photon. Technol. Lett. 18, 7 (2006).

[3] M. D. Craven, S. H. Lim, F. Wu, J. S. Speck, and S. P. DenBaars, Appl. Phys. Lett. 81, 469 (2002).

[4] M. D. Craven, F. Wu, A. Chakraborty, B. Imer, U. K. Mishra, S. P. DenBaars and J. S. Speck, Appl. Phys. Lett. 84, 1281 (2004).

[5] Y. Kato, S. Kitamura, K. Hiramatsu and N. Sawaki, 1994 J. Cryst. Growth 144, 133 (1994).

[6] S. Kitamura, K. Hiramatsu and N. Sawaki, Jpn. J. Appl. Phys. 34, L1184 (1995).

[7] O. H. Nam, M. D. Bremer, B. L. Ward, R. J. Nemanich and R. F. Davis, Jpn. J. Appl. Phys. 36, L532 (1997).

[8] H. Miyake, A. Motogaito, and K. Hiramatsu, Jpn. J. Appl. Phys. 38, L1000 (1999).

[9] K. Hiramatsu, K. Nishiyama, M. Onishi, H. Mizutani, M. Narukawa, A. Motogaito, H. Miyake, Y. Iyechika, and T. Maeda, J. Cryst. Growth 221, 316 (2000).

[10] X. Ni, Ü. Özgür, Y. Fu, N. Biyikli, J. Xie, A. A. Baski, H. Morkoç, and Z. Liliental-Weber,　Appl. Phys. Lett. 89, 262105 (2006).

[11] T. Gühne, Z. Bougrioua, P. Vennéguès, M. Leroux, and M. Albrecht, J. Appl. Phys. 101, 113101 (2007).

[12] B. Imer, F. Wu, J.S. Speck and S.P. DenBaars, J. Cryst. Growth 306, 330 (2007).

[13] B. Imer, F. Wu, M. D. Craven, J. S. Speck, S. P. DenBaars, Jpn. J. Appl. Phys., Vol. 45, 8644 (2006).

[14] C. Netzel, T. Wernicke, U. Zeimer, F. Brunner, M. Weyers and M. Kneissl, J. Crystal Growth 8, 310 (2008).

[15] B. Ma, W. Hu, H. Miyake, and K. Hiramatsu, Appl. Phys. Lett. 95, 121910 (2009).

[16] S. C. Cruz, S. Keller, T. E. Mates, U. K. Mishra, S. P. DenBaars, J. Cryst. Growth 311, 3817 (2009).

Mater. Res. Soc. Symp. Proc. Vol. 1202 © 2010 Materials Research Society   1202-I05-10

Kinetics and Mechanism of Formation of GaN From β-Ga$_2$O$_3$ by NH$_3$

Toshiki Sakai, Hajime Kiyono and Shiro Shimada
Graduate School of Engineering, Hokkaido University, Sapporo 060-8628, Japan

ABSTRACT

Nitridation of β-Ga$_2$O$_3$ to GaN in an atmosphere of NH$_3$/Ar was investigated from the view points of kinetic results by thermogravimetric analysis (TGA) and microstructural observation. TGA and X-ray powder diffraction results showed that the nitridation of Ga$_2$O$_3$ to GaN starts at about 650°C and decomposition of GaN formed occurs from 870°C. Isothermal TGA results showed that the nitridation proceeds linearly with time at 800 – 1000°C. Microstructural observation of the samples nitrided at 800°C showed that fine GaN particles (~50 nm size) deposit on surfaces of Ga$_2$O$_3$ particles at an early stage, and the deposits grow with progress of the nitridation.

INTRODUCTION

Gallium nitride (GaN) has been applied to optoelectronic devices such as light emitting diodes, laser diodes, and ultra-violet photo detector because of its wide-band gap (3.4 eV) [1-3]. There have been various methods for fabrication of GaN. Of these, nitridation of gallium oxide (Ga$_2$O$_3$) to GaN by NH$_3$ gas has advantages due to easy handling and lower cost. In the nitridation, Ga$_2$O$_3$ is reacted with NH$_3$ gas (Ga$_2$O$_3$ + 2NH$_3$ → 2GaN + 3H$_2$O (1)). This method has been used for production of GaN powders, epilayers, and nanotubes [4-6]. Balkaş et al. reported that the reaction (1) dose not occur directly, but proceeds via an intermediate gaseous species Ga$_2$O on the basis of thermodynamic calculation [7]. Jung et al. suggested the nitridation of β-Ga$_2$O$_3$ by NH$_3$ proceeds via intermediates of solid state gallium oxynitride (GaO$_x$N$_y$) [8,9]. Therefore the mechanism of the nitridation has been not clarified. In the present study, the mechanism of nitridation for GaN from Ga$_2$O$_3$ by NH$_3$ gas was investigated on the basis of kinetic results and microstructural observation.

EXPERIMENTAL DETAILS

β-Ga$_2$O$_3$ powder (99.9% purity; Koujyundo Kagaku Co.) and α-alumina powder (99.9% purity; Koujyundo Kagaku Co.) were used in the present work. The as-received β-Ga$_2$O$_3$ powder shows rod-like particle about 1-10 μm long consisting of primary particle of about 0.1 μm and its specific surface area was about 10 m^2/g as measured by the BET method using N$_2$ gas (BELSORP-mini, BEL JAPAN). The β-Ga$_2$O$_3$ powder sintered at 1400°C for 1h in Ar was also used as a starting material. Particle size of the powder was not changed significantly at the sintering temperature of 1400°C but the primary particles were sintered each other and the specific surface area was decreased to about 1 m^2/g.

Weight change was measured by using an electric micro balance (Chan D200). About 30 mg of the starting material (as-received, sintered β-Ga$_2$O$_3$ powder or mixed powder of as-received β-Ga$_2$O$_3$ and α-alumina) were placed in an alumina cell which was hanged with an alumina wire in a fused quartz tube in a vertical furnace. Non-isothermal nitridation was carried out up to 1200°C at a rate of 5°C/min and isothermal nitridation was carried out at a temperature range from 800°C to 1000°C for about 1 h. The nitrided samples were characterized by X-ray powder diffraction (XRD, RINT 2000, RIGAKU), scanning electron microscopy (SEM, JSM-6300F, JEOL), and high-resolution transmission electron microscopy (TEM)-SEM equipment (HD-2000, at the OPEN FACILITY, Hokkaido University Sousei Hall, HITACHI).

RESULTS & DISCUSSION

Thermogravimetrical results

Figure 1 shows the non-isothermal TGA results for nitridation using as-received β-Ga$_2$O$_3$ powder. In Ar atmosphere, no weight change was observed over a range from 20 to 1200 °C. In H$_2$/Ar atmosphere, the weight loss started at about 1000 °C, showing that β-Ga$_2$O$_3$ was decomposed by H$_2$. Decomposed β-Ga$_2$O$_3$ should be in the form of gas species such as Ga$_2$O [10]. In NH$_3$/Ar atmosphere, the weight loss began from 650°C and apparently stopped at 850°C. After a short plateau the weight loss started again from about 1000°C. Figure 2 shows the XRD patterns of as-received β-Ga$_2$O$_3$ and the samples obtained after non-isothermal TGA. The XRD pattern of as-received β-Ga$_2$O$_3$ (Figure 2(a)) exhibits only β-Ga$_2$O$_3$ peaks, and the XRD patterns of sample heated to 870°C and 1200°C (Figure 2(b) and (c)) shows that the sample is changed to wurtzite GaN. These results indicate that nitridation of β-Ga$_2$O$_3$ to GaN by NH$_3$ begins at about 650°C and later weight loss from 1000°C is due to decomposition of GaN.

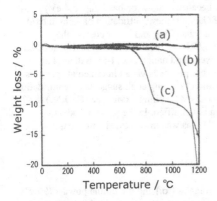

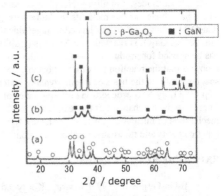

Figure 1. Plots of non-isothermal nitridation in; (a) Ar, (b) H$_2$/Ar and (c) NH$_3$/Ar. The heating rate is 5 °C/min.

Figure 2. XRD patterns of (a) as-received β-Ga$_2$O$_3$ and the sample heated to (b) 870°C, (c) 1200°C.

The results of isothermal TGA in NH_3/Ar atmosphere are shown in Figure 3. In the all temperatures the weight of the sample was decreased linearly with time to -9 ~ -10%. The XRD patterns of the samples obtained after TGA gave only the GaN peaks (Figure 4). An activation energy of nitridation of Ga_2O_3 was 110 kJ mol^{-1}, calculated from the Arrhenius plots of the slopes obtained from the isothermal TG curves (see dashed lines superimposed on the TG curves). The linear kinetics suggest that the nitridation is limited by an interfacial reaction.

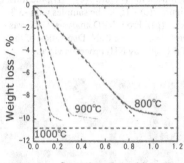

Reaction time / h

Figure 3. Plots of isothermal nitridation in NH_3/Ar. The horizontal dotted line at -10.7% indicates a weight loss corresponding to a completed nitridation of Ga_2O_3.

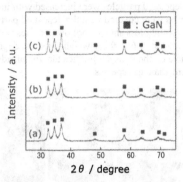

Figure 4. XRD patterns of the samples obtained after isothermal nitridation at; (a) 800°C, (b) 900°C and (c) 1000°C.

Figure 5(a) shows that the weight loss by isothermal TGA at 800°C in different P_{NH3} partial pressures of ammonia. The weight of the sample was decreased linearly with time in all the pressures. The slopes obtained by TG curves were plotted with P_{NH3}. The slope was linearly increased with P_{NH3} (Figure 5(b)). Figure 6 shows the comparison of the isothermal TGA results at 800°C between as-received and sintered β-Ga_2O_3 powders. The TG curves of the sintered sample showed a sigmoid curve with a linear region from -2% to -8%. The slope of the linear region was almost same as that obtained from the isothermal nitridation curves of as-received β-Ga_2O_3 powder.

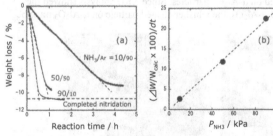

Reaction time / h

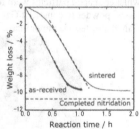

Reaction time / h

Figure 5. The plots of; (a) isothermal nitridation in various P_{NH3} at 800°C, (b) relationship between the nitridation rate constant and P_{NH3}.

Figure 6. Comparison of the isothermal nitridation results at 800°C between as-received and sintered β-Ga_2O_3 powder.

Microstructural observation

Figure 7 shows SEM images of the sintered β-Ga$_2$O$_3$ powders and those isothermally nitrided at 800°C at different weight losses. The sintered β-Ga$_2$O$_3$ powders have a smooth surface (Figure 7(a)). In the samples with -0.3% weight loss (Figure 7(b) and (b')) small particles about ten nanometer in size were deposited on the surface of β-Ga$_2$O$_3$. The number and size of the deposited particles were increased with increasing nitridation (Figure 7(c) ~ (f)); the nitrided samples were covered with the deposited particles (Figure 7(c) ~ (g)). The samples with -5.3 ~ -9.6% weight loss showed hollow particles (Figure 7(e) ~ (g)). From XRD analysis, GaN was observed at -1.6 ~ -9.6% weight loss; the deposited particles must be GaN. Deposition of the particle on the β-Ga$_2$O$_3$ implies that the nitridation of β-Ga$_2$O$_3$ by NH$_3$ proceeds via an intermediate gas species.

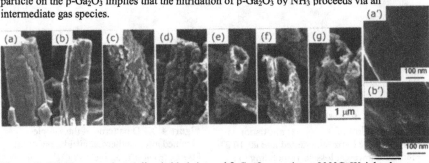

Figure 7. SEM images of partially nitrided sintered β-Ga$_2$O$_3$ powder at 800°C. Weight changes of the partially nitrided samples are (a) 0%, (b) -0.3%, (c) -1.6%, (d) -3.2%, (e) -5.3%, (f) -7.5%, and (g) -9.6%. The picture of (a') and (b') show that magnified image of (a) and (b), respectively.

Isothermal nitridation was carried out using a mixed powder of β-Ga$_2$O$_3$ and α-Al$_2$O$_3$ at 800°C. Figures 8(a) and (b) show SEM images of as-received α-Al$_2$O$_3$ and the mixed powder of β-Ga$_2$O$_3$ and α-Al$_2$O$_3$ heated at 800°C for 1.5h in NH$_3$/Ar (50/50 kPa). As-received α-Al$_2$O$_3$ particles have irregular surface (Figure 8(a)). It is found that the morphology of α-Al$_2$O$_3$ particles is not changed by heating at 800°C in NH$_3$/Ar atmosphere. In the mixed powder (Figure 8(b)), whisker-shaped particles were formed on α-Al$_2$O$_3$ particles. Since XRD analysis showed the mixed powder to consist of GaN and α-Al$_2$O$_3$, the whisker-shaped particle on α-Al$_2$O$_3$ must be GaN deposited via vapor phase.

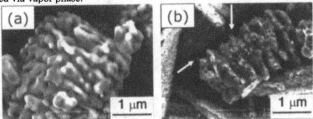

Figure 8. SEM images of (a) as-received α-Al$_2$O$_3$ and (b) nitrided mixed powder (α-Al$_2$O$_3$ and β-Ga$_2$O$_3$) heated at 800°C for 1.5 h in NH$_3$/Ar atmosphere.

Nitridation mechanism

Figure 9 shows a mechanism for nitridation of β-Ga_2O_3 by NH_3 gas, based on the results obtained in the present work. First, β-Ga_2O_3 is reacted with NH_3 gas or gas species (NH_2, NH, H) formed by thermal decomposition of NH_3 to generate gas species such as Ga_2O (Figure 9·1). Second, this intermediate gas species is nitrided by NH_3 gas and deposited as GaN particles on the surface of the β-Ga_2O_3 (Figure 9·2). Since formation of GaN nucleus was difficult on surfaces of β-Ga_2O_3, the nitridation proceeds slower than the subsequent growth stage. Third, GaN particles are grown by deposition of GaN as a result of nitridation of the intermediate gas species with NH_3 on the surface (Figure 9·3). β-Ga_2O_3 is mostly covered with deposited GaN particles and a crust of GaN is formed due to sintering of these GaN particles. However, a GaN crust is so porous that the intermediate gas species could go through the crust without deposition on the inside of the β-Ga_2O_3. As a result, the intermediate gas species continuously vaporizes to the outside of the GaN crust and the inside β-Ga_2O_3 is run out. At a final stage (Figure 9·4), a hollow GaN is formed during nitridation of Ga_2O_3.

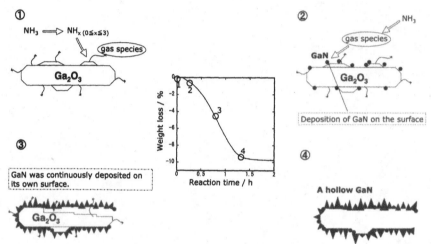

Figure 9. A model for formation of GaN from Ga_2O_3 by NH_3 gas.

CONCLUSIONS

We investigated the formation of GaN from β-Ga_2O_3 with NH_3 gas by TGA and microstructural observation. Non-isothermal TGA results showed that the nitridation starts at about 650°C, followed by decomposition of GaN formed above 1000°C. Isothermal TGA results at 800 – 1000°C showed that the weight loss occurs linearly with time, implying that the nitridation is limited by an interfacial reaction. From microstructural observation, fine GaN particles are deposited on surfaces of Ga_2O_3 particles at an early stage of the nitridation at 800°C of the sintered sample. These results proposed a mechanism of nitridation of β-Ga_2O_3 with NH_3.

REFERENCES

1. R. P. Vaudo, D. Goepfert, T.D. Moustakas, D. M. Beyea, T. J. Frey and K. Meehan, *J.Appl. Phys.*, **79**, 2779-2783 (1996).
2. G. Jacob and D. Bios, *Appl. Phys. Lett.*, **30**, 412-414 (1977).
3. S. J. Pearton and F. Ren, *Advanced Materials*, **12**, 1571 (2000).
4. H.D. Xiao, H.L. Ma, C.S. Xue, J. Ma, F.J. Zong, X.J. Zhang, F. Ji, and W.R. Hu, *Mater. Chem. Phys.*, **88**, 180-184 (2004)
5. H.J. Lee, T.I. Shin, and D.H. Yoon, *Surf. Coat. Tech.*, **202**, 5497-5500 (2008)
6. J. Dinesh, M. Eswaramoorthy, and C. N. R. Rao, *Physical Chemistry C.*, **111**, 510-513 (2007)
7. C.M. Balkaş and R.F. Davis, *J. Am. Ceram. Soc.*, **79**[9] 2309-12 (1996)
8. W. S. Jung, *Mater Lett.*, **60**, 2954-2957 (2006).
9. W. S. Jung, O. H. Han and S. A. Chae, *Mater. Chem Phys.*, **100**, 199-202 (2006).
10. Darryl P. Butt, Youngsoo Park and Thomas N. Taylor, *J. Nucl. Mater.*, **264**, 71-77 (1999)

Mater. Res. Soc. Symp. Proc. Vol. 1202 © 2010 Materials Research Society   1202-I05-09

Praveen Kumar[1,2], Mahesh Kumar[1], Govind[1], B. R. Mehta[2] and S. M. Shivaprasad[3*]

[1]Surface Physics and Nanostructure Group, National Physical Laboratory New Delhi-110012, India

[2]Department of Physics, Indian Institute of Technology, New Delhi-110016, India

[3]Jawaharlal Nehru Centre for Advanced Scientific Research, Jakkur, Bangalore-560 064, India

Abstract:

GaN and related nitride semiconductors have attracted great attention in view of their wide applications in photonics and high temperature & high power electronic devices. Among other issues, reduction of defect densities by forming these interfaces at lower temperature and on novel substrates has been the motivation for several researchers. In the present study ion-induced conversion of Si (111) surface into silicon nitride at room temperature is optimized and used as substrate for the growth of Ga films. These Ga films are again nitrided by optimal N^+ ion bombardment. Experiments have been performed in-situ in an ultra high vacuum chamber equipped with a Ga source and X-ray photoelectron spectrometer (XPS) at base pressure of 2×10^{-10} torr. The energy dependence of the nitridation is carefully performed at constant flux. The results clearly demonstrate the Si-N bond formation after a energy of 2 keV and the formation of GaN layer after 800eV of N_2^+ ion bombardment on Si (111) 7x7 surface and Ga adsorbed silicon nitride surface, respectively. The FWHM and chemical shifts in the core-level spectra of Si(2p), Ga(2p) and N(1s) have been analyzed to probe the interface reactions. The results demonstrate a possible novel and low temperature approach towards the integration of III-nitride & silicon technologies, since silicon nitride bonds can act as barriers to dislocation propagation.

Key Words: XPS, GaN, Silicon Nitride, Si(111)

*Corresponding Author e mail: smsprasad@jncasr.ac.in

INTRODUCTION

Group III-nitride semiconductors based on GaN have received great attention as materials for the realization of blue and green light emitting diodes [1], laser diodes [2] and high-temperature & high power devices [3]. GaN is an attractive material for short-wavelength optoelectronics and high power devices due to its unique properties, including a large direct band-gap and a high critical electrical field. However, the choice of substrate to grow epitaxial GaN is a serious issue. Recently various substrates have been tried to grow wurtzite GaN, such as sapphire [4], 6H-SiC (0001) [5], GaAs (111) [6] and Si (111) [7]. Among these substrates, silicon has become a desirable substrate material for hetero-epitaxy because of its high crystalline quality, large wafer size, low cost, and easier integration with the well-established silicon device technology. However, it is difficult to grow high quality wurtzite GaN on silicon due to the large mismatch in lattice constant (-16.9%), and thermal expansion coefficient mismatch (57%), that results in high dislocation density in the GaN films and thermal diffusion of silicon. These concerns arise due to the tensile stress from the mismatch in thermal expansion coefficients between GaN and silicon during cooling after high temperature growth. Therefore, buffer layers such as AlN, SiC, GaAs, GaN and silicon nitride, have been developed to overcome the problem [8-13]. AlN is by far the most widely employed, but silicon nitride layer formation by nitridation of Si substrate is also an attractive solution [14]. Silicon nitride layer acts as an antisurfactant for GaN growth and improve the crystallinity of the overlayer. The strong silicon nitride bonds can act as barrier to the propagation of dislocation through the film. Thus, the need for a systematic study of the nitridation of Si to form silicon nitride at room temperature, which can be used as a template for the growth of GaN layers, has prompted us to undertake this work. We have recently reported on several surface modification studies on GaAs (001) [15, 16] and Si(111) [17, 18] surfaces, to make these compatible for GaN growth. In this work we present a careful X-ray photoelectron spectroscopy (XPS) study of the nitridation of Si(111)-7x7 reconstructed surface by N_2^+ ions of different energies at room temperature. Further this silicon surface modified into silicon nitride has been used as substrate to grow GaN.

EXPERIMENTAL:

The experiments were performed in a Perkin Elmer (Physical Electronics) ultra high vacuum (UHV) system (model PHI-1257) working at a base pressure of 5 x 10^{-10} Torr. The chamber is equipped with a dual anode Mg K_α (1253.6 eV) and Al K_α (1486.6 eV) X-Ray sources, a high precision sample manipulator and a high resolution hemispherical electron energy analyzer, to perform X-Ray Photoelectron Spectroscopy (XPS). The chamber is also equipped with a differentially pumped N_2^+ ion source with precise control of energy (0-5 keV), fluence and raster area on the sample. The sample, a 10 mm x 10 mm piece cut from a commercial Si(111) wafer is cleaned by the Shiraki process [19] before being introduced into the UHV chamber. The sample was degassed at 600°C for 12 hours followed by repeated flashing up to 1200°C for 5sec by electron beam heating and cooling to RT at a very slow rate of 2°C/sec. The sample temperature was monitored by a W-Re (5%, 25%) thermocouple mounted behind the sample and calibrated by an optical pyrometer. The atomic cleanliness of a sample was ascertained by the absence of carbon and other contaminants on the surface by XPS, and observation of the well ordered 7x7 reconstruction pattern in LEED. Gallium deposition was made from a home made tantalum-Knudsen cell, whose flux rate was controlled by regulating the current to the cell. The experiment involves a controlled bombardment of N_2^+ ions of different energies at a constant flux rate with the sample current of 1.5×10^{-6} Amp being monitored to adjust the exposure time and consequently the estimation of the ion fluence by considering the spot size, raster area and time of exposure.

RESULTS AND DISCUSSION

The growth of GaN was performed in two steps. First the Si(111) surface has been modified by bombardment with different energies of N_2^+ ions, which results in the formation of silicon nitride. Subsequently this nitride surface has been used as substrate to grow GaN epitaxially by subsequent Ga deposition and N_2^+ ion bombardment. The core-level spectra were analyzed to identify the elemental chemical states. The relative percentage elemental compositions are calculated by measuring the area under the core level after suitable background subtraction and using the atomic sensitivity factors [20]. XPS scans of Si(2p), N(1s), and Ga(2p) were used to monitor the elemental composition and chemical shifts due to silicon nitride and GaN formation as a function of N_2^+ ion energy.

1. Nitridation of Si(111)-7x7 reconstructed surface:

The Si(111)-7x7 reconstructed surface has been bombardment by different energies of N_2^+ ion ranging from 0.2keV to 5keV with a fluence of 3×10^{15} ions/cm^2. After nitridation we have recorded the XPS spectra for survey scan (0-1200eV), core levels and Auger transitions, valence band and true secondary electron emission, to evaluate the effects at the interface.

Figure 1 shows the change in Si(2p) core level spectra during nitridation of Si(111) surface. Curve (a) shows Si(2p) for the clean Si(111)-7x7 reconstructed surface. Our survey scan and

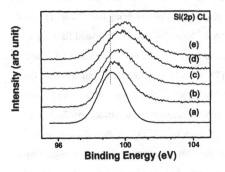

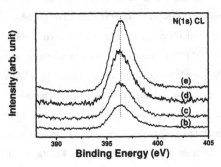

Figure 1: Si(2p) Core level spectra during the nitridation of Si(111)-7x7 surface. Curve (a), (b), (c), (d) and (e) corresponds to clean Si, 2keV, 3keV, 4keV and 5keV N_2^+ ion bombarded samples.

Figure 2: N(1s) Core level spectra during the nitridation of Si(111)-7x7 reconstructed surface. Curve (b), (c), (d) and (e) corresponds to 2keV, 3keV, 4keV and 5keV nitridation.

core level results show that for N_2^+ ions less than 2keV, we do not observe peak of N(1s) core level and any shift in the Si(2p) peak position. This reveals that there is no Si-N reaction for the N_2^+ ion energies less than 2keV.Curve 1 (a) shows the Si(2p) core level spectra for clean Si with position at 99.2eV and FWHM of 1.4eV. Curve (b) shows the Si(2p) core level spectra after 2keV N_2^+ ion bombardment, with a shift of 0.3eV and an increase in its peak width by 0.4eV FWHM. As this surface is bombardment by 2keV N_2^+ions the Si(2p) peak shifts to 99.5 eV with an increase in the FWHM to 1.8eV. The deconvoluted Si(2p) core level spectra, into the Si-Si and S-N related Gaussian components, manifest the formation of silicon nitride after 2keV N_2^+ ion bombardment. Curves (c), (d) and (e) of Fig 1 corresponds to 3keV, 4keV and 5keV N_2^+ ion bombardment of the Si surface. Thus nitridation decreases the peak height of the Si peak shifts it by 1.2 eV towards higher binding energy and its peak width increases. Deconvolution of Si(2p) spectra into Gaussian components shows that as the N_2^+ion energy increases the Si-N related peak while the Si-Si related contribution decreases with in the sampling depth.

244

Figure 2 depicts the evolution of the N(1s) core level spectra with increasing nitridation energy of the Si(111) surface. Curve (b), (c), (d) and (e) correspond to the nitridation by 2keV, 3keV, 4keV and 5keV N_2^+ ion bombardment. Figure shows that N(1s) peak increases in intensity with increase in nitridation energy. We do not observe any shift in the N(1s) peak during the nitridation process, which indicates that all the nitrogen atoms are bonded with silicon to form a stable stoichiometric silicon nitride.

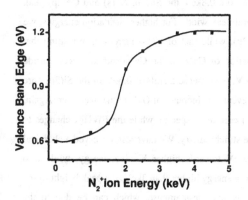

Figure 3 shows the plots the valence band maxima position with respect to the Fermi level, as a function of the N_2^+ ion energy. Figure shows the shift from 0.6eV to the 1.2eV, with the increase in the N_2^+ ion energy, reflecting an increase in the band gap after the nitridation process.

Figure 3: Separation between Valence band maxima position and Fermi level as a function of the N2+ ion energy.

Secondary electron emission spectra show that the work function of the modified surface changes from 4.0eV to 4.5eV. All these results show the formation of silicon nitride on Si(111) surface at room temperature by N_2^+ ion bombardment.

2. GaN growth on modified silicon nitride surface:

Subsequently we used this silicon nitride surface as substrate to grow ion-induced GaN, and discuss this in the following section. Figure 4 shows the core level spectra of Si(2p), N(1s) and Ga(2p) during the Growth of GaN on the modified Si surface. For the growth of GaN, initially

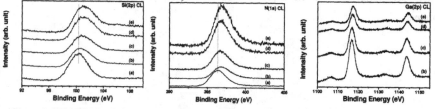

Figure 4: Si(2p), N(1s) and Ga(2p) core level spectra during the growth of GaN on nitrided silicon surface. Curve (a), (b), (c), (d) and (e) corresponds to the modified silicon nitride surface, followed by 2ML Ga adsorbed surface, 0.8keV, 2keV, and 4keV nitridation.

2ML Ga has been adsorbed onto the silicon nitride surface, followed by the bombardment of N_2^+ ions of different energies with a constant fluence of 3×10^{15} ions/cm^2. In figure 4, (a), (b), (c), (d) and (e) curve corresponds to the 2ML Ga overlayer bombarded by 0.8keV, 2keV and 4keV N_2^+ ions, respectively. It has been observed that before 0.8keV the Si(2p), N(1s) and Ga(2p) peak remain unchanged in peak position intensity and peak width. For higher nitridation energies, we observe a decrease in the peak height of Si(2p), while that of N(1s) increases, indicating that 0.8keV is the threshold energy for the formation of GaN on the Ga adsorbed silicon nitride surface. For higher N_2^+ ion energies up to 3keV, we observe a shift of 0.6eV in the Si(2p) core level peak and a change in FWHM of 0.5eV, reveals the increase of GaN in the overlayer. Again, we do not observe any peak shift in the N(1s) core level spectra, while the FWHM changes to 0.4eV, indicating GaN formation, with a stable stoichiometry. We have also observed a chemical shift in the Ga(2p) core level spectra of 0.6eV towards higher binding energy, while also confirming the formation of GaN. For nitridation energy higher than 3keV, the peak height of the Si (2p) peak increases, while Ga(2p) peak becomes less intense, which can be due to the domination of the sputtering of the GaN layers by the high energy N_2^+ ions. All these results strongly suggest the formation of GaN on this modified surface.

CONCLUSIONS

In summary, we have performed careful room temperature experiments of the ion induced formation of silicon nitride on Si(111)-7x7 reconstructed surface. The results show a threshold energy of 2keV of N_2^+ ions for the formation of SiN$_x$ on the Si(111) surface. This modified surface has been used as a substrate to grow GaN. Threshold energy of 0.8keV of N_2^+ ions has been optimized to form GaN on Ga adsorbed silicon nitride surface. The report demonstrates the controlled formation of silicon nitride overlayer on Si, which acts as a template to grow GaN. The strong silicon nitride bonds can act as barriers to the dislocation propagation and result in better GaN overlayers. A structural probe to understand the mechanism in more detail is in progress.

ACKNOWLEDGMENTS

The authors thank Prof C. N. R. Rao and Director NPL for their encouragement and support. One of the authors (P. Kumar) is grateful to the University Grant Commission, India for providing research fellowship for carrying out this work.

REFERENCES

1. Nakamura, *Science* **281**, 956 (1998).

2. R. F. Karlicek, jr , M. Schurman, C. Tran, T. Salagaj, Y. Li and R. A. Stall, *III-Vs Review* **10**, 20 (1997).

3. S. Nakamura, M. Senoh, S.Nagahama, N. Iwasa, T. Yamada, T. Matsushita, Y. Sugimoto and H. Kiyoku, *Appl. Phys. Lett.* **69**, 4056 (1996).

4. Q. Sun, Y. S. Cho, B. H. Kong, H. K. Cho, T. S. Ko, C. D. Yerino, I. Lee and J Han, J. *Cryst. Growth* **311**, 2948 (2009).

5. Q. Xue, Q. K. Xue, R. Z. Bakhtizin, Y. Hasegawa, I. S. T. Tsong, and T. Sakurai and T. Ohno, *Phy. Rev B* **59**,12604 (1999).

6. S. C. Huang, H. Y. Wang, C. J. HSU, J. R. Gong, C. Chiang, S. L. Tu and H. Chang, *J. Mater. Sci. Lett.* **17**,1281 (1998).

7. R. Armitage, Q. Yang, H. Feick, J. Gebauer, E. R. Weber, S. Shinkai and K. Sasaki, *Appl. Phys. Lett.* **81**, 1450 (2002).

8. A. Le. Louarn, S. Vezian, F. Semond and J. Massies, *J. Cryst. Growth* **311**, 3278 (2009).

9. T S Huang, T B Joyce, R T Murray, A J Papworth and P R Chalker, *J. Phys. D: Appl. Phys.* **35**, 620 (2002).

10. O. Kryllouk, T. Anderson, K. C. Kim, *US Patent 7001791* (2006) B2.

11. S. Nakamura, *Jpn. J. Appl. Phys.* **30**, L1705 (1991).

12. K. Ito, Y. Uchida, S. Lee, S. Tsukimoto, Y. Ikemoto, K. Hirata, N. Shibata, and M. Murakami, *Mater. Res. Soc. Symp. Proc.* **916**, 0916-DD04 (2006).

13. C. H. Kuo, S. J. Chang, Y. K. Su, C. K. Wang, J. K. Sheu, T. C. Wen, W. C. Lai, J. M. Tsai and C. C. Lin, *Solid-State Electronics* **47**, 2019 (2003).

14. J. S. Pan, A. T. S. Wee, C. H. A. Huan, H. S. Tan and K. L. Tan, *Vacuum* **47**, 1495 (1996).

15. P. Kumar, M. Kumar, Govind, B.R. Mehta, S.M. Shivaprasad, *App. Surf. Sci. 256,* *(2009) 517.*

16. Praveen Kumar, S. Bhattacharya, Govind, B. R. Mehta, and S. M. Shivaprasad, *J. Nanosci. Nanotechnol.* **9**, 5659 (2009).

17. P. Kumar, L. Nair, S. Bera, B. R. Mehta, S.M. Shivaprasad, *App. Surf. Sci.* **255**, 6802 (2009).

18. P. Kumar, M. Kumar, B.R. Mehta, S.M. Shivaprasad, *Appl. Surf. Sci.* **256**, 480 (2009).

19. Y. Enta, S. Suzuki, S. Kono, T. Sakamoto, *Phys. Rev. B* **39**, 56 (1989).

20. C.D. Wagner, W.M. Riggs, L.E. Davis, J.F. Moulder, G.E. Muilenberg, Handbook of Xray Photoelectron Spectroscopy, Published by: PerkinElmer Corp., 1979.

Mater. Res. Soc. Symp. Proc. Vol. 1202 © 2010 Materials Research Society 1202-I05-16

Low-resistance ohmic contacts to N-face p-GaN for the fabrication of functional devices

Seung Cheol Han,[1] Jae-Kwan Kim,[1] Jun Young Kim,[1] Joon Seop Kwak,[1] Kangho Kim,[2] Jong-Kyu Kim,[3] E. F. Schubert,[3] Kyoung-Kook Kim,[4] and Ji-Myon Lee [1,*]

[1] Department of Materials Science & Metallurgical Engineering, Sunchon National University, Sunchon, Jeonnam 540-742, Korea
[2] Korea Photonics Technology Institute, 5 Cheomdan, Bukgu, Gwangju 500-779, Korea
[3] The Future Chips Constellation, Department of Electrical, Computer, and Systems Engineering, Rensselaer Polytechnic Institute, Troy, New York 12180, USA
[4] Department of Nano-Optical Engineering, Korea Polytechnic University, Siheung 429-793, Korea

ABSTRACT

The electrical properties of Ni-based ohmic contacts N-face p-type GaN are presented. The specific contact resistance of N-face p-GaN exhibits a liner decrease from 1.01 Ω cm^2 to 9.05 × 10^{-3} Ω cm^2 for the as-deposited and the annealed Ni/Au contacts, respectively, with increasing annealing temperature Furthermore, the specific contact resistance could be decreased by four orders of magnitude to 1.03 × 10^{-4} Ω cm^2 as a result of surface treatment using an alcohol-based (NH$_4$)$_2$S solution. The depth profile data measured by the intensity of O1s core peak in the x-ray photoemission spectra showed that the alcohol-based (NH$_4$)$_2$S treatment was effective in removing of the surface oxide layer of GaN. In addition, a Ga 2p core-level peak showed a red-shift of binding energy by 0.3 eV by alcohol-based (NH$_4$)$_2$S treatment, indicating that the surface Fermi level was shifted toward the valence-band edge. Thus, the low ohmic contact behavior observed in our treated sample might be explained in terms of the removal of the oxide layer and reducing the barrier heights by reduced band-bending effect.

INTRODUCTION

GaN-based materials have been successfully used in optical and electronic devices, such as light emitting diodes (LEDs), metal–semiconductor field effect transistors, high electron mobility transistors, and laser diodes.[1-5] The process of high-quality ohmic contacts which have low resistance and thermal stability is of considerable technological importance. In fact, the high contact resistance of p-GaN is one of the major obstacles to the realization of long-lifetime operation of GaN-based optical devices. Therefore, it is important to develop high-quality Ohmic contacts on p-GaN to improve the electrical and optical performance of such devices.
Up to date, the growth of heavily doped p-GaN (>10^{18} cm^{-3}) and the absence of contact metals, which have a work function larger than that of p-GaN (6.5eV),[6] have been realized as major obstacles in the achievement of low-resistance Ohmic contacts to p-GaN. To obtain a low-resistance Ohmic contact to p-type GaN, a variety of surface treatments of p-GaN using solutions such as aqua-regia solution,[7] and (NH$_4$)$_2$S$_x$,[8,9] have been proposed. It has been reported that the

249

surface treatment with a $(NH_4)_2S_x$ solution is especially efficient in removing the native oxide layer on the p-type GaN surface, resulting in a low-resistance Ohmic contact to p-type GaN.[9] In the area of GaAs devices, several studies have reported that the alcohol-based $(NH_4)_2S_x$ solution removes the native oxide on the GaAs surface more efficiently than the normal $(NH_4)_2S_x$ solution due to the low dielectric constant of the alcohol-based $(NH_4)_2S_x$ solution, resulting in a larger improvement in device characteristics.[10,11] Zhilyaev et al.[12] have reported that the photoluminescence intensity of n-type GaN is considerably enhanced as a result of the surface treatment with an alcoholic sulfide solution. Our previous results[13] also showed that, for the case of n-type GaN, the surface treatment with an alcohol-based $(NH_4)_2S_x$ solution removes the insulating layer on the n-type GaN surface more effectively than the normal, $(NH_4)_2S_x$ solution, leading to more enhanced electrical properties. In this letter, we report on an investigation of the effect of an alcohol-based $(NH_4)_2S_x$ solution [t-$C_4H_9OH_1(NH_4)_2S$] for use in the treatment of Ohmic contacts to N-face p-type GaN by observing the oxide layer and the shift of the surface Fermi level, as evidenced by the x-ray photoelectron spectroscopy (XPS) spectra. In addition, the Ga2p core-level spectra result showed that the specific contact resistivity for a sample treated with an alcohol-based $(NH_4)_2S_x$ solution is sharply decreased by four orders of magnitude over that of an untreated sample.

EXPERIMENT

The p-GaN samples were grown on a c-face (0001) sapphire substrate by using a metal organic chemical-vapor deposition system. After the growth of the GaN buffer layer and subsequent undoped GaN layer, 2.0-um-thick p-GaN:Mg was grown. The GaN layers were rapid-thermal annealed at 650 °C for 1 min in N_2 ambient to activate the Mg dopant. Hall measurements showed that the carrier concentration was about 1×10^{17} cm³. Using a KrF excimer laser (248 nm), a laser lift-off process was then performed to separate the sapphire substrate from the GaN layer, followed by HCl:DI (de-ionized) water (=1:2) cleaning for 5 min to remove Ga droplets. The undoped GaN surface was then dry etched to a thickness of 1.7 um for inductively coupled plasma (ICP). The GaN layer was degreased by treatment with trichloroethylene, acetone, and methanol for 5 min at each step, and then rinsed in DI water for 10 min. The samples were first etched in $HCl:H_2O(1:1)$ solution for 1 min and were then dipped into a room temperature alcoholic solution for 5 min, which was composed of 90 vol% tert-butanol (t-C_4H_9OH) and 10 vol% $(NH_4)_2S$ before metal deposition. The residual solution was removed from the sample by rinsing the samples in DI water for 1 min. After the sulfur treatment, conventional photo resist mask was used for the fabrication of the circular transfer length method (CTLM) patterns. The size of the pads was 200 x 200 um² and the spacings between the pads were 10, 20, 40, 60, 80, and 100 um. The Ni (20 nm)/ Au (80 nm) films were then deposited on N-face p-type GaN by electron-beam evaporation. Contact resistivity data were measured at room temperature using a parameter analyzer (HP4145B). To characterize the chemical bonding at the interface between the Ni layer and sulfur-treated p-type GaN, XPS was performed using the Mg $K\alpha$ line (1253.6 eV) as an excitation source in a ultra-high-vacuum system with a base pressure of ~1×10^{-10} Torr.

250

DISCUSSION

Figure 1 (a) show the specific contact resistance of the samples which had been deposited on N-face p-type GaN as a function of annealing temperatures. Specific contact resistances for the samples were determined from a plot of the measured resistance versus the spacings between the CTLM pads.

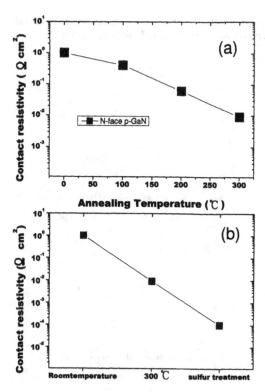

Figure 1. Variation of contact resistivity for Ni(20nm)/Au(80nm) deposited on N-face p-type GaN as (a) a function of annealing temperatures and (b) surface treatement.

The specific contact resistance was determined to be $1.01 \times 10^0 \; \Omega \; cm^2$ for the untreated samples due to the high Schottky barrier height of surface state. As the annealing temperatures were increased to 300 °C, the specific contact resistance was also decreased to the value of 9.1×10^{-3} $\Omega \; cm^2$. However, compared to the value of reported N-face n-type GaN sample and for the purpose of fabricating high efficiency LEDs, the specific contact resistance should be decreased further. We have tried the surface treatment method in order to quench the specific contact resistance and results are shown in Fig. 1 (b). By treating a alcohol-based $(NH_4)_2S$ solution to N-face p-type GaN, the specific contact resistance was decreased by four and by two orders of magnitude for no-treated and 300 °C annealed sample, achieving 1.03×10^{-4} Ω cm^2 for the alcohol based $(NH_4)_2S$ treated samples.

To study the mechanisms for the improvement of the Ohmic characteristics due to this surface treatment, XPS spectra were obtained from the Ni/sulfur-treated N-face p-type GaN sample.

Figure 2 shows the depth-profile data of Ga, Ni, O, C, and N core-level peak from the Ni/GaN for untreated and alcohol-based $(NH_4)_2S$-treated samples, respectively.

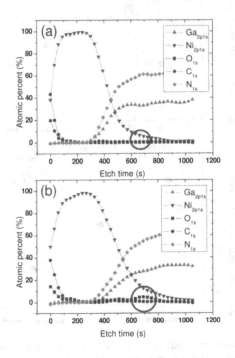

Figure 2. XPS depth profiles of (a) untreated and (b) an alcohol-based $(NH_4)_2S$-treated N-face p-type GaN.

As shown in Fig. 2, the oxide atomic percent for untreated sample is about 7 % at the interface of Ni and GaN. However, atomic percent of O peak at the interface of the samples treated with alcohol-based $(NH_4)_2S$ solution were significantly decreased compared to that of the untreated sample, indicating that the treatment effectively remove the oxide layer at the interface.

It was known that the suitable surface treatment can effectively decrease the Schottky barrier height (SBH). According to the metal semiconductor band theory,[17,18] the effective SBHs can be influenced by the presence of native oxide at the Ni/p-GaN interface, which can be described as,[18]

$$q\phi_b = q\phi_{b0} + 4\pi kT / h(2m\chi)^{1/2}\delta$$

where ϕ_{b0} is the SBH without native oxide. The second term is associated with the presence of an oxide layer, where m is the mean tunneling effective mass of the carriers, χ the mean tunneling barrier for carrier injection form a metal to a semiconductor, and δ the oxide layer thickness. This indicates that the effective removal of native oxide due to the surface treatment (Fig.2) could make a contribution to the observed reduction of the SBHs due to the decreased δ. This result shows that the oxide layer was efficiently removed from the N-face p-GaN surface by the surface treatment with alcohol-based $(NH_4)_2S$ solution. In other words, increased barrier height for hole injection from metal to p-type GaN by formation of oxide layer, leading to a detrimental effect on Ohmic contacts to p-type GaN[14] could be reduced by treatment with the alcohol-based $(NH_4)_2S$ solution. Hence, the improved Ohmic characteristics for samples can be attributed to the removal of the interfacial oxide layer.

Figure 3 shows the Ga 2p core level for the Ni/GaN interface regions of the samples before and after treatment.

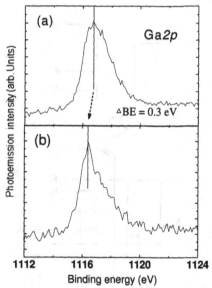

Figure 3. Ga 2p core-level spectra of the Ni/N-face p-type GaN interface for (a) an untreated sample and (b) an alcohol-based $(NH_4)_2S$-treated sample.

It is shown that the Ga 2p core level of the treated sample shifts toward the lower binding energy side by 0.3 eV, compared to that of the untreated sample. The position of surface Fermi level can be determined from the energy position of the Ga-N peaks in the core-level spectra. The difference between the Ga 2p core-level binding energy and valence band maximum in bulk GaN is 1116.7 eV. Therefore, the following equation is used to determine the position of surface Fermi level[19]:

$$E_F = E_{Ga2p} - 1116.7 \ eV$$

Where E_F is surface Fermi level position relative to the valence band maximum and E_{Ga2p} core-level binding energy for Ga-N bonds. This implies that the alcohol-based $(NH_4)_2S$-treated samples causes the surface Fermi level to shift toward the valence-band edge, resulting in a reduction in the band bending in the p-GaN. A previous study[15] reported that sulfur does not bond with nitrogen but bonds with gallium or occupies nitrogen-related vacancies. The S ions in the solution would then bond to the Ga atoms on the N-face GaN sub-surface, producing Ga sulfides and this sulfides are soluble in the sulfur solution.[16] Hence, the simultaneous process of reaction and dissolution can leave a very thin sulfide layer on the sulfur-treated GaN surfaces, leading to the effective removal of surface oxide, formation of Ga vacancies, and protection from the formation of the native oxide on exposure to air prior to metal deposition.

Figure 4 shows the schematic energy band diagram of Ni/p-GaN ohmic contact to N-face p-type GaN.

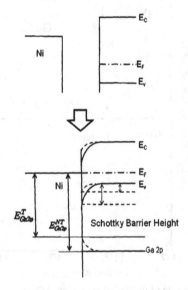

Figure 4. Schematic energy band diagram of Ni/p-GaN ohmic contact.

254

As the binding energy of Ga2p peak of surface treated sample shifted to valence-band, the Fermi energy of the GaN also shifted to valence-band maxima, resulting in the decrease of band bending between metal and GaN interface. According to this, the Schottky barrier height should be decreased as follows;

$$q\phi_p = E_{Ga2p}^T - (E_{Ga2p}^{NT} - E_V)$$

Where $q\phi_p$ is SBH between Ni and p-GaN, and E_{Ga2p}^T, E_{Ga2p}^{NT}, E_V is a binding energy of core-level spectrum for surface treated and non-treated sample, and valence band maxima, respectively. From these results, the reduced band-bending renders the decrease of effective SBH resulting in the enhanced transfer of holes between metal and valence band of GaN, indicating that the alcohol based $(NH_4)_2S$ treatment effectively enhance the ohmic contact properties of Ni/N-face p-type GaN.

CONCLUSIONS

We report on results about electrical properties of Ni-based ohmic contact on N-face p-type GaN. Contact resistance are decreased as the annealing temperature was increased. In order to lower a contact resistance further, alcohol-based $(NH_4)_2S$ surface treatment on N-face p-type GaN were conducted. Compared to the untreated sample, the specific contact resistance was drastically decreased by four orders of magnitude. The oxygen depth-profiling data showed that the alcohol-based $(NH_4)_2S$ treatment is very effective in the removal of the surface oxide layer. The Ga $2p$ core-level peak for the alcohol-based $(NH_4)_2S$-treated sample showed a red-shift by 0.3 eV toward the valence-band edge compared to that for the untreated sample. The drastic improvement in the Ohmic characteristics of the alcohol-based $(NH_4)_2S$-treated sample can be attributed to the effective removal of the surface oxide and the shift of the surface Fermi level toward the valence-band edge, resulting in the lowering of SBH.

REFERENCES

1. S. Nakamura, M. Senoh, N. Iwasa, and S. Nagahama, Jpn. J. Appl. Phys., Part 2 **34**, L797 (1995).
2. S. Nakamura, M. Senoh, N. Iwasa, T. Yamada, T. Matsushita, H. Kiyoku, and Y. Sugimoto, Jpn. J. Appl. Phys., Part 2 **35**, L217 (1996).
3. M. Asif Khan, A. R. Bhattarai, J. N. Kuznia, and D. T. Olson, Appl. Phys. Lett. **63**, 1214 (1993).
4. M. Asif Khan, J. N. Kuznia, A. R. Bhattarai, and D. T. Olson, Appl. Phys. Lett. **62**, 1786 (1993).
5. M. Asif Khan, J. N. Kuznia, D. T. Olson, J. M. Van Hove, M. Blasingame, and L. F. Reitz, Appl. Phys. Lett. **60**, 2917 (1993).
6. V. M. Bermudez, J. Appl. Phys. **80**, 1190 (1996)
7. J. K. Kim, J.-L. Lee, J. W. Lee, H. E. Shin, Y. J. Park, and T. Kim, Appl. Phys. Lett. **73**, 2953 (1998).
8. J.-S. Jang, S.-J. Park, and T.-Y. Seong, J. Vac. Sci. Technol. B **17**, 2667 (1999).

9. J.-L. Lee, J. K. Kim, J. W. Lee, Y. J. Park, and T. Kim, Electrochem. Solid-State Lett. **3**, 53 (2000).
10. V. N. Bessolov, M. V. Lebedev, and D. R. T. Zahn, J. Appl. Phys. **82**, 2640 (1997).
11. V. N. Bessolov, A. F. Ivankov, E. V. Konenkova, and M. V. Lebedev, Tech. Phys. Lett. **21**, 20 (1995).
12. Y. V. Zhilyaev, M. E. Kompan, E. V. Konenkova, and S. D. Raevskii, MRS Internet J. Nitride Semicond. Res. **4S1**, G6.14 (1999)
13. C. Huh, S.-W. Kim, H.-S. Kim, I.-H. Lee, and S.-J. Park, J. Appl. Phys. **87**, 4591 (2000)
14. Z. L. Yuan, X. M. Ding, H. T. Hu, Z. S. Li, J. S. Yang, X. Y. Miao, X. Y. Chen, X. A. Cao, X. Y. Hou, E. D. Lu, S. H. Xu, P. S. Xu, and X. Y. Zhang, Appl. Phys. Lett. **71**, 3081 (1997).
15. Y.-J. Lin, C.-D. Tsai, Y.-T. Lyu, and C.-T. Lee, Appl. Phys. Lett. **77**, 687 (2000).
16. Handbook of Chemistry and Physics, 64th ed., edited by R. C. Weast, M. J. Astle, and W. H. Beyer (CRC, New York, 1985).
17. E. H. Rhoderick and R. H. Williams, *Metal-Semiconductor contacts* (Clarendon, Oxford, U.K. 1998).
18. K. Hattori and Y. Izumi, J. Appl. Phys. **53**, 6906 (1982).
19. G. Landgren, R. Ludeke, Y. Jugnet, J. F. Morar, and F. J. Himpsel, J. Vac. Sci. Technol. B **2**, 351 (1984).

Mater. Res. Soc. Symp. Proc. Vol. 1202 © 2010 Materials Research Society   1202-I09-12

Size Reduction and Rare Earth Doping of GaN Powders Through Ball Milling

Xiaomei Guo[1], Tiju Thomas[2], Kewen K. Li[1], Jifa Qi[3], Yanyun Wang[1], Xuesheng Chen[4], Jingwen Zhang[1], Michael G. Spencer[2], Hua Zhao[5], Yingyin K. Zou[1], Hua Jiang[1], Baldassare Di Bartolo[5]
[1]Boston Applied Technologies, Inc., Woburn, MA
[2]School of Electrical and Computer Engineering, Cornell University, Ithaca, NY
[3]Department of Materials Science and Engineering, Massachusetts Institute of Technology, Cambridge, MA
[4]Department of Physics and Astronomy, Wheaton College, Norton, MA
[5]Department of Physics, Boston College, Chestnut Hill, MA

ABSTRACT

Ball milling of ammonothermally synthesized GaN powders was performed in an ethanol solution for a variety of durations, resulting in average particle sizes of nanometer. The ball milled powders showed an obviously brightened color and improved dispersability, indicating reduced levels of aggregation. X-ray diffraction (XRD) peaks of the ball milled GaN powders were significantly broadened compared to those of the as-synthesized powders. The broadening of the XRD peaks was partially attributed to the reduction in the average particle size, which was confirmed through SEM analyses. On the other hand, rare earth doping of commercial GaN powders was also achieved through a ball mill assisted solid state reaction process. Rare earth salts were mixed with GaN powder by ball milling. The as-milled powders were heat treated under different conditions to facilitate the dopant diffusion. Luminescence properties of the rare earth doped GaN powders at near infrared range were investigated and the results were discussed.

INTRODUCTION

One of the major issues encountered with high power lasers is the management of the heat generated during lasing. Gallium nitride (GaN) possesses high heat conductivity (1.3 W cm^{-1} °C^{-1}), making it a promising candidate for high power applications. However, it is extremely difficult and costly to produce single crystal GaN with sufficient size and optical quality. Polycrystalline laser hosts offer several remarkable advantages over conventional single crystal ones: higher active ions doping concentrations, larger sizes, more complex shapes, and much lower fabrication costs due to shortened fabrication process and high volume production capability [1]. Therefore, exploration in fabricating GaN ceramic laser host has been an intensively researched area in recent years.

Proper particle size and good size uniformity of starting powders are key factors to achieving desirable optical transparency in GaN ceramics since its crystallographic anisotropy effect can be reduced as the crystallite size falls into the submicron range [2].

However, mono-dispersed nanoparticle syntheses of GaN have remained a challenging task. The difficulties confronted by researchers include impurity removal, yield, size uniformity and crystallinity, etc.. Wu et al. have produced GaN powders with high yield and high purity using an ammonothermal process [3]. However, the obtained powders are in micron sizes with an apparently broad size distribution, which typically results in a wide pore size distribution, requiring higher sintering temperatures and longer durations to eliminate the porosity during ceramics consolidation process. Consequently, there would be few chances to obtain a well densified transparent ceramics using such powders.

To improve the ammonothermally synthesized GaN powers for a potential application in laser gain medium, we conducted ball milling experiments and successfully realized size reduction in the powders. In addition, demonstration of rare earth doping in commercial GaN powders through a ball milling assisted solid state reaction process was performed. The luminescence properties of doped GaN powders were investigated.

EXPERIMENTAL DETAILS

The starting GaN powders for ball milling were synthesized using a gas phase ammonothermal method, as described in somewhere else [4]. The as-synthesized powders are dark in color, possibly due to mixed nitrides in the powders. These nitrides can be dissolved away using nitric acid. A roll milling machine (U.S. Stoneware Company, model: 784) was utilized for the ball milling process. Yttria-stabilized zirconia balls with mixed sizes are used with a total ball-to-powder weight ratio ~ 20:1. Ball mill was performed in ethanol. The as received and ball milled powders were examined using Rigaku X-ray diffractometer, Malvern Zetasizer (Nano-ZS) and scanning electron microscopy (LEO 1550 Field Emission SEM).

For luminescence property investigations, powder mixtures of GaN (from Alfa Aesar) and rare earth acetates were ball milled in ethanol for 12 hours. Resulted powders were dried in air, pressed into pellets, and annealed under different conditions. Erbium and ytterbium doped GaN samples were pumped by a 970 nm continuous wave (CW) laser (Boston laser, Inc.) with a maximum power of 3 Watt (working power at 1 Watt) and power density ~2.0 Watt/cm^2. A NanoLog spectrometer from Horiba Jobin Yvon with a FL-1073 detector working at 950 V was used to measure the luminescence emission spectra. A diode laser with emission wavelength centered at 798 nm (LD1-820) was used to pump thulium doped GaN samples. A PbS detector working at 1400 nm to 2200 nm and a photomultiplier (Hamamatsu R1387) with a scanning monochromator (McPherson Model 2051) were used to measure the luminescence emission spectra. All photoluminescence measurements were performed at room temperature.

RESULTS AND DISCUSSION

Size reduction through ball milling process

The as-synthesized GaN powders were rinsed with nitric acid. The as-rinsed powders are brownish gray in color and used for ball milling. After ball milling, the powders showed noticeably brightened color and improved dispersibility, indicating possibly reduced levels of agglomerations. Obvious broadening of the XRD peaks was observed in the milled powders as shown in Figure 1. The peak broadening was attributed to either, or both, the reduction in averaged particle size and/or possible induced lattice strain by the milling action. Significant size reduction was confirmed by SEM observation as shown in Figure 2. Measurements of particle size using a Dynamic Light Scattering (DLS) based Malvern Zetasizer agree well with Scherrer formula based analysis of XRD data and SEM observation. The factors that affect the final powder size include but not limited to the duration, media and the liquid used in the ball milling.

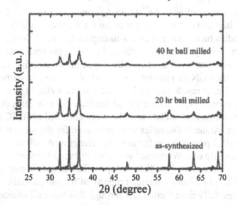

Figure 1. XRD patterns of before and after ball milled GaN powders.

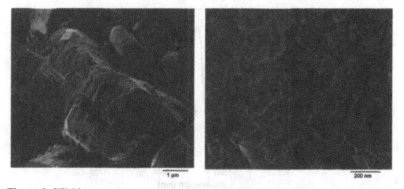

Figure 2. SEM images of ammonothermally synthesized GaN powders before (left) and after (right) 40 hours of ball milling.

The ball millings were performed using a roll milling machine under relatively mild condition instead of a high energy one. Such a low energy soft nano-sizing process can minimize contaminations introduced by the milling media thus can be practically useful. By utilizing such nano-sizing process with pre-determined characteristics, one can control the particle surface morphologies within a wide range, this is especially important in realizing transparent ceramics. Powders with smaller sizes and narrow size distribution enable consolidation at lower temperatures and shorter durations, which can then remarkably restrain the grain and pore growth during the final consolidation stage.

Rare earth doping via ball milling assisted solid state reaction

Due to the large ionic radius of Er^{3+} (1.75 Å) compared to 1.30 Å for Ga^{3+}, the concentration of erbium that can be doped into the GaN lattice could be considered rather small. However, studies have shown that in situ doping can be achieved up to 3-5 at. % with molecular beam epitaxy technology while preserving the optical activation of erbium [5]. Erbium was also ammonothermally doped in situ into the GaN powder and green emission from the GaN:Er powder was observed [6]. An alternative way to realize the doping is by solid state reactions, and ball mill could accelerate mixing and even activate the powder surface [7,8], which would facilitate the rare earth doping into GaN.

Erbium and ytterbium doped GaN samples were annealed at 600°C for 4 hours in an ammonia atmosphere. Annealed samples were measured for the room temperature photoluminescence (PL) properties at infrared wavelengths. A 970 nm centered diode laser with a power density of 2 Watts/cm^2 was used to excite the samples. Shown in Figure 3 is the PL spectrum of a 1 at. % Er^{3+} doped GaN sample at near infrared wavelength range. The photoluminescence property of the sample indicated that the Er^{3+} ions have been successfully doped into GaN through this ball mill assisted solid state reaction route [9].

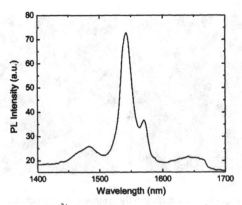

Figure 3. PL spectrum of Er^{3+}-doped GaN powders measured at room temperature.

The PL properties of the doped GaN powders under different annealing conditions were investigated based on the findings that oxygen co-doping could be beneficial to the Er^{3+} emission due to changes in local chemical bonding to a more preferred ionic environment [10]. We re-annealed the Er^{3+} and Er^{3+}/Yb^{3+}(1 at. % and 5 at. % of Ga^{3+}, respectively) doped GaN samples in air at 650°C for 5 hours. Two new samples were added with 1 at. % and 2 at. % Er^{3+} doping for the annealing. The obtained spectra were shown in Figure 4.

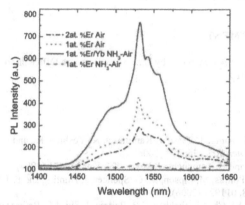

Figure 4. PL spectra of doped GaN powders annealed under different conditions.

One can see that the re-annealed Er^{3+}/Yb^{3+} co-doped sample exhibits the strongest emission peak among the four, and Er^{3+} doped GaN heated in ammonia and then in air exhibited the lowest emission, indicating that the annealing in oxygen-containing atmosphere benefited the photoluminescence emissions more in the Er^{3+}/Yb^{3+} co-doped GaN than in the one doped with Er^{3+} alone. The 2 at. % Er^{3+} doped sample gave lower emission than the one with 1at. % Er^{3+}, suggesting a concentration quenching effect. The samples annealed in air show broadened peaks generally, compared to the ones annealed in ammonia, indicating more varieties of the surrounding conditions of the Er^{3+} ions.

Thulium doped GaN samples were also produced through a similar ball mill assisted solid state reaction process. Mixed GaN powders with 0.5 at. %, 1 at. %, 2 at. % Tm^{3+} were pressed into pellets and annealed for 4 hours at 850°C in a nitrogen atmosphere. A PbS detector that was supposed to work at the Tm emission wavelength (1600 nm to 2000 nm) was used to investigate the luminescence properties. A diode laser with emission wavelength centered at 798 nm was used to pump the samples. Unfortunately, no luminescence signal was detected. This maybe due to the fact that the Tm^{3+} emission is too weak compared to that of erbium or the laser pumping power (~ miliwatts) is not sufficient.

261

CONCLUSIONS

To synthesize size uniformly distributed GaN nanoparticles with proper rare earth ions doping remains a challenging task. In the present work, we successfully reduced the micron-sized ammonothermally synthesized GaN particles to nanometers via a simple ball milling process. Rare earth doping to commercial GaN powders were also realized through ball milling assisted solid state reaction process. These results can be useful for realizing GaN ceramic laser gain medium.

ACKNOWLEDGMENT

This work is supported in part by U.S. Army Research Office through contract No. W911NF-09-C-0012.

REFERENCES

1. A. Ikesue, Y.L. Aung, T. Taira, T. Kamimura, K. Yoshida, K., and G.L. Messing, *Annu. Rev. Mater. Res.,* **36,** 397 (2006).
2. N. Kuramoto and H. Taniguchi, *J. Mater. Sci. Lett.,* **3,** 471 (1984).
3. H. Wu, C. B. Poitras, M. Lipson, M. G. Spencer, J. Hunting and F. J. Di Salvo, *Appl. Phys. Lett.,* **88,** 011921 (2006).
4. T. Thomas, MVS Chandrashekhar, C.B. Poitras, J. Shi, J.C. Reiherzer, F.J. DiSalvo, M. Lipson, and M.G. Spencer in *Rare-Earth Doping of Advanced Materials for Photonic Applications,* edited by V. Dierolf, Y. Fujiwara, U. Hommerich, P. Ruterana, J. Zavada (Mater. Res. Soc. Symp. Proc. **1111,** Warrendale, PA, 2009), paper no.1111-D04-01.
5. P.H. Citrin, P.A. Northrup, R. Birkhahn, and A.J. Steckl, *Appl. Phys. Lett.,* **76,** 2865 (2000).
6. H.Wu, C.B. Poitras, M. Lipson, and M.G. Spencer, J. Hunting and F.J. DiSalvo, *Appl. Phys. Lett.,* **86,** 191918 (2005).
7. C. Suryanarayana, *Progress in Materials Science,* **46,** 1 (2001).
8. K. Sakurai and X. Guo, *Materials Science and Engineering,* **A304-306,** 403 (2001).
9. G. Sun, X. Liu, S.D. Tse, E.E. Brown, U. Hömmerich, S. Trivedi and J.M. Zavada in *Rare-Earth Doping of Advanced Materials for Photonic Applications,* edited by V. Dierolf, Y. Fujiwara, U. Hommerich, P. Ruterana, J. Zavada (Mater. Res. Soc. Symp. Proc. **1111,** Warrendale, PA, 2009), paper no.1111-D02-08.
10. J.M. Zavada, "Rare Earth Impurities in Wide Gap Semiconductors," *Processing of Wide Band Gap Semiconductors,* edited by S.J. Pearton (Elsevier Science Publishers, New York, 2000), p.354.

Mater. Res. Soc. Symp. Proc. Vol. 1202 © 2010 Materials Research Society   1202-I01-04

Selective Area Epitaxy of InGaN/GaN Stripes, Hexagonal Rings, and Triangular Rings for Achieving Green Emission

Wen Feng[1,2], Vladimir V. Kuryatkov[1,2], Dana M. Rosenbladt[1,3], Nenad Stojanovic[1,3], Mahesh Pandikunta[1,2], Sergey A. Nikishin[1,2] and Mark Holtz[1,4]
[1]Nano Tech Center, Texas Tech University, Lubbock, Texas;
[2]Department of Electrical and Computer Engineering, Texas Tech University, Lubbock, Texas;
[3]Department of Mechanical Engineering, Texas Tech University, Lubbock, Texas;
[4]Physics, Texas Tech University, Lubbock, Texas.

ABSTRACT

We report selective area epitaxy of InGaN/GaN micron-scale stripes and rings on patterned (0001) AlN/sapphire. The objective is to elevate indium incorporation for achieving blue and green emission on semi-polar crystal facets. In each case, GaN structures were first produced, and the InGaN quantum wells (QWs) were subsequently grown. The pyramidal InGaN/GaN stripe along the <11-20> direction has uniform CL emission at 500 nm on the smooth {1-101} sidewall and at 550 nm on the narrow ridge. In InGaN/GaN triangular rings, the structures reveal smooth inner and outer sidewall facets falling into a single type of {1-101} planes. All these {1-101} sidewall facets demonstrate similar CL spectra which appear to be the superposition of two peaks at positions 500 nm and 460 nm. Spatially matched striations are observed in the CL intensity images and surface morphologies of the {1-101} sidewall facets. InGaN/GaN hexagonal rings are comprised of {11-22} and {21-33} facets on inner sidewalls, and {1-101} facets on outer sidewalls. Distinct CL spectra with peak wavelengths as long as 500 nm are observed for these diverse sidewall facets of the hexagonal rings.

INTRODUCTION

III-nitrides are important materials for blue and green solid state light emitters. Selective area epitaxy (SAE) technique has wide latitude in both composition and layer thickness through control of the mask design and growth conditions [1], and has been widely employed for the growth of InGaN/GaN materials due to several factors. First, lateral epitaxial overgrowth over the mask during SAE reduces the vertical propagation of threading dislocations in III-nitrides to a low level of 10^5 cm^{-2} [2]. Secondly, various nano- and microstructures can be formed through SAE of III-nitrides. These microstructures have been shown to be advantageous for fabricating light emitters with high extraction efficiency [3]. Furthermore, the sidewalls of microstructures generated by SAE are usually semi- or non-polar planes possessing reduced internal fields in comparison to the (0001)-plane [4].

We investigate SAE of InGaN/GaN structures with three geometrical shapes. We exploit the advantages of SAE to fabricate InGaN/GaN stripes along with triangular and hexagonal rings. Single type of sidewall facets are obtained on triangular rings to get uniform light emission while diverse sidewall facets and light emission exist on hexagonal rings. Stripes with high indium content are found to produce green light emission with good wavelength uniformity.

EXPERIMENT

AlN/sapphire of (0001) plane was used as substrates. A 100-nm thick SiO_2 layer was deposited on the substrates as mask by plasma enhanced chemical vapor deposition at 350 °C. Three different sets of SiO_2 mask patterns were produced by photolithography and wet etching. For the SAE of InGaN/GaN stripes, mask stripes were arranged along <11-20> direction, and the window width was set at 2 μm. Triangular ring openings were prepared for the SAE of triangular rings according to our observations on the SAE of InGaN/GaN stripes. The edges of the triangular ring openings were also arranged along <11-20> direction. Edge length of outer triangle was set at 30 μm, while edge length of inner triangle was 12 μm. Circle-like ring openings were produced for the SAE of hexagonal rings. The inner and outer diameters of the openings were 18 μm and 20 μm, respectively. The patterned substrates were then introduced into a vertical MOVPE reactor.

Bulk GaN was first deposited on the patterned substrate. Growth temperature and pressure were 1060 °C and 100 Torr, respectively. NH_3 flux of 1000 SCCM (SCCM denotes cubic centimeter per minute at standard temperature and pressure) and trimethylgallium (TMGa) flux of 17.5 SCCM were supplied with H_2 carrier gas. Following this, three pairs of $In_xGa_{1-x}N/In_yGa_{1-y}N$ QWs were grown on the facets of the bulk GaN using N_2 carrier gas at 500 Torr. The QWs were sandwiched between two $In_yGa_{1-y}N$ cladding layers. The well layers, $In_xGa_{1-x}N$, were grown at 830 °C, while the barrier/cladding layers were grown at 870 °C. The growth time for well layers was 120 seconds with NH_3 flux of 700 SCCM, TMGa flux of 1.0 SCCM, and trimethylindium (TMIn) flux of 100 SCCM. The growth time for barrier layers was 330 seconds with NH_3 flux of 700 SCCM, TMGa flux of 1.0 SCCM, and TMIn flux of 30 SCCM. Epitaxy was terminated with a thin GaN cover layer. In addition to microstructures, all growth was carried out on reference AlN/sapphire substrates without the SiO_2 mask during the same MOVPE run. The luminescence of the reference QW peaks at 450 nm. Based on X-ray diffraction, the well and barrier thickness (In compositions) of the reference sample are estimated to be 2 nm (x=0.20) and 4 nm (y=0.12), respectively. Profile and morphologies of the InGaN/GaN triangular microrings were mainly investigated using scanning electron microscope (SEM) images, as well as atomic force microscopy and interferometric microscopy. Optical properties of the samples were examined at room temperature using SEM-based CL system at beam energy of 7 keV with a spot diameter ~ 15 nm. A cooled CCD was used for CL spectroscopy and spectral imaging. CL intensity imaging was carried out at select wavelengths with a photomultiplier detector. More details have been described elsewhere [5, 6].

DISCUSSION

The SEM images in Fig. 1 show that growth was restricted to the openings in the SiO_2 mask for all three geometrical shapes investigated. Figure 1 (a) shows the SEM image of one completed pyramidal stripe. Smooth and flat {1-101} sidewall facets were obtained for the stripe, which is consistent with previous report [7]. Stripes oriented along the <11-20> direction are usually inclined and bound by semi-polar {1-101} sidewall facets, while stripes oriented along <1-100> can lead to semi-polar {11-22} or non-polar {11-20} sidewall facets[8]. Figure 1 (b) shows the image of representative InGaN/GaN triangular ring. The triangular ring exhibits smooth surfaces on all sidewall facets, denoted as A, B, C, and D. All these sidewall facets have an angle of ~ 62° to the (0001) plane and fall into the same family plane of {1-101}. This

264

identification is consistent with above observations of {1-101} sidewall facets for InGaN/GaN stripes oriented along the <11-20> direction.

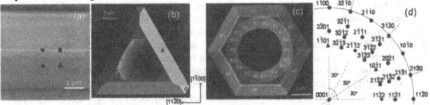

, Fig. 1. SEM images of selectively grown InGaN/GaN (a) stripes, (b) triangular rings, and (c) hexagonal rings. (d) Equivalent stereographic (0001) projection of hexagonal structures.

Figure 1(c) shows the more complicated structure of the InGaN/GaN hexagonal ring. The outer shape is hexagonal while the inner boundary is composed of a series of sidewall facets. Upon close inspection, the outer and inner sidewalls are defined by multiple inclined facets. The wall width at the narrowest point of the hexagonal ring is 6.4 μm, corresponding to lateral epitaxial overgrowth of 5.4 μm. The six outer sidewall facets of the hexagonal ring are denoted S1 through S6. Eighteen sidewall facets are observed on the inner sidewall denoted R1 through R12 and P1 through P6. Facets given common designations belong to the same family. The equivalent stereographic (0001) projection diagram of hexagonal structures, Fig. 1(d), is useful for identifying the crystal facets since it corresponds to the angular relationship between the lattice planes and directions. The facets on the outer sidewall have an angle of ~ 62° to the (0001) plane. Based on the stereographic projection, the S facets are identified as equivalent semi-polar {1-101} planes, corresponding to the triangles in Fig. 1(d). This identification is consistent with prior experiments in which pyramidal stripes were grown by SAE with orientation parallel to the <11-20> direction. The R and P facets are found to be inclined at an angle of ~ 58° to the (0001) plane. The P facets are identified as six equivalent semi-polar {11-22} planes, corresponding to the diamonds in Fig. 1(d). The {11-22} P facets have been observed for InGaN/GaN stripes oriented along the <1-100> direction. Based on Fig. 1 (d), the R facets are best identified as semi-polar {21-33} planes (squares).

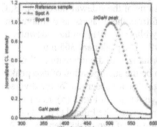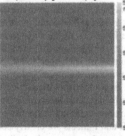

Fig. 2. (color online) (a) CL spectra of InGaN/GaN stripe. (b) Corresponding CL wavelength mapping across one full stripe.

Local CL spectra from the sidewall (spot A), ridge (spot B), and from the reference sample are plotted in Fig. 2(a). The spectrum from the reference region has maximum at wavelength 450 nm. The local CL spectrum from sidewall position A has emission peak red shifted to 500 nm. The CL spectrum from the ridge position B has maximum near 550 nm, redshifted by ~ 100 nm relative to the reference region. Figure 2 (b) shows the CL spectrum peak-wavelength mapping

265

of the stripe. The CL peak wavelength is uniform across the {1-101} sidewall facet, ranging between 500 and 510 nm. Upon reaching the ridge the CL peak abruptly shifts to ~ 550 nm.

The redshifts observed in Fig. 2(a) are attributed to thickness enhancement and indium enrichment in the selectively grown InGaN MQW stripe. Both effects originate from additional source supply due to migration effect from the masked region and lateral vapor diffusion effect during SAE. The migration lengths of the group III atoms on the mask, and their diffusion lengths in the vapor phase, are ordered according to In > Ga > Al. Therefore, indium enrichment during SAE occurs due to the enhanced supply of indium species on the stripe facets. The inclined {1-101} facet shows uniform CL spectra, indicating that QW growth on {1-101} surface is uniform. This can be interpreted as verification that the surface diffusion length is larger than the distance between the sidewall base and ridge so that InGaN forms on the inclined {1-101} facet under locally homogeneous conditions. Indium species arriving at the ridge, while undergoing surface migration, incorporate where these facets join, possibly due to the presence of a diffusion barrier. In addition, faster InGaN growth rate on the (0001) plane is well known, and the presence of a (0001) facet at the ridge may contribute to efficient indium incorporation. These factors combined produce the large redshift seen at the ridge in Fig. 2.

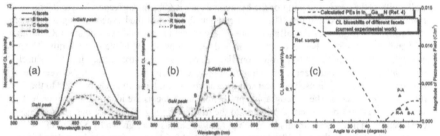

Fig. 3. (color online) (a) CL spectra from different sidewall facets of InGaN/GaN triangular ring. (b) CL spectra from different sidewall facets of InGaN/GaN hexagonal ring. (c) Blueshift CL (left axis) of InGaN QW on planes with different angles to c-plane. Curves are theoretical calculations of the magnitude of piezoelectric field (right axis).

For InGaN/GaN triangular rings, despite different dimensions of the sidewall facets, A, B, C, and D, overall consistency is observed for the spectra obtained from these facets, as displayed in Fig. 3(a). This is attributed to the formation of the same family plane {1-101} on the sidewalls and the existence of the equivalent conditions during the growth. The peak near 365 nm corresponds to the GaN underlying layer; each spectrum is normalized to the intensity of this feature. The InGaN QW emission in each spectrum appears to be the superposition of two peaks at positions 500 nm and 460 nm. In comparison, distinct CL spectra are observed for the diverse sidewall facets S, R, and P of the InGaN/GaN hexagonal ring, as shown in Fig. 3 (b). Each spectrum has a peak near 365 nm corresponding to the GaN underlying the InGaN QWs. The InGaN-related spectra from the R and P facets are clearly composed of two peaks, denoted as A and B. The InGaN QW bands from each spectrum, including for the S facet, are adequately fit using two peaks with Gaussian shape. Despite differences between spectra from different facet types, all spectra obtained from the six S facets exhibit excellent consistency, as do the measurements from respective twelve R and six P surfaces. This is reasonable, since identical facets are formed under equivalent conditions.

To analyze the internal piezoelectric field, CL spectra were measured under different SEM incident electron beam current. Triangles in Fig. 3 (c) show the CL blueshifts of InGaN QW on the hexagonal ring facets with different angles to c-plane. In each case, the observed blue shift for the A peaks is small. The B peak shift (not shown) is also gradual, < 0.05 meV/pA in S, R, and P. The corresponding blueshift observed for the (0001) reference sample is much larger. The blueshifts are due to excitation-induced screening of the internal fields when carrier density is increased, which are expected to be larger for QWs with greater initial fields. The dashed curve is calculated internal piezoelectric fields (magnitude) as a function of the tilt angle for InGaN QW planes with 15% indium composition[4]. The overall CL blueshift trend agrees with the dependence of the internal piezoelectric fields on the angles of the facet to the c-plane. The smaller blueshift observed for emission from the S, P, and R facets confirms that the internal piezoelectric field is reduced from the maximum (0001) case, as expected for semi-polar QWs.

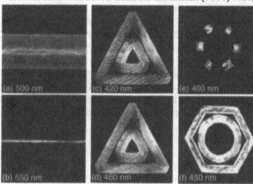

Fig. 4. Monochromatic CL intensity images over InGaN/GaN stripes (a) and (b), triangular rings (c) and (d), hexagonal rings (e) and (f).

To check for uniformity in optical properties of the InGaN/GaN microstructures, monochromatic CL intensity images were measured, as shown in Fig. 4. For the stripes, emission at 500 nm is seen from the entire sidewall facets in Fig. 4(a), supporting our conclusion that MQW growth is homogeneous on the {1-101} facets. As displayed in Fig. 4(b), emission at wavelength 550 nm originates along the entire ridge of the stripe with consistent intensity. This emission is notably absent from the side facets. The monochromatic CL images allow us to estimate the width of the 550 nm emission region to be ~ 250 nm. As for the triangular ring, at a given wavelength, similar features are observed for all different sidewall facets, A through D. The intensities at different wavelengths are consistent with the CL spectra in Fig. 3(a). At 460 nm, strong emission is seen from all sidewall facets of the microring. This agrees with the peak position of the CL spectrum. Emissions at 420 nm become a little weaker, since the light corresponds to the two shoulders of the spectra. An interesting phenomenon observed in Fig. 4(c) is the striations evident in the CL intensity images, especially at 420 nm. These striations exist on all sidewall facets of the triangular ring. The spatial distance between two adjacent light striations is about 1.4 µm, and does not appear to depend on wavelength. Upon rotating the ring by 90° in the substrate plane, striations in the CL intensity images were found to rotate with the sample verifying that they are not an artifact of the CL setup. In high magnification SEM images, morphological striations (not shown) were observed on the sidewall facets of the triangular rings, and match well the light striations. In Fig. 4(e), P facets of the hexagonal rings emit at 400 nm; light at this wavelength is notably absent from all other sidewall facets. This agrees with the presence of a peak in the CL spectrum of P facets at this wavelength and low intensity in the spectra from S and R sidewall facets in Fig. 3(b). At 450 nm, strong and nearly uniform emission intensity is seen from all sidewall facets of the hexagonal rings.

SUMMARY

InGaN/GaN micro-scale structures of different geometries were selectively grown on patterned (0001) AlN/sapphire by MOVPE. For InGaN/GaN triangular rings, each edge of the ring can be regarded as one stripe oriented along the <11-20> direction; hence the structures reveal a single type of inner and outer sidewall facets of {1-101} plane. Although all these {1-101} sidewall facets demonstrate similar CL spectra with two peaks at positions 500 nm and 460 nm, spatially matched striations are observed in the CL intensity images and surface morphologies, which indicates the closed triangular ring structure created a different growth environment for InGaN/GaN. In comparison, InGaN/GaN hexagonal rings created preferential {1-101} facets on outer sidewalls, and {11-22} and {21-33} facets on inner sidewalls. The CL spectra obtained from the same type facets exhibit excellent consistency; however, distinct CL spectra are seen from different type sidewall facets of the hexagonal rings due to dissimilar properties of different type sidewall facets. In pyramidal InGaN/GaN stripes along <11-20> direction, uniform CL emission was observed at 500 nm on the smooth {1-101} sidewall and at 550 nm on the narrow ridge. These wavelength shifts relative to the CL spectrum peak (450 nm) from the reference sample are attributed to thickness enhancement and indium enrichment in selective MOVPE.

ACKNOWLEDGMENTS

This was partly supported by the National Science Foundation (ECS-0609416), U.S. Army CERDEC (W15P7T-07-D-P040), and J. F Maddox Foundation.

REFERENCES

[1]W. Feng, J.Q. Pan, H. Yang, L.P. Hou, F. Zhou, L.J. Zhao, H.L. Zhu, W. Wang, *J. Phys. D* **39**, 2330 (2006).
[2]T.S. Zheleva, O.H. Nam, W.M. Ashmawi, J.D. Griffin, R.F. Davis, *J. Cryst. Growth* **222**, 706 (2001).
[3]H.W. Choi, C.W. Jeon, C. Liu, I.M. Watson, M.D. Dawson, P.R. Edwards, R.W. Martin, S. Tripathy, S.J. Chua, *Appl. Phys. Lett.* **86**, 021101 (2005).
[4]M. Feneberg, K. Thonke, J. Phys.: Condens. Matter 19, 403201 (2007).
[5]W. Feng, V.V. Kuryatkov, A. Chandolu, D.Y. Song, M. Pandikunta, S.A. Nikishin, M. Holtz, *J. Appl. Phys.* **104**, (2008).
[6]W. Feng, V.V. Kuryatkov, D.M. Rosenbladt, N. Stojanovic, S.A. Nikishin, M. Holtz, *J. Appl. Phys.***105**, (2009).
[7]R. Paszkiewicz, *Optica Applicata* **32**, 503 (2002).
[8]M. Ueda, T. Kondou, K. Hayashi, M. Funato, Y. Kawakami, Y. Narukawa, T. Mukai, *Appl. Phys. Lett.* **90**, 171907 (2007).

Mater. Res. Soc. Symp. Proc. Vol. 1202 © 2010 Materials Research Society   1202-I05-20

Mohammad A. Ebdah[1], Martin E. Kordesch[1], Andre Anders[2], and Wojciech M. Jadwisienczak[3].
[1]Department of Physics and Astronomy, Ohio University, Athens, OH 45701, U.S.A.
[2]Lawrence Berkeley National Laboratory, Berkeley, CA 94720, U.S.A.
[3]School of EECS, Ohio University, Athens OH, 45701, U.S.A.

ABSTRACT

In this work, europium implanted $In_xGa_{1-x}N$/GaN SL with a fixed well/barrier thickness ratio grown by metal-organic chemical-vapor deposition (MOCVD) on GaN/(0001) sapphire substrate were investigated. The as-grown and Eu^{3+} ion implanted $In_xGa_{1-x}N$/GaN SLs were annealed at different temperatures ranging from 600°C to 950°C in nitrogen ambient. The quality of the SL interfaces in undoped and implanted structures has been investigated by X-ray diffraction (XRD) at room temperature. The characteristic satellite peaks of SLs were measured for the (0002) reflection up to the second order in the symmetric Bragg geometry. The XRD simulation spectrum of the as-grown SL agrees well with the experimental results. The simulation results show 6 at.% of Indium in the $In_xGa_{1-x}N$ well sub-layers, with thicknesses of 2.4 and 3.3 nm for single $In_xGa_{1-x}N$ well and GaN barrier, respectively. It was observed that annealing the undoped SL does not significantly affect the interfacial quality of the superstructure, whereas, the Eu^{3+} ion implanted $In_xGa_{1-x}N$/GaN SL undergo partial induced degradation. Annealing the implanted SLs shows a gradual improvement of the multilayer periodicity and a reduction of the induced degradation with increasing the annealing temperature as indicated by the XRD spectra.

INTRODUCTION

Growth, structural, and optical characterization of rare earth (RE) ion doped low dimensional III-nitride (III-N) structures such as multiple quantum wells (QWs) and superlattice (SLs) have received recently increasing interest due to the technological applications of such nanostructures [1-4]. InGaN films have been extensively grown by organometallic vapor phase epitaxy (OMVPE) [5-7], molecular beam epitaxy (MBE) [8], metal organic chemical vapor deposition (MOCVD) [9], and radio frequency (RF) magnetron sputtering [10]. In these films, the indium content influences the structural as well as the optical properties of InGaN sub-layers. With the bandgap of end members ($E_g(InN) = 0.8$ eV, and $E_g(GaN) = 3.4$ eV), the band edge of the InGaN alloy can be engineered by varying the indium at.% (x parameter) in the $In_xGa_{1-x}N$ alloy to covers a broad spectral range. Accordingly, the structural design of $In_xGa_{1-x}N$/GaN SL structures is directly related to the desired optical properties. The strain, critical thickness, and phase separation are important factors in stabilizing InGaN on GaN sub-layers. Furthermore, the progress in III-N materials growth creates an opportunity to investigate the sensitization of the RE ions emission when doped into low dimensional quantum structures such as SLs. The most challenging obstacle on the path to achieving the full potential of RE-doped III-N single epilayers is the low radiative quantum efficiency of RE ions in these materials. In general, the RE radiative quantum efficiency strongly depends on the carrier mediated energy transfer processes, which have to compete with fast nonradiative recombination channels abundant in III-N hosts. It was theoretically shown that the excitation of $4f$ electrons near the interface of heterostructures is more effective than such an excitation in the bulk semiconductors [11, 12]. It was also experimentally demonstrated that the presence of quantum structures strongly affects the carrier localization and enhance the radiative emission from RE ions [13, 14].

In this paper, we use XRD and its analysis to determine the indium at.% in $In_xGa_{1-x}N/GaN$ SLs, and investigate the structure and interface quality of thermally treated undoped and europium implanted InGaN/GaN SL structures. In such SL structures, the Fourier components of the composition modulation in direct space is related to the profile of satellite peaks measured by XRD. In this paper, the performed analysis demonstrates that XRD is a powerful nondestructive technique for the structural and compositional characterization of undoped and implanted SL structures.

EXPERIMENT

InGaN/GaN multilayer structures were grown by a metal-organic chemical-vapor deposition (MOCVD) reactor by *Veeco Instruments Inc.* The InGaN/GaN multilayer structure consists of 25 periods of 2.4 nm InGaN well/3.3 nm GaN barrier structure, as determined by the post growth analysis, grown on a buffer layer of 1.0 μm undoped GaN/2.5 μm n-type GaN/(0001) sapphire substrate. The SL was caped with a 3 nm GaN cap layer. Figure 1(a) shows a schematic illustration of the grown SL structure. The designated indium at.% from the growth conditions is 6 %. The SL was nominally undoped. Doping was achieved by implantation with the Eu^{3+} ion beam normal to the surface suing acceleration energy of 150 keV at room temperature. The Eu^{3+} ions dose was ~5.5×10^{15} cm^{-2}. The as-grown and Eu^{3+} ion implanted InGaN/GaN SLs were subjected to isochronal thermal annealing in quartz tube furnace at different temperatures ranging from 600 °C to 950 °C in nitrogen ambient for 3 minutes. The interface quality of all SLs and the structure recovery from the induced implantation damage were characterized by XRD using an X-ray diffractometer [*Rigaku Geigerflex, 2000W*] with (1.54 Å) $Cu\,K\alpha_1$ and (1.544 Å) $Cu\,K\alpha_2$ radiations. The SL characteristic satellite peaks were measured in the vicinity of the (0002) reflection of GaN buffer layer in the symmetric Bragg geometry using the (θ-2θ) scan mode. Simulation of the XRD for the as-grown SL structure was done based on fully strained InGaN sub-layers model using *X'Pert Epitaxy* software v.4.2 [15]. Simulation of the Eu^{3+} ion profile depth in the SL was made using the SRIM-2008 software [16].

RESULTS AND DISCUSSION

For a SL of hexagonal alternating layers of material A and B, with layer thicknesses of d_A and d_B, respectively, the angular positions, θ_0, of the SL fundamental Bragg peaks {SL(0)s} are given by the Bragg formula

$$2D\sin\theta_0 = m\lambda, \tag{1}$$

where m is the index of the SL average atomic plane, λ is the X-ray wavelength, and D is the average lattice constant of the SL in the growth direction [0001]. D is given by

$$D = (C_A\,d_A + C_B\,d_B)/(d_A+d_B) = (C_A\,d_A + C_B\,d_B)/\Lambda \tag{2}$$

where C_A and C_B are the atomic lattice constants in the [0001] direction of material A and B, respectively, and Λ is the period of the SL. For the (0002) reflection, $m = 2$. The XRD pattern of the SL is a group of the zero-th order peaks {SL(0)s} and the secondary satellite peaks {SL(n), n = ± integer $\neq 0$} around each SL(0). The positions of SL(n)s around the m-th SL(0) are given by the modified Bragg formula:

270

$$2\sin(\theta_n) = \lambda(m/D + n/\Lambda) \qquad (3).$$

Thus, the period, of the SL can be written in terms of two satellite peaks (i and j) [17]

$$\Lambda = d_A + d_B = (n_i - n_j)\,(\lambda/2)/(\sin\theta_i - \sin\theta_j) \qquad (4).$$

According to the Vegard's law, the out-of-plane relaxed lattice constants of a ternary alloy and the binary members are related linearly through the composition at.%, x [18]. Therefore, in the case of $In_xGa_{1-x}N$, we have that

$$C_{InGaN} = x\,C_{InN} + (1-x)\,C_{GaN} \qquad (5).$$

By letting A = GaN, and B = $In_xGa_{1-x}N$, the indium at.% can be calculated from the (0002) XRD spectra by combining Eqs. (2)-(5), and by using the lattice constants $C_{InN} = 5.702$ Å, $C_{GaN} = 5.1851$ Å, respectively [19, 20]. However, in order to calculate x and due to the correlation between x and d_{InGaN}, it is necessary to use a well known value of d_{InGaN} (or d_{GaN}) obtained from the growth conditions, or simulation methods. If $In_xGa_{1-x}N$ is biaxially strained in the (0001) plane to the GaN substrate of in-plane lattice parameter a_{GaN}, then the in-plane lattice constants are matched resulting in distorting the out-of-plane lattice constant of the $In_xGa_{1-x}N$ alloy layer according to [21]

$$(C^*_{InGaN} - C_{InGaN})/C_{InGaN} = -[2v/(1-v)]\,[\,(a_{GaN} - a_{InGaN})/a_{InGaN}\,] \qquad (6),$$

where v is Poisson's ratio of $In_xGa_{1-x}N$, and C^*_{InGaN} is the strained out-of-plane lattice constant of $In_xGa_{1-x}N$. Therefore, in the case of fully strained InGaN, the lattice constant $C_A = C_{InGaN}$ in Eq. (2) must be replaced with the distorted lattice constant $C^*_A = C^*_{InGaN}$. The calculated value of C^*_{InGaN} from Eq. (2) is then used to obtain the relaxed constant C_{InGaN} using Eq. (6), which is needed for calculating the indium at.%, x, from Vegard's law (Eq. (5)). The Possion's ratio for $In_xGa_{1-x}N$ alloy is obtained by linear interpolation of those of InN and GaN ($v_{InN} = 0.291$, $v_{GaN} = 0.203$) [22]. InGaN sub-layers are fully strained (pseudomorphic) on GaN when less than the critical layer thickness (CLT) [23,24]. The CLT for InGaN films grown on GaN substrates has

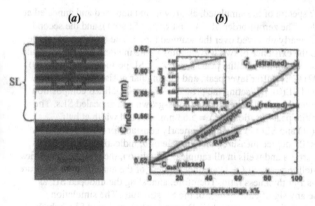

Figure 1: (a) Schematic diagram of the InGaN/GaN SL structure. (b) The lattice constant of InGaN alloy as a function of x for both relaxed and fully strained states. The inset shows the linear and non-linear relation between C_{InGaN} and x in the two cases, respectively.

been investigated and reported before [23]. For $x = 6$ at.%, the CLT is less than 100 nm, which is two orders of magnitude higher than the thickness of InGaN sublayers in our SL samples.

(a) *(b)*

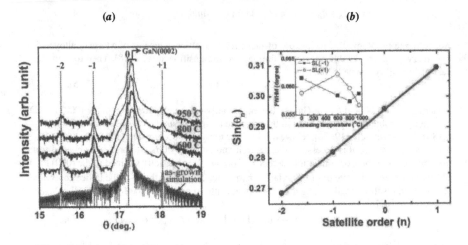

Figure 2: (a) The (0002) XRD spectra of the simulation (bottom), as-grown, and undoped and annealed at 600, 800, and 900 ℃ SLs. (b) The sine function of the satellite peaks as a function of the satellite order. The red line represents fitting sin (θ_n) to Eqs. (3) with the aid of Eqs. (2)-(6). The fitting result gives $x = 5.98$ at.%, using the simulation results of $\Lambda = 5.7$ nm, and $d_{InGaN} = 2.4$ nm, respectively. The inset shows the FWHM of the SL(-1) and SL(+1) satellite peaks as a function of the annealing temperature.

Figure 2(a) shows the XRD spectra of the simulated, as-grown, and undoped and annealed at 600, 800, and 900 ℃ SL samples. The zeroth-order SL(0), first-orders SLs(±1), and the second-order SL(-2) satellite peaks are clearly observed over the scanned range. The (0002) peak of the buffer GaN layer is split due to the existence of two wavelengths in the incident radiations, while such splitting is not clearly observed in the satellite peaks since $\lambda \ll$ SL period [25]. The SL(0) peak is resolved from the (0002) GaN buffer layer peak, and the angular positions of satellite peaks are well distributed around SL(0), indicating the presence of a well-defined composition modulations of InGaN/GaN along the [0001] direction of the as-grown and annealed SLs. The calculated period from the satellite peak positions is $\Lambda = 5.63$ nm. The full width at half maximum (FWHM) of the SL(-1) and SL(+1) satellite are nearly constant for all undoped SLs with a variation of less than 6×10^{-3} degree measured on the scale of θ, indicating the same interface quality between the barriers and wells in all samples. In addition, the relative intensities of the positive and negative satellite peaks SL(± 1) are independent of the annealing temperature, indicating a constant well to barrier thickness ratio. Therefore, annealing the undoped SL, as specified above, does not cause any significant change of the SL structure. The simulation spectrum is in good agreement with the experimental XRD spectra of all undoped SLs, which gives thicknesses of a single InGaN well and GaN barrier of 2.4 and 3.3 nm, respectively, and

272

thus a period of 5.7 nm. The simulation results yields x = 6 at.% in the $In_xGa_{1-x}N$ well. On the other hand, Fig.1(b) shows the sine function of the satellite peaks as a function of the satellite order fitted to Eqs. (3). The fitting result gives x = 5.98 at.%, using the simulation values of Λ = 5.7 nm, and d_{InGaN} = 2.4 nm. This result is in excellent agreement with the simulation result. For x = 6 at.% the lattice constant of $In_xGa_{1-x}N$ layer can be directly obtained from Fig.1(b), C^*_{InGaN} = 5.233 Å, with a total strain normal to the growth direction [0001] of +0.33 % with respect to the relaxed layer structure assuming the GaN layers fully relaxed. Calculation of the average lattice constant of the SL structure from the angular position of SL(0) gives D = 5.205 Å, which is in excellent agreement with the value obtained from Eq. (2), D = 5.206 Å.

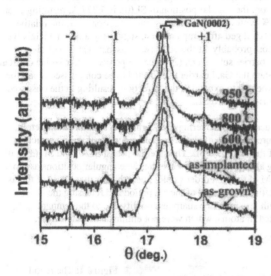

Figure 3: (from bottom to top) The (0002) XRD spectra of the as-grown SL, Eu^{3+} ions implanted SL, and SL:Eu^{3+} SLs annealed at 600, 800, and 900 °C SLs, respectively. The vertical dashed lines represent the positions of the satellite peaks in as-grown SL.

It is known that the quality of RE-implanted SL interfaces is reduced due to the ion implantation induced defects, strain, and/or diffusional intermixing of the sub-layer materials. For layers that are deeply enough far away from the surface of the SL structure and due to the lack of Eu^{3+} to penetrate to that depth, the lattice mismatch of adjacent layers is absorbed by elastic deformation, with InGaN fully strained to the GaN sub-layers without forming misfit dislocations [26], similar to the strain in the as-grown sample. On the other hand, layers closer to the surface of the SL structure suffer more displacement defects/strain caused by implantation that tends to expand the surface area due to the formation of point defects [27], and hence adds an additional strain to these layers [28,29]. This results in a non-uniform implantation induced strain along the depth of the SL structure that broadens the satellite peaks [30]. In the near-surface region, the implantation induced expansion is not restricted along the normal to the surface direction [0001], while expansion parallel to the plane of growth is constrained due to the strain to the underlying undamaged layers, resulting in a total implantation induced stress over the area, and thus bending the stack of layers along the [0001] direction. Moreover, the diffusional intermixing of the sub-layer materials, mainly caused by migrating the In atoms from the wells to the barriers, adds another contribution to the satellite peaks broadening that can not be restored by annealing. It can be seen from Fig.3 that implantation causes a kind of degradation to the SL structure. For the as-implanted sample, the satellite peaks are broader with less intensity than those of the as-grown sample, and SL(-1) satellite peak is shifted to lower angles while the position of SL(+1) remained unchanged. This indicates a less sharp interface quality between the implanted barriers and wells, and defines a non-uniform period along the [0001] direction.

The implantation induced stress is obviously manifested by the observed shift of the satellite peak SL(0) in the as-implanted SL with respect to the as-grown SL. The average lattice constant of the as-implanted SL, estimated from the angular position of SL(0), is 5.227 Å, resulting in an implantation induced strain of +0.4% with respect to the as-grown SL. Moreover, the relative intensities of SL(-1) and SL(+1) has changed after implantation, suggesting a minor InGaN well thickness modification., which is most probably attributed to two reasons: (1) the partial migration of In from the well to the barrier sub-layers, (2) since the perpendicular lattice constant of GaN is less than that of InGaN, then the GaN barrier is expected to be compressed along the [0001] direction, while InGaN is expected to expand in this direction, resulting in the observed positive perpendicular strain in the as-implanted SL.

Annealing at 600 °C does not succeed in recovering the original SL structure completely, while annealing at higher temperatures provide a relatively good recovery of the SL structure with the SL period equals to that of the as-grown SL. The satellite peaks, SL(0) and SL (±1), of the as-implanted SL after annealing at 800 and 950 °C have the same angular positions as those of the as-grown SL, with the FWHM of the first order satellite peak higher than those of the as-grown SL by (0.02° - 0.03°) , indicating a recovery of the SL period and releasing the implantation induced-strain, but with less interface sharpness quality due to the remnant implantation defects as well migrated In atoms, which were not eliminated by annealing.

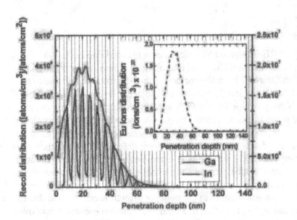

Figure 4: The recoil distribution of the In and Ga atoms generated using SRIM software. The vertical lines indicate the interface boundaries between the sub-layers. The inset shows the depth profile for the Eu^{3+} ions implanted in InGaN/GaN (2.4nm/3.3nm) SL.

Figure 4 shows the recoil distribution of the In and Ga atoms and the depth profile for the Eu^{3+} ions implanted in InGaN/GaN (2.4nm/3.3nm) SL generated using SRIM software. The ion energy and dose used for the simulation were the same as the experimental ones used for implantation. The vertical lines indicate the interface boundaries between the sub-layers of the SL. It is obvious that the distribution of the recoiled In and Ga atoms shows a semi-oscillatory behavior over the depth range close to the SL surface, which is a result of compositional modulation of InGaN/GaN. The depth profile of the Eu^{3+} ions does not show such behavior, which is attributed to the density ratio of InGaN/GaN (6.1896/6.15 ~ 1), and thus having an equal probability of stopping an Eu^{3+} ions in either the InGaN or GaN sublayers. The peaks of the In and Ga atoms occur very close to the sub-layers interfaces indicating a migration/intermixing of composition for adjacent layers, as predicted by the XRD results. The Eu^{3+} depth distribution profile shows a peak at ~ 30 nm from the SL surface. The Eu^{3+} ion density reaches 7.5×10^{-4} of its maximum value at depth ~80 nm. The SL sub-layers at depth >80 nm are more protected from the implantation induced strain with period thickness closer to the as-grown SL.

CONCLUSIONS

We have studied the structural properties of InGaN/GaN and Eu^{3+} ion implanted InGaN/GaN SLs. Calculations of the In at.%, and the sub-layers thicknesses were accomplished in the framework of a multi-layered strained structure. It was found that implantation-induced damage does not destroy the SL structure completely, and significant structure recovery is attained by thermal annealing. Strictly, the implantation induced strain on the SL was highly released by annealed at enough high temperature. Further work is in progress to investigate optical properties of the Eu-implanted SL and the energy transfer from the SL structure to the $4f$-shell system of the Eu^{3+} ion and comparing obtained results with a reference Eu-implanted GaN and InGaN epilayers.

ACKNOWLEDGMENT

The project was partially supported by the Ohio University 1804 Fund grant received by W.M.J. A.A. acknowledges support by the U. S. Department of Energy under Contract No. DE-AC02-05CH11231. The authors acknowledge the assistance of Mr. Jingzhou Wang in this project.

REFERENCES

1. A. Steckl, and J. Zavada, *MRS Bull.* **24**, 33 (1999).
2. H. Lozykowski, W. Jadwisienczak, and I. Brown, *Appl. Phys. Lett.* **74**, 1129 (1999).
3. A. Steckl, J. Park, and J. Zavada, *Materials Today* **10**, 20 (2007).
4. R. Dahal, C. Ugolini, J. Lin, H. Jiang, and J. Zavada, *Appl. Phys. Lett.* **93**, 033502 (2008).
5. S. Nakamura, T. Mukai, and M. Senoh, *Appl. Phys. Lett.* **64**, 1687 (1994).
6. H. Amano, T. Tanaka, Y. Kunii, S. T. Kim, and I. Akasaki, *Appl. Phys. Lett.* **64**, 1377 (1994).
7. M. Asif Khan, S. Krishnankutty, R. A. Skogman, J. N. Kuznia, D. T. Olson, and T. George, *Appl. Phys. Lett.* **65**, 520 (1994).
8. R. Singh, D. Doppalapudi, T. D. Moustakas, and L. T. Romano, *Appl. Phys. Lett.* **70**, 1089 (1997).
9. N. A. El-Masry, E. L. Piner, S. X. Liu, and S. M. Bedair, *Appl. Phys. Lett.* **72**, 40 (1998).
10. M. A. Ebdah, D. R. Hoy, and M. E. Kordesch, *Mat. Res. Soc. Symp. Proc.* **1151**, SS03-05, (2009).

11. O. Gusev, J. Prineas, E. Lindmark, M. Bresler, G. Khitrova, H. Gibbs, I. Yassievich, B. Zakharchenya, and V. Masterov, *J. Appl. Phys.* **82**, 1815 (1997).

12. V. Masterov, and L. Gerchikov, *Semiconductors* **33**, 616 (1999).

13. M. A. Ebdah, W. M. Jadwisienczak, M. E. Kordesch, S. Ramadan, H. Morkoc, and A. Anders, *Mat.Res. Soc. Symp. Proc.* **1111**, D04-12 (2009).

14. H. J. Lozykowski, and W.M. Jadwisienczak, J. Han, I.G. Brown, *Appl. Phys. Lett.* **77**, 767 (2000).

15. *X'Pert Epitaxy* software v.4.2 from *PANalytical B. V.*, Netherlands, http://www.panalytical.com.

16. *SRIM-2008* software, http://www.srim.org.

17. W. Li, P. Bergman, I. Ivanov, W. Ni, H. Amano, and I. Akasa, *Appl. Phys. Lett.* **69**, 22 (1996).

18. L. Vegard, *Z. Phys.* **5**, 17, (1921).

19. J. Dyck, K. Kash, C. Hayman, A. Argoitia, M. Grossner, J. Angus, and Wei-Lie Zhou, *J. Mater. Res.* **14**, 2411 (1999).

20. H. Angerer *et al*, *Appl. Phys. Lett.* **71**, 1504 (1997).

21. M. Schuster, P. O. Gervais, B. Jobst, W. Hosler, R. Averbeck, H. Riechert, A. Iberi, and R. Stommer, *J. Phys. D* **32**, A56 (1999).

22. A. Wright, *J. Appl. Phys.* **82**, 2833 (1997).

23. C. A. Parker, J. C. Roberts, S. M. Bedair, M. J. Reed, S. X. Liu, and N. A. El-Masry, *Appl. Phys. Lett.* **75**, 2776 (1999).

24. M. E. Vickers, M. J. Kappers, T. M. Smeeton, E. J. Thrush, J. S. Barnard and C. J. Humphreys, *J. Appl. Phys. Lett.* **94**, 1565 (2003).

25. C. Jin, B. Zhang, Z. Ling, J. Wang, X. Hou, Y. Segawa, X. Wang, *J. Appl. Phys.* **81**, 8, (1997).

26. J. W. Matthews and A. E. Blakeslee, *J. Cryst. Growth* **27**, 118 (1974).

27. R. E. Whan and G. W. Arnold, *Appl. Phys. Lett.* **17**, 378 (1970).

28. E. P. EerNisse, *Appl. Phys. Lett.* **18**, 581 (1971); also E. P. EerNisse, Sandia Report SC-RR-710424 (1971).

29. D. R. Myers, P. L. Gourely, and P. S. Peercy, *J. Appl. Phys.* **54**, 5032 (1983).

30. D. R. Myers, S. T. Picraux, B. L. Doyle, G. W. Arnold, and R. M. Biefeld, *J. Appl. Phys.* **60**, 3631 (1986).

Mater. Res. Soc. Symp. Proc. Vol. 1202 © 2010 Materials Research Society   1202-I05-21

Growth temperature - phase stability relation in In$_{1-x}$Ga$_x$N epilayers grown by high-pressure CVD

G. Durkaya[1], M. Alevli[1], M. Buegler[1,2], R. Atalay[1], S. Gamage[1], M. Kaiser[2], R. Kirste[2], A. Hoffmann[2], M. Jamil[3], I. Ferguson[3] and N. Dietz[1]

[1] Department of Physics & Astronomy, Georgia State University, Atlanta, GA 30303
[2] Technical University Berlin, Institute of Solid State Physics, Berlin, Germany
[3] School of ECE, University of North Carolina at Charlotte, Charlotte, NC 28223

ABSTRACT

The influence of the growth temperature on the phase stability and composition of single-phase In$_{1-x}$Ga$_x$N epilayers has been studied. The In$_{1-x}$Ga$_x$N epilayers were grown by high-pressure Chemical Vapor Deposition with nominally composition of x = 0.6 at a reactor pressure of 15 bar at various growth temperatures. The layers were analyzed by x-ray diffraction, optical transmission spectroscopy, atomic force microscopy, and Raman spectroscopy. The results showed that a growth temperature of 925 °C led to the best single phase InGaN layers with the smoothest surface and smallest grain areas.

INTRODUCTION

The ternary In$_{1-x}$Ga$_x$N alloy system attracts significant attention due to its unique physical properties such as direct band-gap, high carrier mobility, and strong chemical bonding[1,2]. The optical band gap of the In$_{1-x}$Ga$_x$N alloy system can be tuned from ultraviolet (E$_g^{GaN}$=3.4 eV) to near-infrared (E$_g^{InN}$=0.7eV), spanning over more than 80% of the solar spectrum. This is of interest for the development of high-efficiency monolithic multijunction photovoltaic solar cells based on In$_{1-x}$Ga$_x$N / Ga$_{1-x}$In$_x$N heterostructures. However, the growth of the In$_{1-x}$Ga$_x$N alloys and heterostructures is a challenge due to the lower disassociation temperature of InN compared to that of GaN. A further challenge is the large difference between the lattice constants of the binaries InN and GaN[3] (~11%), which may induce lattice strain[4] and contribute to a potential solid-phase miscibility gap in the ternary In$_{1-x}$Ga$_x$N system[5]. These facts contribute to the reported compositional inhomogeneity observed in InGaN layers[6-11], which reduces the device efficiencies of InGaN based optoelectronic structures.

The phase stability of InGaN epilayers has been studied for different growth temperatures with different growth techniques[7,10,11]. For instance, InGaN layers grown by RF-MBE show a linear correlation between gallium incorporation with increased growth temperature between 600°C - 700°C[10]. MOVPE grown InGaN layers exhibited a similar behavior in the temperature range between 700°C - 850°C[12]. Pantha et al.[7] reported the growth of single phase InGaN layers by MOCVD and observed a decreased indium incorporation with increasing growth temperature from 600°C - 750°C. This research effort explores the potential of high-pressure Chemical Vapor Deposition (HPCVD) to improve the phase stability in InGaN layers, utilizing high pressures nitrogen gas to stabilize the In$_{1-x}$Ga$_x$N growth surface and effectively suppressing the thermal decomposition process above the growth surface[12,13]. This contribution focuses on the characterization of a set of single-phase In$_{1-x}$Ga$_x$N layers that were grown under identical growth conditions, varying the growth temperature only. X-Ray

Diffraction (XRD), Optical Transmission Spectroscopy (OTS), and Raman spectroscopy were used to analyze the structural and optical properties of the epilayers. The findings were linked to the surface morphological properties of the $In_{1-x}Ga_xN$ layers.

EXPERIMENTAL DETAILS

The $In_{1-x}Ga_xN$ epilayers analyzed were grown by HPCVD on ~5 μm thick GaN/c-plane sapphire templates. Trimethylindium (TMI), Trimethylgallium (TMG) and ammonia (NH$_3$) precursors were used to provide active indium, gallium and nitrogen fragments respectively to the growth surface. As depicted in Figure 1, the precursors were provided to the growth surface via temporally controlled precursor pulses, which are embedded into the nitrogen main carrier gas stream. NH$_3$ and (TMI, TMG) injection times were 2.0 sec and 0.8 sec, respectively. The pulse separations between TMI/TMG - ammonia and between ammonia - TMI/TMG were set to 1.4 sec and 2.2 sec, respectively. The $In_{1-x}Ga_xN$ layers were grown at a reactor pressure of 15 bar, a nitrogen (N$_2$) main carrier gas flow of 12 slm (standard liters per minute), a group V-III molar precursor ratio of 1500, and a group III composition set value of x = 0.6. All experimental parameters were kept constant except the growth temperature, which was varied between 910°C and 960°C.

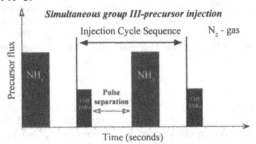

Simultaneous group III-precursor injection

Injection Cycle Sequence   N$_2$ - gas

Figure 1.

The pulsed injection sequence employed for $In_{1-x}Ga_xN$ growth

XRD experiments were carried out utilizing an X˙Pert PRO MPD (Philips) 4-circle diffractometer with a monochromatic X-ray (CuKα) source. XRD spectra were analyzed by Gaussian curve fitting to determine the position and Full Width Half Maximum (FWHM) of the (0002) Bragg reflex. The position and corresponding miller indices of this Bragg reflex were evaluated together to calculate the lattice parameter 'c' of the $In_{1-x}Ga_xN$ layers[3]. The composition of the $In_{1-x}Ga_xN$ epilayers were estimated using Vegard's law, which assumes a linear dependence of the ternary lattice parameters and their binaries alloys GaN and InN, respectively. Neglecting further any interfacial strain effects on XRD spectra, the lattice parameter 'c' can be expressed as

$$c_o^{In_{1-x}Ga_xN} = x \cdot c_o^{GaN} + \left(1-x\right) \cdot c_o^{InN} \qquad (1)$$

In order to analyze the behavior of the absorption edge of $In_{1-x}Ga_xN$ layers for different growth temperatures, optical transmission experiments were carried out at room temperature using a UV-VIS-NIR spectrometer. The acquired optical transmission spectra were corrected for detector, monochromator and light source characteristics and normalized to the growth templates used. The optical absorption spectra (OAS) of the layers were calculated from the optical

278

transmission spectra using Beer–Lambert's law in order to estimate the optical absorption edge of the $In_{1-x}Ga_xN$ alloys. The surface morphology of the layers was analyzed by Atomic Force Microscopy (AFM) using a 'XE 100 Park Systems' AFM in non-contact mode. The AFM tips used in the AFM experiments had a resonance frequency of 300 kHz and a spring constant of 45N/m. The phonon modes of the $In_{1-x}Ga_xN$ layers were studied by Raman spectroscopy in back-scattering geometry (z(xx)z̲) using an excitation wavelength of 532nm. The Raman spectra were analyzed using a Lorentzian peak fitting algorithm in order to obtain peak positions and FWHM's values for the phonon modes.

RESULTS AND DISCUSSION

Figure 2a shows 2Θ-ω scans for the $In_{1-x}Ga_xN$ layers with a nominally composition value x=0.60 grown at growth temperatures ranging from 910°C to 960°C. All epilayers exhibit single $In_{1-x}Ga_xN$ (0002) Bragg reflexes, indicating no macroscopic observable phase separations. The line-shapes and peak positions of the $In_{1-x}Ga_xN$ (0002) Bragg reflexes show a strong dependency with growth temperature. The line-shape and peak position analysis of these Bragg reflexes are summarized in Table 1. The $In_{1-x}Ga_xN$ layer grown at 925°C exhibited the most pronounced $In_{1-x}Ga_xN$ (0002) Bragg reflex. The estimated InGaN composition as a function of growth temperature is depicted in Figure 2b. The red dashed line in Figure 2b marks the experimental set-point, defined by the set values for the precursors TMI and TMG in the gas phase. Figure 2b indicates a nonlinear correlation between the molar group III-ratio in the gas phase and bulk layer, with a closest match for 925°C for which the highest indium incorporation is observed.

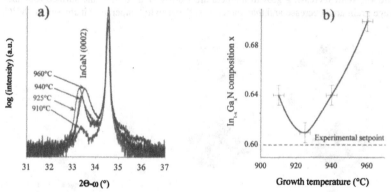

Figure 2. a) XRD patterns and **b)** estimated composition for $In_{1-x}Ga_xN$ epilayers grown on GaN/c-sapphire templates with different growth temperatures varying from 910°C to 960°C

The optical absorption spectra obtained for the $In_{1-x}Ga_xN$ epilayers grown at different temperatures are shown in Figure 3a. As shown, the optical absorption edge changes as a function of the growth temperature. To quantify the absorption edge, a linear slope fit of the curves was used to obtain the intercept point with the energy axis. The calculated intercept values are plotted in Figure 3b as a function of the growth temperature. The estimate shows that the absorption-edge follows the indium composition behavior shown in Figure 2b. The highest indium incorporation is observed for a growth temperature of 925°C with decreasing indium content as the growth temperature increases to 960°C.

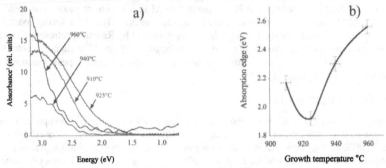

Figure 3. **a)** OAS spectra and **b)** calculated absorption edge of In₁₋ₓGaₓN layers grown at different growth temperatures from 910 °C to 960 °C

Figure 4.a, b, c and, d depict 2μm x 2μm AFM images of the In₁₋ₓGaₓN epilayers grown at 910°C, 925°C, 940°C, and 960°C, respectively. Statistical analysis techniques were used to calculate the surface roughnesses and average grain areas of these AFM images. The correlation of surface roughness and average grain area as function of growth temperature are shown in the Figure 4e. With increasing growth temperature from 910°C to 925° the surface roughness and average grain area decrease and than increase as the growth temperature increases to 960°C.

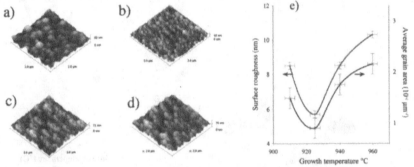

Figure 4. (a-d) 2μm x 2μm AFM images of In₁₋ₓGaₓN layers grown at **a)** 910°C, **b)** 925°C, **c)** 940°C, **d)** 960°C , **e)** Surface roughness and average grain area as a function of the growth temperature.

The Raman spectra for the In₁₋ₓGaₓN epilayers are shown in Figure 5a in the frequency region for the E₂(high) and A₁(LO) phonon modes. In wurtzite InN and GaN, the Raman phonon modes allowed in z(xx)z̄ geometry along (0001) direction are A₁(LO) and E₂(high[14]. The most distinct phonon mode observed in the Raman spectra is the A₁(LO) mode, while the E₂(high) mode is present, but not distinct enough to be statistically analyzed. The A₁(LO) Raman line was fitted using two Lorentzian oscillators side by side. The curve fitting results are summarized in Table 1 and variations in the peak positions as a function of growth temperature is depicted in

Figure 5b. The two oscillator contribution are denoted as low - and high energy $A_1(LO)$ oscillator, respectively. Also drawn in is the $In_{1-x}Ga_xN$ $A_1(LO)$ mode position as function of composition x, assuming a linear behavior of the $A_1(LO)$ mode between the binaries. The fitted high energy $A_1(LO)$ oscillator peak corresponds to the major phase according to the peak area analysis as shown in Table 1 and its position follows very closely a single $A_1(LO)$ mode behavior. The origin and behavior of low energy $A_1(LO)$ oscillator will need further studies.

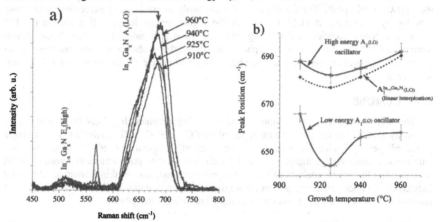

Figure 5. a) Raman spectra of $E_2(high)$ and $A_1(LO)$ region for $In_{1-x}Ga_xN$ layers grown with different growth temperatures, b) $A_1(LO)$ phonon peak positions for the $In_{1-x}Ga_xN$ epilayers.

	Growth temperature	910°C	925°C	940°C	960°C
XRD (0002)	2-θ (in deg)	33.36	33.31	33.38	33.56
	FWHM (arcsec)	2160	1210	1510	2110
	Estimated composition x	0.64	0.61	0.64	0.70
OAS	Absorption edge (eV)	2.2	1.9	2.3	2.6
AFM	Surface roughness (nm)	8.5	5.7	8.6	10.3
	Ave. grain area (10^{-2} μm²)	1.5	0.9	1.8	2.2
RAMAN	Low energy side position (cm⁻¹)	666	644	656	658
	Low energy side FWHM (cm⁻¹)	60	43	49	47
	High energy side position (cm⁻¹)	688	682	685	693
	High energy side FWHM (cm⁻¹)	33	38	32	34
	Area ratio (High energy peak / Low energy peak)	1.5	1.6	1.5	1.8

Table 1. Summary of the results obtained from XRD, AFM, OAS, and Raman spectroscopy analyzing $In_{1-x}Ga_xN$ layers with a nominal set value of x = 0.6, as a function of growth temperature.

The XRD analysis suggests that – under the set of chosen growth conditions - at the growth temperature of 925°C the highest amount of indium was incorporated in the $In_{1-x}Ga_xN$ (x=0.61) epilayers. Smaller and high growth temperatures led the less indium incorporation and reduced structural quality. The XRD (0002) Bragg reflex from the $In_{1-x}Ga_xN$ layer grown at 925°C showed an optimum, but is at least a factor 3 to 5 too high for device quality material. The same

In$_{1-x}$Ga$_x$N layer, showed the lowest optical absorption edge, the smoothest surface roughness and the smallest average grain area. The A$_1$(LO) phonon line in Raman spectra had to be fitted using two contributions, where the high energy oscillator position follows closely a single A$_1$(LO) mode behavior with composition x. The low energy oscillator in the A$_1$(LO) peak might be due to microscopic compositional fluctuations that are not observable in XRD spectra, however, a further analysis is required to understand the origin and behavior of low energy contribution to the A$_1$(LO) mode peak. The results from Raman analysis are in good agreement with those reported by Hernandez et al.,[15] in which the higher indium incorporation shifted the LO phonons to lower energies, while the spectral line-shape is broadened. Although the growth temperature 925°C exhibits the highest phase stability, further studies in intermediate steps of the investigated growth temperatures are needed in order to realize the peculiarity of this growth temperature.

CONCLUSIONS

We have studied the compositional variations in single-phase In$_{1-x}$Ga$_x$N (x=0.6) epilayers grown by HPCVD in the temperature range of 910°C - 960°C. XRD analysis revealed single phase epilayers with a local structural optimum at growth temperature of 925°C, which coincide with the smoothest surface morphology and the smallest average grain area as analyzed by AFM analysis. The line shape analysis of the A$_1$(LO) phonon mode observed in Raman spectroscopy indicates the highest strain component for a growth temperature of 925°C, at which the indium incorporation was closest to the set molar group III ratio in the gas phase.

ACKNOWLEDGMENTS

We gratefully acknowledge the support by AFOSR grant# FA9550-07-1-0345 and GSU-REP.

REFERENCES

1. H. Morkoc, *Nitride semiconductors and devices*, Springer-Verlag, (1999).
2. Wu, J., W. Walukiewicz, K. M. Yu, W. Shan, J. W. Ager III, E. E. Haller, H. Lu and W. J. Schaff, W. K. Metzger, S. Kurtz, J. of Appl. Phys. 94(10): 6477 6482. (2003).
3. F.K. Yam, Z. Hassan, Superlatt. and Microstr., 43 1–23 (2008).
4. S. Yu. Karpov, MRS Internet J. Nitride Semicond. 3, 16 (1998).
5. I. Ho, G.B. Stringfellow, Appl. Phys. Lett. 69, 2701-2703 (1996).
6. R. Singh, D. Doppalapudi, T. D. Moustakas, and L. T. Romano, Appl. Phys. Lett. 70, 1089 (1997)
7. B. N. Pantha, J. Li, J. Y. Lin, and H. X. Jiang, Appl. Phys. Lett. 93, 182107 (2008)
8. N. A. El-Masry, E. L. Piner, S. X. Liu, and S. M. Bedair, Appl. Phys. Lett. 72, 40 (1998).
9. C.A. Chang, C. F. Shin, N. C. Chen, T. Y. Lin, and K. S. Liu, Appl. Phys. Lett. 85, 6131, (2004)
10. K. Kushi, H. Sasamoto, D. Sugihara, S. Nakamura, A. Kikuchi and K. Kishino, Mater. Sci. Eng. B, 59 (1999).
11. W. Van der Stricht, I. Moerman, P. Demeester, L. Considine, E.J. Thrush and J.A. Crawley, MRS Internet J. Nitride Semicond. Res. 2 (1997).
12. N. Dietz, in "*III-Nitrides Semiconductor Materials*", Ed. Z.C. Feng, Imperial College Press, ISBN 1-86094-636-4, pp. 203-235 (2006).
13. J. McChesney, P. M. Bridenbaugh, and P. B. O'Connor, Mater. Res. Bull. 5, 783 (1970).
14. V. Y. Davydov and A. A. Klochikhin, Semiconductors 38, 861-898 (2004).
15. S. Hernández, R. Cuscó, D. Pastor, L. Artús, K. P. O'Donnell, R. W. Martin, I. M. Watson, Y. Nanishi, and E. Calleja. J. of Appl. Phys. 97 013511 (2005)

Mater. Res. Soc. Symp. Proc. Vol. 1202 © 2010 Materials Research Society   1202-I05-25

Lattice Parameter Variation in ScGaN Alloy Thin Films on MgO(001)
Grown by RF Plasma Molecular Beam Epitaxy

Costel Constantin[1], Jeongihm Pak[2], Kangkang Wang[2], Abhijit Chinchore[2], Meng Shi[2], and Arthur R. Smith[2]

[1]Department of Physics, Seton Hall University, South Orange, NJ 07079

[2]Nanoscale & Quantum Phenomena Institute, Department of Physics and Astronomy, Ohio University, Athens, OH 45701

ABSTRACT

We present the structural and surface characterization of the alloy formation of scandium gallium nitride $Sc_xGa_{1-x}N(001)/MgO(001)$ grown by radio-frequency molecular beam epitaxy over the Sc range of x = 0-100%. In-plane diffraction measurements show a clear face-centered cubic surface structure with single-crystalline epitaxial type of growth mode for all x; a diffuse/distinct transition in the surface structure occurs at near x = 0.5. This is consistent with out-of-plane diffraction measurements which show a linear variation of perpendicular lattice constant $a_\perp$ for x = 0 to 0.5, after which $a_\perp$ becomes approximately constant. The x = 0.5 transition is interpreted as being related to the cross-over from zinc-blende to rock-salt structure.

INTRODUCTION

The technological achievement of light emitting diodes (LEDs) made out of wurtzite GaN (w-GaN) spurred much interest in related III-nitrides such as aluminium nitride (AlN) and and indium nitride (InN). Wurtzite GaN has a band gap of 3.4 eV and emits invisible, highly energetic ultraviolet light, but when some of the gallium atoms are substituted by indium atoms, highly efficient violet, or blue LEDs can be obtained. However, alloying w-GaN with AlN or InN has shown a decrease in device efficiency as the emission wavelength is shifted towards the near infrared-end of the visible spectrum. A very good alternate to w-GaN is the cubic GaN (c-GaN) which has several interesting properties such as a direct wide band gap of $E_D^{c-GaN} = 3.2\,eV$, a tetrahedral zinc-blende bonding structure with a lattice constant of $a_{c-GaN} = 0.452$ nm [1, 2]. On the other hand, scandium nitride (ScN) is a semiconductor with a direct band gap of $E_D^{ScN} = 2.15\,eV$ and a lower indirect band gap of $E_I^{ScN} = 0.9\,eV$ [3-7]. The most stable crystal structure of ScN observed experimentally so far is the rock-salt structure with a relaxed lattice constant $a_{ScN} = 0.4501$ nm [6]. It can be noticed that there is ~ 0.2 % lattice constant mismatch between c-GaN and ScN, and also the different bandgaps [i.e., c-GaN (3.2 eV), and ScN (2.15 eV)] makes ScGaN alloying and the growth of c-GaN/ScN and ScN/c-GaN extremely appealing. We previously reported the growth of $Sc_xGa_{1-x}N$ on wurtzite GaN, and we found that for both low x and high x, alloy-type behavior is observed. For x ≥ 0.54, rock-salt structure is found; for small x up to 0.17, an anisotropic expansion of the ScGaN lattice is observed which is interpreted in terms of local lattice distortions of the wurtzite structure in the vicinity of Sc_{Ga} substitutional sites in which there is a decrease of the N-Sc-N bond angle. This tendency toward flattening of the wurtzite bilayer is consistent with a predicted h-ScN phase [8-10].

So far there are no experimental reports regarding the growth of the ScGaN on substrates with a cubic symmetry [i.e., MgO(001)]. In this article we explore the growth of ScGaN on MgO(001) substrates by radio-frequency plasma-assisted molecular beam epitaxy (rf-MBE). We grew a set of five films of $Sc_xGa_{1-x}N(001)/MgO(001)$ with x = 0-1 in increments of 0.25. The films are measured *in-situ* with reflection high-energy electron diffraction (RHEED) which has a 20 keV electron beam, and *ex-situ* by x-ray diffraction (XRD) which has Cu $K_{\alpha 1}$ and $K_{\alpha 2}$ x-rays.

EXPERIMENT

The growth experiments are performed in a homemade radio-frequency plasma molecular beam epitaxy (rf-plasma MBE) utilizing gallium and scandium effusion cells. The N_2 gas was delivered through an rf-plasma source; the N_2 flow rate and rf-plasma power were maintained constant throughout the entire growth at 1.1 sccm ($P_{chamber} = 9 \times 10^{-6}$ Torr) and 500 Watts, respectively. Growth was performed on MgO(001) substrates which were ultrasonically cleaned with acetone and isopropanol, after which they were loaded into the MBE chamber and prepared by heating and nitridation at a temperature of ~ 900°C for 30 minutes, while keeping the N_2 flow rate to 1.1 sccm and the rf-plasma power at 500 W. The ScGaN films were then grown at a substrate temperature of ~ 900°C and a flux ratio $r = J_{Sc} / (J_{Sc} + J_{Ga})$ and to a thickness of 80 – 495 nm. The Sc and Ga fluxes were $f_{Sc} = 0 – 2.2 \times 10^{14} /(cm^2$ sec), and $f_{Ga} = 0 – 5.8 \times 10^{14} /(cm^2$ sec). The sum of metal fluxes, $f_{Sc} + f_{Ga}$, for each film was within the range $2.2 – 7.7 \times 10^{14} /(cm^2$ sec).

RESULTS AND DISCUSSIONS

Figure 1(a) – (j) show RHEED patterns of $Sc_xGa_{1-x}N(001)/MgO(001)$ with x = 0-1. Figure 1(α) and 1(β) show generic RHEED patterns for an MgO(001) substrate. As can be seen from Fig. 1, RHEED shows that the $Sc_xGa_{1-x}N(001)$ growth is epitaxial for all values of x. The ScGaN film has a clearly crystalline structure at each value of x, and there is a cube-on-cube epitaxial relationship with the substrate, as $[110]_{ScGaN} \parallel [110]_{MgO}$ and $[100]_{ScGaN} \parallel [100]_{MgO}$. Despite the fact that GaN is zinc-blende and ScN is rock-salt, both have fcc crystal structure. This can be considered the reason why the RHEED patterns at all values of x along a given azimuth (α or β) display the same reciprocal space symmetry. From the RHEED patterns of Fig. 1, one feature stands out clearly as a function of x. Namely, as x increases, the RHEED pattern becomes noticeably brighter at around x = 0.5. This may be interpreted as an effect of the surface structure and surface stoichiometry. Whereas GaN(001) growth proceeds well under Ga-rich conditions with excess Ga adatoms, leading to a partly diffusive RHEED pattern, ScN(001) grows smoothly under either Sc-rich or N-rich conditions without accumulating more than 1 monolayer of Sc adatoms, thus leading to a more distinctive RHEED pattern [2, 5, 6]. These growth modes are closely tied to the underlying crystal structure of the surface, zinc-blende vs. rock-salt. It is therefore clear that as x increases, the growth transitions from zinc-blende structure at small x to rock-salt structure at large x. The transition point appears to be in the vicinity of x = 0.5, at which point the RHEED pattern becomes noticeably brighter compared to at x = 0.25.

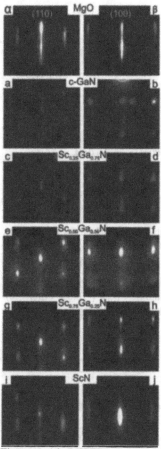

Figure 1 (α) RHEED pattern of MgO(001) substrate along [110]; (β) RHEED pattern of MgO(001) substrate along [100]; (a) – (j) RHEED patterns for $Sc_xGa_{1-x}N(001)/MgO(001)$ with $x = 0 - 1$; images (a), (c), (e), (g), and (i) are taken along same azimuth as (α); images (b), (d), (f), (h), and (j) are taken along same azimuth as (β).

Shown in Fig. 2 are RHEED patterns of ScGaN growth on sapphire(0001) [10]. For these values of x = 0.54 and 0.78, the RHEED pattern is similar to that of ScN(111) along [1-10] but with also the development of a ring-like structure, indicating polycrystalline ScGaN (111)-oriented grains. Comparing this to the growth at x = 0.5 and 0.75 on MgO(001) [see Fig. 1(e) and 1(g)], we see that the growth on MgO with (001) orientation is of significantly better quality in that it does not develop such polycrystallinity. This is clearly an advantage of the (001)-oriented growth.

Figure 2 (a) and (b) RHEED patterns of $Sc_xGa_{1-x}N$/sapphire(0001) with $x = 0.54$, and 0.78, respectively (image reproduced from Ref. 10).

In Fig. 3 we present a portion of the XRD data (39° - 41°) for c-GaN, $Sc_{0.25}Ga_{0.75}N$, $Sc_{0.50}Ga_{0.50}N$, $Sc_{0.75}Ga_{0.25}N$, and ScN films. The XRD data presented in Fig. 3 was calibrated using the MgO 002 peak of the substrate (not shown here) with $2\theta = 42.94°$ which gives a lattice constant of $a_{MgO} = 0.4213$ nm. In Fig. 3, one can observe that there are peak shifts. For example, the 002 peak shifts to the right by ~ 0.07° from $2\theta = 39.81(6)°$ (c-GaN) to $2\theta = 39.88(6)°$ ($Sc_{0.25}Ga_{0.75}N$). As the Sc concentration is increased to 50% the peak shifts to $2\theta = 39.96(6)°$ ($Sc_{0.50}Ga_{0.50}N$ peak). At 75%, the 002 peak is at $2\theta = 39.95(2)°$ ($Sc_{0.75}Ga_{0.25}N$ peak). Finally, the ScN peak is found at $2\theta = 39.96(3)°$. It is worth noting that the peak for $Sc_{0.75}Ga_{0.25}N$ film is very close to the ScN peak (~ 0.01° difference). This small difference between $Sc_{0.75}Ga_{0.25}N$ and ScN clearly indicates that the crystal structure preferred by $Sc_{0.75}Ga_{0.25}N$ is rock-salt – the same as ScN.

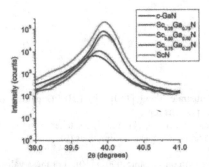

Figure 3 XRD data plots of c-GaN, $Sc_{0.25}Ga_{0.75}N$, $Sc_{0.50}Ga_{0.50}N$, $Sc_{0.75}Ga_{0.25}N$, and ScN on MgO(001) in the range $2\theta = 39 - 41°$. The peaks shown are the 002 peaks.

We calculated the out-of-plane ($a_\perp$) lattice constant using Bragg's law. The XRD data was first calibrated to the MgO peak as mentioned above. We used a value of $\lambda = 0.1542$ nm for the average wavelength of our Cu $K_{\alpha1}$ and $K_{\alpha2}$ x-rays. The $a_\perp$ and layer thickness values are presented in Table I.

286

Table I Layer thickness (t_{ScGaN}), and out-of-plane ($a_\perp$) lattice constant values for $Sc_xGa_{1-x}N$ at various x compositions ($0 \leq x \leq 1$).

Composition x	t_{ScGaN}(nm)	a_{OUT}(nm)
0	187	0.4529
0.25	80	0.4521
0.50	280	0.4512
0.75	430	0.4514
1	495	0.4513

The evolution of the $a_\perp$ lattice constants as a function of Sc concentration x can be observed in Fig. 4. Our measured $a_\perp$ lattice constants for c-GaN (0.4529 nm), and ScN (0.4513 nm) are in very good agreement with previously reported value of 0.4530 nm [2], and 0.4501 nm [6], respectively. It is interesting to note that the $a_\perp$ lattice constant values for c-GaN, $Sc_{0.25}Ga_{0.75}N$, and $Sc_{0.50}Ga_{0.50}N$ decrease linearly with x (from 0.4529 to 0.4512 nm). This linear behavior clearly suggests that $Sc_{0.25}Ga_{0.75}N$ has the same zinc blende crystal structure as c-GaN. The $Sc_{0.50}Ga_{0.50}N$ film seems to be close to a transition point, as also observed in Fig. 1. As the scandium concentration is increased from 50% to 75 % and then to 100 %, $a_\perp$ is fairly constant with x, which suggests rocksalt structure over most of this range. So far, our preliminary results agree reasonably well with recent theoretical calculations of Zerroug et al. [11] who predicted that for Sc concentrations of 0%, 25%, and 50%, the zinc blende structure is more favorable than the rocksalt structure, whereas for Sc concentrations of 75% and 100%, the rocksalt structure is the most stable configuration with respect to zinc blende structure. Our measured lattice constants of zinc-blende GaN (0.4529 nm), and rock-salt ScN (0.4513 nm), do not agree very well with the corresponding 0.455 nm and 0.454 nm lattice constants obtained by Zerroug et al. Furthermore, the point at x = 0.5 cannot be concluded from our experiment to correspond to zinc-blende structure, as the RHEED data suggests the possibility of rock-salt.

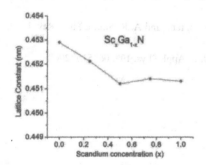

Figure 4 Out-of-plane ($a_\perp$) lattice constants as a function of Sc concentration x.

CONCLUSIONS

ScGaN has been grown with a range of stoichiometry from x = 0 (ZB-GaN) to x = 1 (RS-ScN). The results indicate that the range of x up to ~ 0.5 has ZB structure, whereas the range beyond 0.5 has RS structure. The exact location of the transition point seems to be near x = 0.5. Further studies are in progress to explore the in-plane lattice parameter variations as a function of x and to correlate this with the out-of-plane variations described here.

ACKNOWLEDGMENTS

Funding for this study was provided by the Department of Energy, Office of Basic Energy Sciences (Grant No. DE-FG02-06ER46317) and by the National Science Foundation (Grant No. 0730257).

REFERENCES
[1] M. B. Haider, R. Yang, C. Constantin, E. Lu, A. R. Smith, and H. A. H. Al-Brithen, J. Appl. Phys. **100(08)**, 083516 (2006).
[2] H. A. AL-Brithen, R. Yang, M. B. Haider, C. Constantin, E. Lu, A. R. Smith, N. Sandler, P. Ordejon, Phys. Rev. Letters **95**, 146102 (2005).
[3] W. R. Lambrecht1, Phys. Rev. B **62**, 13538 (2000).
[4] C. Stampfl, W. Mannstadt, R. Asahi, and A. J. Freeman, Phys. Rev. B **63**, 155106 (2001).
[5] H. A. Al-Brithen, E. M. Trifan, D. C. Ingram, A. R. Smith, and D. Gall, J. Cryst. Growth **242**, 345 (2002).
[6] A. R. Smith, H. A. H. Al-Brithen, D. C. Ingram, and D. Gall, J. Appl. Phys. **90**, 1809 (2001).
[7] D. Gall, I. Petrov, L. D. Madsen, J. E. Sundgren, and J. E. Greene, Vac. Sci. Technol. **A 16**, 2411 (1998).
[8] C. Constantin, M. B. Haider, D. Ingram, A. R. Smith, N. Sandler, K. Sun, and P. Ordejón, J. Appl. Phys. **98(12)**, 123501 (2005).
[9] C. Constantin, M. B. Haider, D. Ingram, and A. R. Smith, Appl. Phys. Lett. **85(26)**, 6371 (2004).

[10] C. Constantin, H. Al-Brithen, M. B. Haider, D. Ingram, and A. R. Smith, Phys. Rev. B **70**, 193309 (2004).

[11] S. Zerroug, F. Ali Sahraoui, and N. Bouarissa, J. of Appl. Phys. **103**, 063510 (2008).

Mater. Res. Soc. Symp. Proc. Vol. 1202 © 2010 Materials Research Society   1202-I05-24

Materials study of the competing group-V element incorporation process in dilute-nitride films

Wendy L. Sarney and Stefan P. Svensson
U.S. Army Research Laboratory, Sensors & Electron Devices Directorate,
Adelphi, MD 20783, U.S.A.

ABSTRACT

The incorporation of small amounts of N into III-V antimonide-containing semiconductor alloys allows a drastic expansion of available wavelengths for infrared (IR) detector applications. Quaternary films containing three group-V elements can be lattice matched to the most prevalent substrates for IR applications, such as InAs, GaAs, and GaSb. It is not trivial to incorporate N while maintaining the high crystalline quality required for IR devices. Current materials characterization studies of dilute-nitride films consisting of more than two group-V elements has yielded conflicting information related to their competing behavior and the extent of N incorporation. Due to challenges related to light-element microanalysis for many characterization techniques, and the small concentrations of N involved, it is difficult to quantify the amount of N incorporated into dilute-nitride films. In this study, we use transmission electron microscopy (TEM), energy-dispersive spectroscopy (EDS), and x-ray diffraction (XRD) to study the incorporation behaviors of the competing group-V elements in InAsSbN films.

INTRODUCTION

Dilute nitride semiconductors contain a small amount of N (typically less than 5%) incorporated into a III-V compound. Adding small concentrations of N dramatically expands the range of accessible wavelengths, as shown in Fig. 1. The green region shows the range of wavelengths available with the common III-V semiconductors, and the blue region shows the additional wavelengths that can be obtained by incorporating N.

Early work in the dilute-nitrides was related to alloys that were lattice matched to GaAs for the purpose of reaching the 1.3-1.6 μm wavelength, which is suitable for optoelectronic

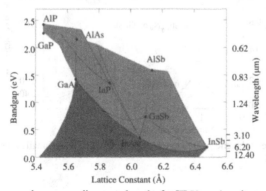

Figure 1: Bandgaps and corresponding wavelengths for III-V semiconductors

289

telecommunication applications [1]. There is recent interest in materials that are lattice matched to GaSb or InSb for IR detector applications [2-3]. Alloys based on GaAs, GaSb, or InAs do not naturally have the bandgaps needed for long-wave IR (LWIR) detector applications requiring operating wavelengths greater than 5 μm. The bandgap for GaAs, for instance, is 1.42 eV, which is in the near-infrared (NIR) range. For this reason, quantum structures such as quantum well IR photodetectors (QWIPs) and type-II strained-layer superlattices are grown to take advantage of confinement effects and to induce an effective bandgap in the desired range. The disadvantage of such structures is that they require many interfaces which must be atomically precise, and the quality of such films may be fundamentally limited.

The dilute nitrides offer the possibility of direct-bandgap detector materials, such as InAsSbN and GaInSbN, with wavelengths that can reach LWIR for N concentrations of approximately 4 %. The calculated bandgap as a function of N composition is shown in Figure 2 [4]. It is noteworthy that both alloys require similar amounts of N, even though the InAs-based material starts with a considerably lower bandgap. Studies find that the optical properties of InAsN are far superior to GaSbN [4-5], and therefore InAsSbN is the focus of our current experiments.

There are drastically conflicting reports on the exact incorporation behaviors of three group-V elements in the limited studies of this material system. Ma, et al., reports that for GaAsSbN single films grown on GaAs substrates, N incorporation is invariant of Sb flux and that N incorporation increases with the N_2 flow rate at the expense of As incorporation [6]. Wicaksono, et al., observed that the presence of Sb flux promotes N incorporation at the expense of Sb incorporation [7]. Zhuang, et al., report that for InAsSbN single films on InAs: (a) N incorporation increases by 2.5 times in presence of Sb, (b) Sb incorporation is enhanced 1.5 times by presence of N, and (c) both Sb and N incorporate at the expense of As [8].

The wide range of results highlights the difficulty in determining the N mole fraction, which is a non-trivial endeavor due to the difficulty of light element analysis and the small concentrations involved. Other studies of N and Sb incorporations compare two reference samples containing a single ternary layer having two group-V elements (such as InAsSb and InAsN) with a third sample containing a single, bulk (~ 1 μm thick) quaternary layer having three group-V elements (such as InAsSbN) [6]. Many characterization studies rely on plan-view

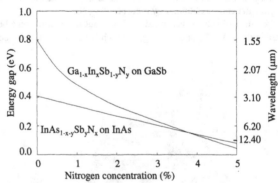

Figure 2: Change of the energy gap due to the addition of Sb and small concentrations of N to InAs and GaSb.

290

electron probe micro-analysis (EPMA) for chemical analysis [6, 8]. This technique uses what is basically a scanning electron microscope (SEM) outfitted with chemical analysis capabilities that include wavelength-dispersive spectrometry (WDS) and energy-dispersive spectroscopy (EDS). EPMA has the advantage of being a non-destructive technique with little sample preparation needed, especially as compared with transmission electron microscopy (TEM). The WDS has better energy resolution (around 10 eV) than the EDS, and therefore allows much better x-ray peak discrimination (peak overlap is a problem in EDS). The major disadvantage is that the EMPA samples a large (~1 μm) area, making it impossible to examine a sample consisting of thin multilayers such as those examined in this study. If the Sb is surface-segregating, as claimed by numerous papers in the literature [5,9], then an EPMA probe of the surface will give an inflated value for the average Sb incorporated in the film.

Our methodology examines thin InAsSb, InAsN, and InAsSbN layers within a single multilayer sample rather than comparing different single-layer bulk samples. We use x-ray diffraction (XRD) to determine the group-V mole fractions in the InAsN and InAsSb layers and TEM/EDS to calculate the As/Sb composition ratio for the InAsSb and the InAsSbN layers. For the alloys studied in this experiment, the identifying peaks are not closely located on the spectrum. Since peak overlap is not a problem, the use of EDS is feasible. The small TEM probe size allows us to separately examine each layer of the film at several locations (laterally and along the depth). Therefore, we can verify that the compositions are uniform throughout the layer thickness and that there are no lateral composition modulations. The combination of the TEM/EDS and XRD data allows the determination of the three group-V mole fractions in the InAsSbN layer. This approach has the following advantages: (a) a single-sample short deposition sequence avoids growth stability and reproducibility issues, (b) the layers are sufficiently thin to avoid relaxation (allowing straightforward XRD analysis), and (c) cross-sectional TEM/EDS avoids complications due to surface accumulation effects. Using this method we confirmed the enhanced incorporation of N in the presence of Sb at the expense of As.

EXPERIMENT

A series of four-layer test structures consisting of InAs/InAsSb/InAsN/InAsSbN on nominally undoped InAs substrates (Figure 3) were grown by solid-source molecular beam epitaxy (MBE). Nitrogen was introduced into the growth chamber using an rf plasma gas source.

Samples were examined with high resolution XRD, TEM, and EDS. Cross sectional TEM samples were prepared by tripod polishing and Ar ion milling, and examined with a JEOL 2010F

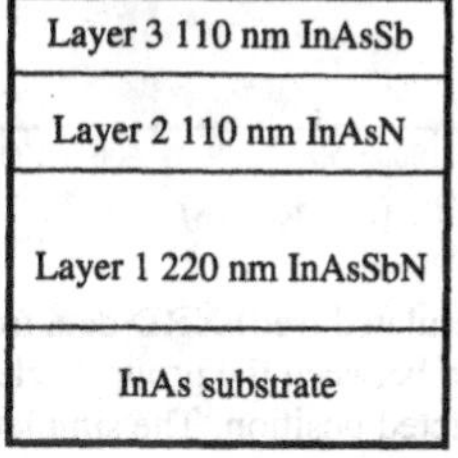

| Layer 4 11 nm InAs |
| Layer 3 110 nm InAsSb |
| Layer 2 110 nm InAsN |
| Layer 1 220 nm InAsSbN |
| InAs substrate |

Figure 3: Schematic of the four layer test structure used in our study.

TEM operated at 200 keV. The minimum probe size is 0.5 nm. The EDS used is an EDAX with a Si-Li x-ray detector and a SUTW (super ultra thin-window). The EDAX spatial resolution is approximately 1 nm and the detector resolution is 133.7 eV at Mn Kα.

RESULTS AND DISCUSSION

Since the lattice constant is affected by the addition of both N and Sb into the InAs lattice, XRD data alone cannot be used for unambiguous determination of the composition. At least one more chemical characterization method must be used. The EDS/TEM technique and the EPMA technique are both unable to detect the lightest elements (below Z=4, Be). Although both instruments can in theory detect N, neither is sensitive enough to accurately give quantitative information for the small N compositions in the dilute-nitrides. Therefore, we restricted the EDS analysis to provide concentration ratios of As/Sb in the InAsSb and the InAsSbN layers. This information in combination with the XRD peak positions allows us to uniquely determine the mole fractions of the three group-V elements.

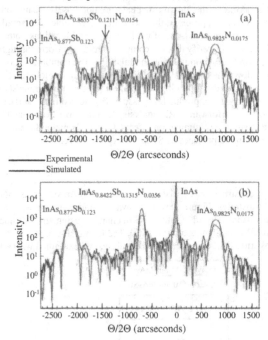

Figure 4: Experimental (black) and simulated (red) XRD data from the test structure. The simulation in (a) assumes no interaction between the group-V elements. Note how the real InAsSbN peak is shifted from the predicted position. The simulation in (b) uses the chemical information from the EDS analysis. Note the enhanced N concentration in the InAsSbN layer.

292

As expected, the InAsSb and InAsN layers were easily identified and characterized with XRD, as shown for one sample in Figure 4. The simulated location of the InAsSbN XRD peak for the simple case of the three group-V elements adding up without competing interactions (red line, Figure 4a) is approximately -1450 arc sec. Our experimental data shows that the peak's actual location is significantly shifted from the predicted position to approximately -700 arcsec (black line, Figure 4a). The simulated data in Figure 4b uses the chemical information obtained from the EDS analysis. Note that the N concentration in the InAsSbN layer is greater than that in (a) the InAsN layer and (b) the predicted concentration for the case of no competing group-V interactions.

Our data for four samples (A-D) are shown in the ternary composition plot in Figure 5. The As and Sb mole fractions in the InAsSb layer and the As and N mole fractions in the InAsN layer were measured by XRD, and are denoted by the star and circle data points, respectively. The dashed black line represents the compositions that produce InAsSbN films that lattice match to InAs. The predicted absorption wavelengths are also indicated along the lattice match line. The triangles denote the data points corresponding to the experimentally measured InAsSbN mole fractions. As shown by the ternary plot, we find that the change in the As mole fraction decreased by up to 3.5% in all but one case where it remained unchanged. The Sb mole fraction increased by a range of 3.3-9.7% in all but one case where it decreased by 11%. This one sample with a decrease in Sb mole fraction was the same sample where the As mole fraction was unchanged. The N mole fraction increases in all cases, ranging from a 56% to a 131% increase. TEM images (not shown here) show that the incorporation of N did not lead to crystal defects. Our XRD and

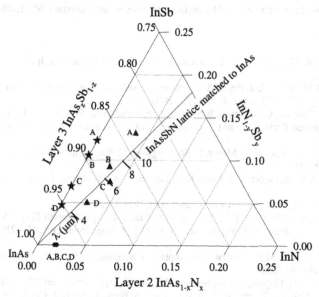

Figure 5: Ternary plot of the composition data

293

EDS results also agree with past studies that have definitively shown that the N incorporation rate is not affected by the nature of the group-III flux [10].

CONCLUSIONS

A direct-gap, bulk, III-V material with sufficiently small bandgap may be the ideal, low-cost material for LWIR applications. Calculations show that the addition of small concentrations of N to III-V semiconductors, such as InAs and GaSb, may lead to dilute-nitride films having the wavelengths needed to reach the LWIR range (8-12 μm) [4]. Since there are commercial applications for III-V semiconductors, there exists the potential to leverage commercial fabrication lines. Therefore costs should be low compared with detectors based on HgCdTe.

In this study, we investigated the competing group-V incorporation process for InAsSbN films grown on InAs substrates. Our characterization methodology examines thin InAsSb, InAsN, and InAsSbN layers within a single multilayer sample with EDS and TEM/XRD. We find that the presence of Sb enhances the extent of N incorporation at the expense of As. At this time, we do not have an explanation for the mechanism that drives the enhanced N incorporation in the presence of Sb. Past studies by other groups indicate a possible surfactant effect [8,9,11], but this cannot be proved conclusively using the method in this particular study.

ACKNOWLEDGMENTS

We acknowledge support by the Missile Defense Agency under contract W911NF0720086.

REFERENCES

[1] M. Kondow, K. Uomi, A. Niwa, T. Kitatani, S. Watahiki and Y. Yazawa, Jpn. J. Appl. Phys. **35**, 1273 (1996).
[2] T. Ashley, T.M. Burke, G.J. Pryce, A.R. Adams, A. Andreev, B.N. Murdin, E.P. O'Reilly, C.R. Pidgeon, Solid State Electron. **47** (3), 387 (2003).
[3] T. Ashley, L. Buckle, G.W. Smith, B.N. Murdin, P.H. Jefferson, F.J. Louis, T.D. Veal, C.F. McConville, Infrared Technology and Applications XXXII, Proceedings of the SPIE **6206**, 62060L (2006).
[4] S.P. Svensson, G. Belenky, J. Meyer, L. Shterengas and I. Vurgaftman, submitted to J. Vac. Sci. Technol., B 2009.
[5] Q. Zhuang, A.M. R. Godenir, A. Krier, K. T. Lai, and S. K. Haywood, J. Appl. Phys. **103**, 063520 (2008).
[6] T.C. Ma, Y.T Lin, H.H. Lin, J. Cryst. Growth **310**, 2854 (2008).
[7] S.W. Wicaksono, S.F. Yoon, K.H. Tan and W.K. Cheah, J. Cryst. Growth **274**, 355 (2005).
[8] Q. Zhuang, A. Godenir, A. Krier, G. Tsai, and H. H. Lin, Appl. Phys. Lett. **93**, 121903 (2008).
[9] H.D. Sun, S. Calvez, M.D. Dawson, J.A. Gupta, G.I. Sproule, X. Wu, Z.R. Wasilewski, Appl. Phys. Lett. **87**, 181908 (2005).
[10] J.C. Harmand, G. Ungaro, L. Largeau, and G. Le Roux, Appl. Phys. Lett. **77**, 2482 (2000).
[11] K. Volz, V. Gambin, W. Ha, M. Wistey, H. Yuen, S. Bank, J. Harris, J. Cryst. Growth **251**, 360 (2003).

AUTHOR INDEX

SUBJECT INDEX